Lecture Notes in Computer Science 14835

Founding Editors

Gerhard Goos
Juris Hartmanis

The series Lecture Notes in Computer Science (LNCS), including its subseries Lecture Notes in Artificial Intelligence (LNAI) and Lecture Notes in Bioinformatics (LNBI), has established itself as a medium for the publication of new developments in computer science and information technology research, teaching, and education.

LNCS enjoys close cooperation with the computer science R & D community, the series counts many renowned academics among its volume editors and paper authors, and collaborates with prestigious societies. Its mission is to serve this international community by providing an invaluable service, mainly focused on the publication of conference and workshop proceedings and postproceedings. LNCS commenced publication in 1973.

Leonardo Franco · Clélia de Mulatier ·
Maciej Paszynski · Valeria V. Krzhizhanovskaya ·
Jack J. Dongarra · Peter M. A. Sloot
Editors

Computational Science – ICCS 2024

24th International Conference
Malaga, Spain, July 2–4, 2024
Proceedings, Part IV

 Springer

Editors

Leonardo Franco (iD)
University of Malaga
Malaga, Spain

Clélia de Mulatier (iD)
University of Amsterdam
Amsterdam, The Netherlands

Maciej Paszynski (iD)
AGH University of Science and Technology
Krakow, Poland

Valeria V. Krzhizhanovskaya (iD)
University of Amsterdam
Amsterdam, The Netherlands

Jack J. Dongarra (iD)
University of Tennessee
Knoxville, TN, USA

Peter M. A. Sloot (iD)
University of Amsterdam
Amsterdam, The Netherlands

ISSN 0302-9743   ISSN 1611-3349 (electronic)
Lecture Notes in Computer Science
ISBN 978-3-031-63771-1   ISBN 978-3-031-63772-8 (eBook)
https://doi.org/10.1007/978-3-031-63772-8

This Springer imprint is published by the registered company Springer Nature Switzerland AG
The registered company address is: Gewerbestrasse 11, 6330 Cham, Switzerland

If disposing of this product, please recycle the paper.

Preface

Welcome to the proceedings of the 24th International Conference on Computational Science (https://www.iccs-meeting.org/iccs2024/), held on July 2–4, 2024 at the University of Málaga, Spain.

In keeping with the new normal of our times, ICCS featured both in-person and online sessions. Although the challenges of such a hybrid format are manifold, we have always tried our best to keep the ICCS community as dynamic, creative, and productive as possible. We are proud to present the proceedings you are reading as a result.

ICCS 2024 was jointly organized by the University of Málaga, the University of Amsterdam, and the University of Tennessee.

Facing the Mediterranean in Spain's Costa del Sol, Málaga is the country's sixth-largest city, and a major hub for finance, tourism, and technology in the region.

The University of Málaga (Universidad de Málaga, UMA) is a modern, public university, offering 63 degrees and 120 postgraduate degrees. Close to 40,000 students study at UMA, taught by 2500 lecturers, distributed over 81 departments and 19 centers. The UMA has 278 research groups, which are involved in 80 national projects and 30 European and international projects. ICCS took place at the Teatinos Campus, home to the School of Computer Science and Engineering (ETSI Informática), which is a pioneer in its field and offers the widest range of IT-related subjects in the region of Andalusia.

The International Conference on Computational Science is an annual conference that brings together researchers and scientists from mathematics and computer science as basic computing disciplines, as well as researchers from various application areas who are pioneering computational methods in sciences such as physics, chemistry, life sciences, engineering, arts, and the humanities, to discuss problems and solutions in the area, identify new issues, and shape future directions for research.

The ICCS proceedings series have become a primary intellectual resource for computational science researchers, defining and advancing the state of the art in this field.

We are proud to note that this 24th edition, with 17 tracks (16 thematic tracks and one main track) and close to 300 participants, has kept to the tradition and high standards of previous editions.

The theme for 2024, "Computational Science: Guiding the Way Towards a Sustainable Society", highlights the role of Computational Science in assisting multidisciplinary research on sustainable solutions. This conference was a unique event focusing on recent developments in scalable scientific algorithms; advanced software tools; computational grids; advanced numerical methods; and novel application areas. These innovative novel models, algorithms, and tools drive new science through efficient application in physical systems, computational and systems biology, environmental systems, finance, and others.

ICCS is well known for its excellent lineup of keynote speakers. The keynotes for 2024 were:

- David Abramson, University of Queensland, Australia
- Manuel Castro Díaz, University of Málaga, Spain
- Jiří Mikyška, Czech Technical University in Prague, Czechia
- Takemasa Miyoshi, RIKEN, Japan
- Coral Calero Muñoz, University of Castilla-La Mancha, Spain
- Petra Ritter, Berlin Institute of Health & Charité University Hospital Berlin, Germany

This year we had 430 submissions (152 to the main track and 278 to the thematic tracks). In the main track, 51 full papers were accepted (33.5%); in the thematic tracks, 104 full papers (37.4%). The higher acceptance rate in the thematic tracks is explained by their particular nature, whereby track organizers personally invite many experts in the field to participate. Each submission received at least 2 single-blind reviews (2.6 reviews per paper on average).

ICCS relies strongly on our thematic track organizers' vital contributions to attract high-quality papers in many subject areas. We would like to thank all committee members from the main and thematic tracks for their contribution to ensuring a high standard for the accepted papers. We would also like to thank Springer, Elsevier, and Intellegibilis for their support. Finally, we appreciate all the local organizing committee members for their hard work in preparing this conference.

We hope the attendees enjoyed the conference, whether virtually or in person.

July 2024

Leonardo Franco
Clélia de Mulatier
Maciej Paszynski
Valeria V. Krzhizhanovskaya
Jack J. Dongarra
Peter M. A. Sloot

Organization

Conference Chairs

General Chair

Valeria Krzhizhanovskaya   University of Amsterdam, The Netherlands

Main Track Chair

Clélia de Mulatier   University of Amsterdam, The Netherlands

Thematic Tracks Chair

Maciej Paszynski   AGH University of Krakow, Poland

Thematic Tracks Vice Chair

Michael Harold Lees   University of Amsterdam, The Netherlands

Scientific Chairs

Peter M. A. Sloot   University of Amsterdam, The Netherlands
Jack Dongarra   University of Tennessee, USA

Local Organizing Committee

Leonardo Franco (Chair)   University of Malaga, Spain
Francisco Ortega-Zamorano   University of Malaga, Spain
Francisco J. Moreno-Barea   University of Malaga, Spain
José L. Subirats-Contreras   University of Malaga, Spain

Thematic Tracks and Organizers

Advances in High-Performance Computational Earth Sciences: Numerical Methods, Frameworks & Applications (IHPCES)

Takashi Shimokawabe	University of Tokyo, Japan
Kohei Fujita	University of Tokyo, Japan
Dominik Bartuschat	FAU Erlangen-Nürnberg, Germany

Artificial Intelligence and High-Performance Computing for Advanced Simulations (AIHPC4AS)

Maciej Paszynski	AGH University of Krakow, Poland

Biomedical and Bioinformatics Challenges for Computer Science (BBC)

Mario Cannataro	University Magna Graecia of Catanzaro, Italy
Giuseppe Agapito	University Magna Graecia of Catanzaro, Italy
Mauro Castelli	Universidade Nova de Lisboa, Portugal
Riccardo Dondi	University of Bergamo, Italy
Rodrigo Weber dos Santos	Federal University of Juiz de Fora, Brazil
Italo Zoppis	University of Milano-Bicocca, Italy

Computational Diplomacy and Policy (CoDiP)

Roland Bouffanais	University of Geneva, Switzerland
Michael Lees	University of Amsterdam, The Netherlands
Brian Castellani	Durham University, UK

Computational Health (CompHealth)

Sergey Kovalchuk	Huawei, Russia
Georgiy Bobashev	RTI International, USA
Anastasia Angelopoulou	University of Westminster, UK
Jude Hemanth	Karunya University, India

Computational Optimization, Modelling, and Simulation (COMS)

Xin-She Yang   Middlesex University London, UK
Slawomir Koziel   Reykjavik University, Iceland
Leifur Leifsson   Purdue University, USA

Generative AI and Large Language Models (LLMs) in Advancing Computational Medicine (CMGAI)

Ahmed Abdeen Hamed   State University of New York at Binghamton,
                             USA
Qiao Jin   National Institutes of Health, USA
Xindong Wu   Hefei University of Technology, China
Byung Lee   University of Vermont, USA
Zhiyong Lu   National Institutes of Health, USA
Karin Verspoor   RMIT University, Australia
Christopher Savoie   Zapata AI, USA

Machine Learning and Data Assimilation for Dynamical Systems (MLDADS)

Rossella Arcucci   Imperial College London, UK
Cesar Quilodran-Casas   Imperial College London, UK

Multiscale Modelling and Simulation (MMS)

Derek Groen   Brunel University London, UK
Diana Suleimenova   Brunel University London, UK

Network Models and Analysis: From Foundations to Artificial Intelligence (NMAI)

Marianna Milano   Università Magna Graecia of Catanzaro, Italy
Giuseppe Agapito   University Magna Graecia of Catanzaro, Italy
Pietro Cinaglia   University Magna Graecia of Catanzaro, Italy
Chiara Zucco   University Magna Graecia of Catanzaro, Italy

Numerical Algorithms and Computer Arithmetic for Computational Science (NACA)

Pawel Gepner   Warsaw Technical University, Poland
Ewa Deelman   University of Southern California, Marina del
                                Rey, USA
Hatem Ltaief   KAUST, Saudi Arabia

Quantum Computing (QCW)

Katarzyna Rycerz   AGH University of Krakow, Poland
Marian Bubak   Sano and AGH University of Krakow, Poland

Simulations of Flow and Transport: Modeling, Algorithms, and Computation (SOFTMAC)

Shuyu Sun   King Abdullah University of Science and
                                Technology, Saudi Arabia
Jingfa Li   Beijing Institute of Petrochemical Technology,
                                China
James Liu   Colorado State University, USA

Smart Systems: Bringing Together Computer Vision, Sensor Networks and Artificial Intelligence (SmartSys)

Pedro Cardoso   University of Algarve, Portugal
João Rodrigues   University of Algarve, Portugal
Jânio Monteiro   University of Algarve, Portugal
Roberto Lam   University of Algarve, Portugal

Solving Problems with Uncertainties (SPU)

Vassil Alexandrov   Hartree Centre – STFC, UK
Aneta Karaivanova   IICT – Bulgarian Academy of Science, Bulgaria

Teaching Computational Science (WTCS)

Evguenia Alexandrova	Hartree Centre – STFC, UK
Tseden Taddese	UK Research and Innovation, UK

Reviewers

Ahmed Abdelgawad	Central Michigan University, USA
Samaneh Abolpour Mofrad	Imperial College London, UK
Tesfamariam Mulugeta Abuhay	Queen's University, Canada
Giuseppe Agapito	University of Catanzaro, Italy
Elisabete Alberdi	University of the Basque Country, Spain
Luis Alexandre	UBI and NOVA LINCS, Portugal
Vassil Alexandrov	Hartree Centre – STFC, UK
Evguenia Alexandrova	Hartree Centre – STFC, UK
Julen Alvarez-Aramberri	Basque Center for Applied Mathematics, Spain
Domingos Alves	Ribeirão Preto Medical School, University of São Paulo, Brazil
Sergey Alyaev	NORCE, Norway
Anastasia Anagnostou	Brunel University London, UK
Anastasia Angelopoulou	University of Westminster, UK
Rossella Arcucci	Imperial College London, UK
Emanouil Atanasov	IICT – Bulgarian Academy of Sciences, Bulgaria
Krzysztof Banaś	AGH University of Krakow, Poland
Luca Barillaro	Magna Graecia University of Catanzaro, Italy
Dominik Bartuschat	FAU Erlangen-Nürnberg, Germany
Pouria Behnodfaur	Curtin University, Australia
Jörn Behrens	University of Hamburg, Germany
Adrian Bekasiewicz	Gdansk University of Technology, Poland
Gebrail Bekdas	Istanbul University, Turkey
Mehmet Ali Belen	Iskenderun Technical University, Turkey
Stefano Beretta	San Raffaele Telethon Institute for Gene Therapy, Italy
Anabela Moreira Bernardino	Polytechnic Institute of Leiria, Portugal
Eugénia Bernardino	Polytechnic Institute of Leiria, Portugal
Daniel Berrar	Tokyo Institute of Technology, Japan
Piotr Biskupski	IBM, Poland
Georgiy Bobashev	RTI International, USA
Carlos Bordons	University of Seville, Spain
Bartosz Bosak	PSNC, Poland
Lorella Bottino	University Magna Graecia of Catanzaro, Italy

Roland Bouffanais	University of Geneva, Switzerland
Marian Bubak	Sano and AGH University of Krakow, Poland
Aleksander Byrski	AGH University of Krakow, Poland
Cristiano Cabrita	Universidade do Algarve, Portugal
Xing Cai	Simula Research Laboratory, Norway
Carlos Calafate	Universitat Politècnica de València, Spain
Victor Calo	Curtin University, Australia
Mario Cannataro	University Magna Graecia of Catanzaro, Italy
Karol Capała	AGH University of Krakow, Poland
Pedro J. S. Cardoso	Universidade do Algarve, Portugal
Eddy Caron	ENS-Lyon/Inria/LIP, France
Stefano Casarin	Houston Methodist Hospital, USA
Brian Castellani	Durham University, UK
Mauro Castelli	Universidade Nova de Lisboa, Portugal
Nicholas Chancellor	Durham University, UK
Thierry Chaussalet	University of Westminster, UK
Sibo Cheng	Imperial College London, UK
Lock-Yue Chew	Nanyang Technological University, Singapore
Pastrello Chiara	Krembil Research Institute, Canada
Su-Fong Chien	MIMOS Berhad, Malaysia
Marta Chinnici	enea, Italy
Bastien Chopard	University of Geneva, Switzerland
Maciej Ciesielski	University of Massachusetts, USA
Pietro Cinaglia	University of Catanzaro, Italy
Noelia Correia	Universidade do Algarve, Portugal
Adriano Cortes	University of Rio de Janeiro, Brazil
Ana Cortes	Universitat Autònoma de Barcelona, Spain
Enrique Costa-Montenegro	Universidad de Vigo, Spain
David Coster	Max Planck Institute for Plasma Physics, Germany
Carlos Cotta	University of Málaga, Spain
Peter Coveney	University College London, UK
Alex Crimi	AGH University of Krakow, Poland
Daan Crommelin	CWI Amsterdam, The Netherlands
Attila Csikasz-Nagy	King's College London, UK/Pázmány Péter Catholic University, Hungary
Javier Cuenca	University of Murcia, Spain
António Cunha	UTAD, Portugal
Pawel Czarnul	Gdansk University of Technology, Poland
Pasqua D'Ambra	IAC-CNR, Italy
Alberto D'Onofrio	University of Trieste, Italy
Lisandro Dalcin	KAUST, Saudi Arabia

Bhaskar Dasgupta	University of Illinois at Chicago, USA
Clélia de Mulatier	University of Amsterdam, The Netherlands
Ewa Deelman	University of Southern California, Marina del Rey, USA
Quanling Deng	Australian National University, Australia
Eric Dignum	University of Amsterdam, The Netherlands
Riccardo Dondi	University of Bergamo, Italy
Rafal Drezewski	AGH University of Krakow, Poland
Simon Driscoll	University of Reading, UK
Hans du Buf	University of the Algarve, Portugal
Vitor Duarte	Universidade NOVA de Lisboa, Portugal
Jacek Długopolski	AGH University of Krakow, Poland
Wouter Edeling	Vrije Universiteit Amsterdam, The Netherlands
Nahid Emad	University of Paris Saclay, France
Christian Engelmann	ORNL, USA
August Ernstsson	Linköping University, Sweden
Aniello Esposito	Hewlett Packard Enterprise, Switzerland
Roberto R. Expósito	Universidade da Coruna, Spain
Hongwei Fan	Imperial College London, UK
Tamer Fandy	University of Charleston, USA
Giuseppe Fedele	University of Calabria, Italy
Christos Filelis-Papadopoulos	Democritus University of Thrace, Greece
Alberto Freitas	University of Porto, Portugal
Ruy Freitas Reis	Universidade Federal de Juiz de Fora, Brazil
Kohei Fujita	University of Tokyo, Japan
Takeshi Fukaya	Hokkaido University, Japan
Wlodzimierz Funika	AGH University of Krakow, Poland
Takashi Furumura	University of Tokyo, Japan
Teresa Galvão	University of Porto, Portugal
Luis Garcia-Castillo	Carlos III University of Madrid, Spain
Bartłomiej Gardas	Institute of Theoretical and Applied Informatics, Polish Academy of Sciences, Poland
Victoria Garibay	University of Amsterdam, The Netherlands
Frédéric Gava	Paris-East Créteil University, France
Piotr Gawron	Nicolaus Copernicus Astronomical Centre, Polish Academy of Sciences, Poland
Bernhard Geiger	Know-Center GmbH, Austria
Pawel Gepner	Warsaw Technical University, Poland
Alex Gerbessiotis	NJIT, USA
Maziar Ghorbani	Brunel University London, UK
Konstantinos Giannoutakis	University of Macedonia, Greece
Alfonso Gijón	University of Granada, Spain

Jorge González-Domínguez	Universidade da Coruña, Spain
Alexandrino Gonçalves	CIIC – ESTG – Polytechnic University of Leiria, Portugal
Yuriy Gorbachev	Soft-Impact LLC, Russia
Pawel Gorecki	University of Warsaw, Poland
Michael Gowanlock	Northern Arizona University, USA
George Gravvanis	Democritus University of Thrace, Greece
Derek Groen	Brunel University London, UK
Loïc Guégan	UiT the Arctic University of Norway, Norway
Tobias Guggemos	University of Vienna, Austria
Serge Guillas	University College London, UK
Manish Gupta	Harish-Chandra Research Institute, India
Piotr Gurgul	SnapChat, Switzerland
Oscar Gustafsson	Linköping University, Sweden
Ahmed Abdeen Hamed	State University of New York at Binghamton, USA
Laura Harbach	Brunel University London, UK
Agus Hartoyo	TU Kaiserslautern, Germany
Ali Hashemian	Basque Center for Applied Mathematics, Spain
Mohamed Hassan	Virginia Tech, USA
Alexander Heinecke	Intel Parallel Computing Lab, USA
Jude Hemanth	Karunya University, India
Aochi Hideo	BRGM, France
Alfons Hoekstra	University of Amsterdam, The Netherlands
George Holt	UK Research and Innovation, UK
Maximilian Höb	Leibniz-Rechenzentrum der Bayerischen Akademie der Wissenschaften, Germany
Huda Ibeid	Intel Corporation, USA
Alireza Jahani	Brunel University London, UK
Jiří Jaroš	Brno University of Technology, Czechia
Qiao Jin	National Institutes of Health, USA
Zhong Jin	Computer Network Information Center, Chinese Academy of Sciences, China
David Johnson	Uppsala University, Sweden
Eleda Johnson	Imperial College London, UK
Piotr Kalita	Jagiellonian University, Poland
Drona Kandhai	University of Amsterdam, The Netherlands
Aneta Karaivanova	IICT-Bulgarian Academy of Science, Bulgaria
Sven Karbach	University of Amsterdam, The Netherlands
Takahiro Katagiri	Nagoya University, Japan
Haruo Kobayashi	Gunma University, Japan
Marcel Koch	KIT, Germany

Harald Koestler	University of Erlangen-Nuremberg, Germany
Georgy Kopanitsa	Tomsk Polytechnic University, Russia
Sotiris Kotsiantis	University of Patras, Greece
Remous-Aris Koutsiamanis	IMT Atlantique/DAPI, STACK (LS2N/Inria), France
Sergey Kovalchuk	Huawei, Russia
Slawomir Koziel	Reykjavik University, Iceland
Ronald Kriemann	MPI MIS Leipzig, Germany
Valeria Krzhizhanovskaya	University of Amsterdam, The Netherlands
Sebastian Kuckuk	Friedrich-Alexander-Universität Erlangen-Nürnberg, Germany
Michael Kuhn	Otto von Guericke University Magdeburg, Germany
Ryszard Kukulski	Institute of Theoretical and Applied Informatics, Polish Academy of Sciences, Poland
Krzysztof Kurowski	PSNC, Poland
Marcin Kuta	AGH University of Krakow, Poland
Marcin Łoś	AGH University of Krakow, Poland
Roberto Lam	Universidade do Algarve, Portugal
Tomasz Lamża	ACK Cyfronet, Poland
Ilaria Lazzaro	Università degli studi Magna Graecia di Catanzaro, Italy
Paola Lecca	Free University of Bozen-Bolzano, Italy
Byung Lee	University of Vermont, USA
Mike Lees	University of Amsterdam, The Netherlands
Leifur Leifsson	Purdue University, USA
Kenneth Leiter	U.S. Army Research Laboratory, USA
Paulina Lewandowska	IT4Innovations National Supercomputing Center, Czechia
Jingfa Li	Beijing Institute of Petrochemical Technology, China
Siyi Li	Imperial College London, UK
Che Liu	Imperial College London, UK
James Liu	Colorado State University, USA
Zhao Liu	National Supercomputing Center in Wuxi, China
Marcelo Lobosco	UFJF, Brazil
Jay F. Lofstead	Sandia National Laboratories, USA
Chu Kiong Loo	University of Malaya, Malaysia
Stephane Louise	CEA, LIST, France
Frédéric Loulergue	University of Orléans, INSA CVL, LIFO EA 4022, France
Hatem Ltaief	KAUST, Saudi Arabia
Zhiyong Lu	National Institutes of Health, USA

Fernando Nobrega Santos	University of Amsterdam, The Netherlands
Joseph O'Connor	University of Edinburgh, UK
Frederike Oetker	University of Amsterdam, The Netherlands
Arianna Olivelli	Imperial College London, UK
Ángel Omella	Basque Center for Applied Mathematics, Spain
Kenji Ono	Kyushu University, Japan
Hiroyuki Ootomo	Tokyo Institute of Technology, Japan
Eneko Osaba	TECNALIA Research & Innovation, Spain
George Papadimitriou	University of Southern California, USA
Nikela Papadopoulou	University of Glasgow, UK
Marcin Paprzycki	IBS PAN and WSM, Poland
David Pardo	Basque Center for Applied Mathematics, Spain
Anna Paszynska	Jagiellonian University, Poland
Maciej Paszynski	AGH University of Krakow, Poland
Łukasz Pawela	Institute of Theoretical and Applied Informatics, Polish Academy of Sciences, Poland
Giulia Pederzani	Universiteit van Amsterdam, The Netherlands
Alberto Perez de Alba Ortiz	University of Amsterdam, The Netherlands
Dana Petcu	West University of Timisoara, Romania
Beáta Petrovski	University of Oslo, Norway
Frank Phillipson	TNO, The Netherlands
Eugenio Piasini	International School for Advanced Studies (SISSA), Italy
Juan C. Pichel	Universidade de Santiago de Compostela, Spain
Anna Pietrenko-Dabrowska	Gdansk University of Technology, Poland
Armando Pinho	University of Aveiro, Portugal
Pietro Pinoli	Politecnico di Milano, Italy
Yuri Pirola	Università degli Studi di Milano-Bicocca, Italy
Ollie Pitts	Imperial College London, UK
Robert Platt	Imperial College London, UK
Dirk Pleiter	KTH/Forschungszentrum Jülich, Germany
Paweł Poczekajło	Koszalin University of Technology, Poland
Cristina Portalés Ricart	Universidad de Valencia, Spain
Simon Portegies Zwart	Leiden University, The Netherlands
Anna Procopio	Università Magna Graecia di Catanzaro, Italy
Ela Pustulka-Hunt	FHNW Olten, Switzerland
Marcin Płodzień	ICFO, Spain
Ubaid Qadri	Hartree Centre – STFC, UK
Rick Quax	University of Amsterdam, The Netherlands
Cesar Quilodran Casas	Imperial College London, UK
Andrianirina Rakotoharisoa	Imperial College London, UK
Celia Ramos	University of the Algarve, Portugal

Robin Richardson	Netherlands eScience Center, The Netherlands
Sophie Robert	University of Orléans, France
João Rodrigues	Universidade do Algarve, Portugal
Daniel Rodriguez	University of Alcalá, Spain
Marcin Rogowski	Saudi Aramco, Saudi Arabia
Sergio Rojas	Pontifical Catholic University of Valparaiso, Chile
Diego Romano	ICAR-CNR, Italy
Albert Romkes	South Dakota School of Mines and Technology, USA
Juan Ruiz	University of Buenos Aires, Argentina
Tomasz Rybotycki	IBS PAN, CAMK PAN, AGH, Poland
Katarzyna Rycerz	AGH University of Krakow, Poland
Grażyna Ślusarczyk	Jagiellonian University, Poland
Emre Sahin	Science and Technology Facilities Council, UK
Ozlem Salehi	Özyeğin University, Turkey
Ayşin Sancı	Altinay, Turkey
Christopher Savoie	Zapata Computing, USA
Ileana Scarpino	University "Magna Graecia" of Catanzaro, Italy
Robert Schaefer	AGH University of Krakow, Poland
Ulf D. Schiller	University of Delaware, USA
Bertil Schmidt	University of Mainz, Germany
Karen Scholz	Fraunhofer MEVIS, Germany
Martin Schreiber	Université Grenoble Alpes, France
Paulina Sepúlveda-Salas	Pontifical Catholic University of Valparaiso, Chile
Marzia Settino	Università Magna Graecia di Catanzaro, Italy
Mostafa Shahriari	Basque Center for Applied Mathematics, Spain
Takashi Shimokawabe	University of Tokyo, Japan
Alexander Shukhman	Orenburg State University, Russia
Marcin Sieniek	Google, USA
Joaquim Silva	Nova School of Science and Technology – NOVA LINCS, Portugal
Mateusz Sitko	AGH University of Krakow, Poland
Haozhen Situ	South China Agricultural University, China
Leszek Siwik	AGH University of Krakow, Poland
Peter Sloot	University of Amsterdam, The Netherlands
Oskar Slowik	Center for Theoretical Physics PAS, Poland
Sucha Smanchat	King Mongkut's University of Technology North Bangkok, Thailand
Alexander Smirnovsky	SPbPU, Russia
Maciej Smołka	AGH University of Krakow, Poland
Isabel Sofia	Instituto Politécnico de Beja, Portugal
Robert Staszewski	University College Dublin, Ireland

Magdalena Stobińska	University of Warsaw, Poland
Tomasz Stopa	IBM, Poland
Achim Streit	KIT, Germany
Barbara Strug	Jagiellonian University, Poland
Diana Suleimenova	Brunel University London, UK
Shuyu Sun	King Abdullah University of Science and Technology, Saudi Arabia
Martin Swain	Aberystwyth University, UK
Renata G. Słota	AGH University of Krakow, Poland
Tseden Taddese	UK Research and Innovation, UK
Ryszard Tadeusiewicz	AGH University of Krakow, Poland
Claude Tadonki	Mines ParisTech/CRI – Centre de Recherche en Informatique, France
Daisuke Takahashi	University of Tsukuba, Japan
Osamu Tatebe	University of Tsukuba, Japan
Michela Taufer	University of Tennessee, USA
Andrei Tchernykh	CICESE, Mexico
Kasim Terzic	University of St Andrews, UK
Jannis Teunissen	KU Leuven, Belgium
Sue Thorne	Hartree Centre – STFC, UK
Ed Threlfall	United Kingdom Atomic Energy Authority, UK
Vinod Tipparaju	AMD, USA
Pawel Topa	AGH University of Krakow, Poland
Paolo Trunfio	University of Calabria, Italy
Ola Tørudbakken	Meta, Norway
Carlos Uriarte	University of the Basque Country, BCAM – Basque Center for Applied Mathematics, Spain
Eirik Valseth	University of Life Sciences & Simula, Norway
Rein van den Boomgaard	University of Amsterdam, The Netherlands
Vítor V. Vasconcelos	University of Amsterdam, The Netherlands
Aleksandra Vatian	ITMO University, Russia
Francesc Verdugo	Vrije Universiteit Amsterdam, The Netherlands
Karin Verspoor	RMIT University, Australia
Salvatore Vitabile	University of Palermo, Italy
Milana Vuckovic	European Centre for Medium-Range Weather Forecasts, UK
Kun Wang	Imperial College London, UK
Peng Wang	NVIDIA, China
Rodrigo Weber dos Santos	Federal University of Juiz de Fora, Brazil
Markus Wenzel	Fraunhofer Institute for Digital Medicine MEVIS, Germany

Contents – Part IV

Computational Health

Biomedical and Bioinformatics Challenges for Computer Science

Exploiting Medical-Expert Knowledge Via a Novel Memetic Algorithm for the Inference of Gene Regulatory Networks

Adrián Segura-Ortiz[1]([✉])[iD], José García-Nieto[1,2][iD], and José F. Aldana-Montes[1,2][iD]

[1] ITIS Software and Department Lenguajes y Ciencias de la Computación, University of Málaga, Málaga, Spain
{adrianseor.99,jnieto}@uma.es
[2] Biomedical Research Institute of Málaga (IBIMA), Málaga, Spain

Abstract. This study introduces an innovative memetic algorithm for optimizing the consensus of well-adapted techniques for the inference of gene regulation networks. Building on the methodology of a previous proposal (GENECI), this research adds a local search phase that incorporates prior knowledge about gene interactions, thereby enhancing the optimization process under the influence of domain expert. The algorithm focuses on the evaluation of candidate solutions through a detailed evolutionary process, where known gene interactions guide the evolution of such solutions (individuals). This approach was subjected to rigorous testing using benchmarks from editions 3 and 4 of the DREAM challenges and the yeast network of IRMA, demonstrating a significant improvement in accuracy compared to previous related approaches. The results highlight the effectiveness of the algorithm, even when only 5% of the known interactions are used as a reference. This advancement represents a significant step in the inference of gene regulation networks, providing a more precise and adaptable tool for genomic research.

Keywords: Memetic Algorithm · Gene Regulatory Networks · Optimization · Bioinformatics

1 Introduction

In the field of computational biology, the inference of gene regulatory networks (GRNs) has become an indispensable mean to comprehend the mechanisms governing gene expression and their implications in various areas of biomedical research. These networks, which are crucial for understanding biological processes at the molecular level, provide a valuable perspective in the study of diseases [14,30] and in the development of genetic therapies [29,36].

However, despite significant advances in this field, the accurate inference of GRNs remains a considerable challenge [15,23,34,42]. There are two main difficulties. The first is the inherent complexity of biological systems [28]. The second

L. Franco et al. (Eds.): ICCS 2024, LNCS 14835, pp. 3–17, 2024.
https://doi.org/10.1007/978-3-031-63772-8_1

is the limitations related to the quantity and quality of empirically validated data [6], which are also difficult to properly incorporate into existing methodologies to improve the accuracy of the results. There is a clear need to take advantage of the knowledge that the medical expert and the literature can bring to the partial construction of networks through a priori known interactions.

In response to these challenges, this research proposes an advanced methodology that extends the previous work carried out in GENECI [33]. GENECI has proven effective in addressing the complexity and diversity of networks through the clever consensus of various techniques. Building upon this solid foundation that addresses the first challenge, an additional stage has been integrated to tackle the second drawback, focusing on maximizing the use of known information. This has been approached by designing an adapted additional local search phase, which incorporates prior knowledge about genetic interactions to guide the optimization process, thus allowing for greater precision in the inference of GRNs, through the injection of domain experts' knowledge.

In this domain, it is common for experts to have partial knowledge or hypotheses about specific genetic interactions. This research focuses on the importance of integrating such knowledge into the inference of GRNs. The experimentation in this work is based on the idea of refining and testing the proposal on generic benchmarks with the intention of subsequently validating its application in real-world problems where the complete solution is unknown. This has been conducted by means of a well-grounded set of benchmarks, including DREAM challenges [26] (specifically their 3rd and 4th editions) and the yeast network of IRMA [4]. Results have shown that the application of this approach introduces significant improvements in the inference of GRNs even when a minimal amount of information is used.

This article is organized as follows. The state of the art in this field is presented in Sect. 2, followed by a detailed description of the approach and methodology in Sect. 3. Subsequently, the experimentation of this study is presented in Sect. 4. Conclusions and future lines of work are discussed in Sect. 5.

2   Related Work

The inference of gene regulatory networks from expression data is a well-studied challenge in computational biology. The literature has explored multiple approaches, including probabilistic graphical models [35], ordinary differential equations (ODEs) [11,16,37], and machine learning techniques such as neural networks [10,12,20,39]. Integrative methods combining different omics data types have also been explored [41], along with causality-based approaches [8] and works related to mutual information [38]. The diversity of approaches has led to a wide range of computational techniques aimed at inferring GRNs. Among them, notable for their accuracy and popularity in the literature are ARACNE [25], C3NET [1], CLR [7], GENIE3 [17], and GRNBOOST2 [27].

In the field of genetic regulatory network inference, seeking a consensus among the results of multiple techniques has been a prominent trend. The

DREAM challenge [26] was a significant turning point, demonstrating that combining results from various techniques produces more accurate solutions than individual methods alone [24]. This revelation spurred the exploration of diverse approaches to achieve consensus, such as the analysis of topological features [19], graph mining [18], and evolutionary algorithms [9,31].

Recent advances reveal novel strategies for achieving consensus among inference techniques, although they still lack a robust methodology tailored to real-world biological networks. EnGRaiN [2] approaches consensus from a mathematical perspective without considering the biological context, while GReNaDIne [32] considers a limited number of techniques with a simple consensus procedure. The challenge of building a weighted and optimal consensus from a set of techniques, taking into account the biological nature of the problem, was addressed in GENECI [33], with results demonstrating a significant improvement in the accuracy of inferred networks.

In the biomedical field, the adoption of memetic algorithms has gained significant traction, demonstrating their versatility and effectiveness in several applications [3,5,13,21,22,40]. These algorithms, which combine intensive local search with global evolutionary strategies, have been successfully applied to solve complex problems in this domain. For instance, in [13] the optimization of PPI (Protein-Protein Interactions) network alignment considers both topological structure and sequence similarities, surpassing existing methods in accuracy. Additionally, in protein structure prediction, memetic algorithms have been designed using knowledge from databases to guide the search towards similar native structures, showing promising results comparable to reference prediction methods [5,21]. In the field of cancer diagnosis, the application of memetic algorithms has demonstrated to enhance the selection of relevant genes by combining local and global search techniques to identify discriminant genes with precision [3].

Finally, the memetic approach has also reached the focus of this work, the reconstruction of GRNs. In [22], an innovative approach is proposed to learn parameters of Recurrent Neural Networks (RNN) and develop an LASSO (Least Absolute Shrinkage and Selection Operator) based framework for the effective reconstruction of GRNs. This method demonstrates superior ability to handle the complexity and sparsity of relationships in real GRNs, outperforming other RNN learning algorithms in large-scale network reconstruction. More recently, in [40], a memetic algorithm is proposed for inferring sparse GRNs using Maximum Entropy Probability Models (MEPMs). This approach addresses the problem from a multi-objective optimization perspective, considering maximum entropy and MEPM constraints as separate objectives.

Given the statistical rigor demonstrated by the GENECI proposal in its results and considering the validity that the memetic approach has shown in biomedical domain problems, it is more than justified to introduce this approach to address the specific problem of reaching a consensus among several inference techniques for the reconstruction of GRNs.

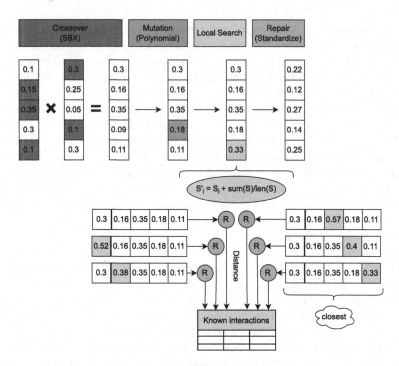

Fig. 1. Succession of phases within the evolutionary process. Individuals are crossed through simulated binary crossover and subsequently subjected to polynomial mutation. Following this, the local search begins where several variations of the individual (encoding a given solution) are compared to select the one whose consensus network is closest to the known interactions. Finally, the individuals are repaired to resume their representation in the form of a weight vector.

3   Proposed Approach

In this article, a memetic algorithm is proposed to optimize the consensus of different techniques for the inference of gene regulation networks. This is based on our previous proposal where an evolutionary process drives this optimization based on the quality and topological characteristics of the networks [33]. This tool has been complemented with a local search phase to guide the optimization process, thanks to prior knowledge of certain gene interactions in the network. This additional phase is located and exemplified in Fig. 1. For a more technical analysis, the pseudocode is set out in Algorithm 1.

The set of candidates subjected to local search is iteratively explored in a loop spanning the length of the individual (line 3 in Algorithm 1). This set comprises the individual provided by the previous phase without any modification (case $i = -1$ in Algorithm 1) and each of the variations resulting from granting an additional vote to each technique (case $i \neq -1$ in Algorithm 1). In other words, the first variation will correspond to adding an additional vote of confidence to the first technique, quantified as the value of one vote in the case that the

Algorithm 1. Main code of the local search phase

Input Individual sol, Known interactions involved in distance calculation ref.
Output Improved individual $resSol$.

```
 1: resSol ← copyOf (sol)
 2: minDistance ← inf
 3: for i in (−1, len (sol)) do
 4:     tmpSol ← copyOf (sol)
 5:     if i ≠ −1 then
 6:         tmpSol[i] += sum (sol) /len (sol)
 7:     RepairSolution (tmpSol)
 8:     net ← GetNetwork(tmpSol)
 9:     distance ← Distance (net, ref)
10:     if distance < minDistance then
11:         minDistance ← distance
12:         resSol ← tmpSol
13: return resSol
```

system is not weighted (case $i = 0$ and line 6 in Algorithm 1). The exact formula for calculating the new value of the technique in the vector is explained and exemplified in Fig. 1.

After generating the candidates, they are repaired and the consensus network derived from each of them is constructed (lines 7 and 8 in Algorithm 1). Finally, the distance of their confidence levels from the known interactions in the network is measured (line 9 in Algorithm 1). The known interactions will usually be assigned a confidence level equal to 1 in the comparison file. However, if the medical researcher wishes to assign a certain probability to their knowledge, any other value between 0 and 1 is accepted. This means that knowledge of a non-existent interaction could also be reflected, but this case is less common.

If the distance is less than the recorded minimum, the current one becomes the new minimum and the best solution is replaced by the current one (lines 10–12 in Algorithm 1). At the end of the loop, the solution with the smallest distance to the reference is returned (line 13 in Algorithm 1).

The distance is calculated as a simple summation of the absolute value differences between the value of the known interactions (usually 1) and the confidence levels assigned by the consensus network for these interactions. However, the possibility that the set of known interactions is a poorly distributed sample that always favors the same technique during the consensus, has been considered. To mitigate this possibility, an additional parameter has been added that defines the interactions that participate in the calculation of the distance on each iteration.

This parameter is exemplified in Fig. 2 by covering its three possible values, namely: the option *all* is contemplated, in which all the known interactions participate in all local searches; the option *some* in which a randomly chosen subset of them participates on each occasion; and finally the option *one* in which only one of the known interactions chosen randomly is used on each local search.

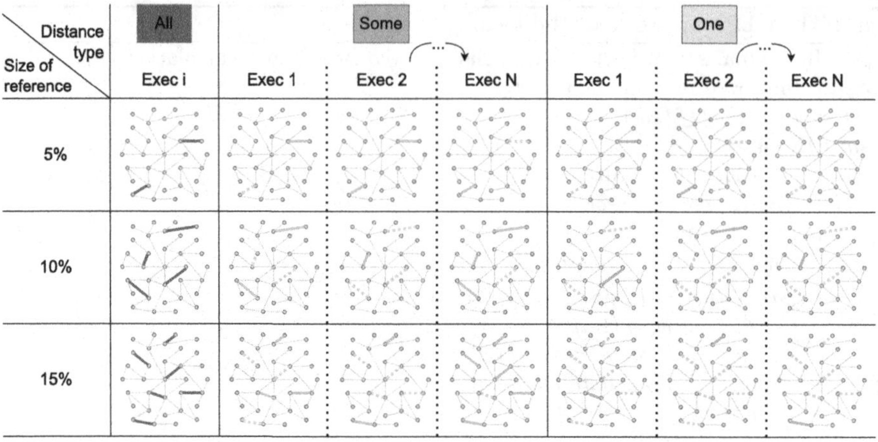

Fig. 2. Examples of interactions involved in the distance calculation in different executions based on the proportion of the gold standard extracted as a set of "known by the expert" interactions (rows) and the type of distance (columns). The case of extracting 5%, 10%, and 15% of the gold standard for the distance types *all*, *some*, and *one* respectively, is shown. As can be observed, all executions take the same reference in the case of *all*, while for *some* and *one*, there is a certain random component that causes differences on each local search.

This local search phase aims at breaking the limitations imposed by GENECI in its aggregate term *Quality*, where techniques whose confidence levels are quite consistent with the remaining ones are somewhat rewarded. Although the consistency of confidence values can increase the reliability of a technique, this strategy sometimes lets certain peculiar interactions that are only inferred by a small subset of techniques slip away. The local search allows for the utilization of prior information to the inference of the network to identify these cases and redirect the evolution of the individuals. It is evident that both strategies are interdependent and must coexist in the evolutionary process, as exceeding the use of previously known information could provoke overfitting.

4   Experimentation

The experimentation addressed in this study employs the academic benchmarks provided by the DREAM challenges [26] (specifically their 3rd and 4th editions) and the yeast network of IRMA [4]. DREAM challenges focused on subnetworks associated with *Escherichia coli* (E. coli) and *Saccharomyces cerevisiae* (yeast) organisms, and includes networks with sizes ranging from 10 to 100 genes. IRMA network comprises 5 genes (CBF1, GAL4, SWI5, GAL80, and ASH1) and encompasses 6 regulatory interactions. These interactions lead to the creation of both "switch on" and "switch off" versions of the network, achieved by cultivating cells in either galactose or glucose conditions, respectively. All these networks were also part of the experimentation of GENECI and constitute a total of 27 inference cases. The known interactions of these networks that will guide the evolutionary process have been defined from their gold standards (known solutions information). Specifically, 5% of these references have been extracted for each execution.

The accuracy of the results will be calculated using the AUROC and AUPR metrics, which were set by the DREAM challenges themselves for their competition and make it possible to compare these results with other studies in the literature. Other metrics such as F1-Score and MCC are not considered, as the use of the chosen benchmark standards is deemed sufficient to cover this study.

This section presents the parameter configuration of the proposed method and the subsequent rigorous comparison with regard to GENECI.

4.1 Parameter Settings

Given that this proposal partially follows the evolutionary process of GENECI, which is in fact common in standard EA settings, it has been decided to keep as much as possible the parameter setting that was configured in the experimentation of its corresponding article, hence allowing a fair comparison. Therefore, the default settings of simulated binary crossover (with a probability of 0.9), polynomial mutation (with a probability of $1/n$, where n is the number of techniques to be consolidated), and repair based on vector standardization have been established. However, for the additional phase proposed in this work, it remains to determine the probability with which the local search is carried out (which is independent of the crossover and mutation probability) and the way the information from the known interactions is used for the calculation of the distance.

To find the most suitable values for these two parameters, all possible combinations between their values have been considered. For the probability of the local search, the candidate values 0.1, 0.25, 0.4, and 0.55, have been defined. And for the type of distance, the already discussed options of *all*, *some*, and *one*.

Each combination of parameters has been tested with 15 independent executions for each network considered in this study. Afterwards, the performance of each solution was calculated using the AUROC and AUPR metrics with regard to the gold standards. For each network and combination of parameters, the median of their precision values was extracted, which finally allowed the calculation of a Friedman statistical ranking with Holm's non-parametric tests.

The results are shown in Table 1 for the AUPR metric and in Table 2 for the AUROC metric. It can be seen how the winning combination for both cases is the one that always takes into account all the known interactions in the distance calculation and with a higher probability of local search. That is, the combination that employs to a greater extent the external information provided. However, rigorous statistical significance cannot be attributed to this victory since only in one case does it meet the established threshold of $p < 0.05$.

A point to consider regarding the lack of statistical significance is that academic problems have a relatively small network size, sometimes around 10 nodes. This causes the difference between taking all or only a subset of interactions for distance calculation to rely on a couple of interactions, which does not allow for

Table 1. Friedman mean rank with Holm's adjusted p values (0.05) for AUPR. Several distance (D) and local search probability (P) configurations are compared based on the AUPR metric. For this purpose, 15 independent runs of each configuration were performed and the median of them (Median) was rescued. After running Friedman's statistical ranking (second column), the winner (highlighted in bold with $*$) is taken as a reference to measure statistical significance against the rest using Holm's nonparametric tests (third column).

AUPR		
Algorithm	$Friedman's Rank$	$Holm's Adj - p$
*Median D-all P-0.55	**4.88889**	-
Median D-one P-0.25	5.90741	0.725979
Median D-one P-0.1	5.96296	0.725979
Median D-all P-0.25	6.03704	0.725979
Median D-some P-0.4	6.24074	0.673303
Median D-all P-0.4	6.53704	0.465230
Median D-some P-0.1	6.62963	0.456477
Median D-one P-0.55	6.75926	0.396552
Median D-some P-0.25	6.90741	0.341178
Median D-one P-0.4	6.92593	0.341178
Median D-some P-0.55	7.00000	0.314504
Median D-all P-0.1	8.20370	0.008033

Table 2. Friedman mean rank with Holm's adjusted p values (0.05) for AUROC. The procedure and nomenclature are identical to those in Table 1.

AUROC		
Algorithm	$Friedman's Rank$	$Holm's Adj - p$
*Median D-all P-0.55	**5.53704**	-
Median D-some P-0.1	6.24074	1.99405
Median D-one P-0.25	6.42593	1.99405
Median D-one P-0.4	6.42593	1.99405
Median D-all P-0.25	6.46296	1.99405
Median D-one P-0.55	6.46296	1.99405
Median D-one P-0.1	6.59259	1.99405
Median D-all P-0.1	6.61111	1.99405
Median D-all P-0.4	6.74074	1.99405
Median D-some P-0.4	6.77778	1.99405
Median D-some P-0.25	6.79630	1.99405
Median D-some P-0.55	6.92593	1.72664

a significant statistical conclusion. However, there is an observable trend towards providing more accurate solutions when the available information is maximized simultaneously through probability and the method of distance calculation.

Regarding the other combinations, another factor that cannot be measured and may have affected the results should be taken into account, granting better precision to combinations with less use of information and worsening the results of others that made greater use of it. In each execution, to form the set of known interactions, a random 5% of the network's gold standard was extracted. Although the number of reference interactions was the same in all executions, their informational value is not necessarily equivalent. That is, the knowledge about the existence of certain interactions may be more valuable than that of others. This is an unpredictable and inevitable fact, since eliminating randomness and establishing fixed reference relationships could bias the results even more.

In the context of the academic networks employed in this study, it is logical to consider extending the winning combination and adding a higher probability of local search to further improve precision levels. However, it should be noted that in such academic problems, the temporal expression levels are simulated from a predefined set of interactions, which ultimately represents the gold standard of the problem. This means that whenever known interactions are added from this gold standard, information from the optimal solution is being shared. This is not the case with real-world networks, and even less so with networks that are intended to be inferred (e.g. in vivo experiments that are not performed yet). In other words, in the cases for which this proposal is intended, the information provided could form part of a good solution known to the domain expert, i.e. a set of interactions that effectively provides a logical explanation of what happens to the gene expression levels during the experiment. However, this may not be the only possible explanation, and there may be other similar alternatives that fit the scenario better. If such information is consistently favored with high probability, it could disturb the direction in which the population evolves during the algorithm execution. Nevertheless, keeping these interactions in mind regularly can bring the population closer to a high-potential zone without condemning the evolution to a possible local minimum.

Given that the optimal solution for these real-world networks intended to be inferred is unknown, the deviation that can be caused by overusing local search could be critical. Therefore, in this case, the most intelligent stance is caution rather than blindly parameterizing in full this proposal based on simulated problems without this broader perspective.

Furthermore, even in academic data where the information injected into the local search is part of the optimal solution to the problem, there is a certain risk that a poorly distributed sample of known interactions may end up diverting the evolution of individuals. The deterioration that these cases can cause to the accuracy of the results increases with the probability of local search. Therefore, once again, setting certain limits is a good practice to maintain a balance that ensures the proposal's security.

Therefore, despite the lack of rigorous statistical significance, the combination of distance *all* and probability 0.55 is chosen as the winner, as it has obtained the first position in the ranking for both precision metrics.

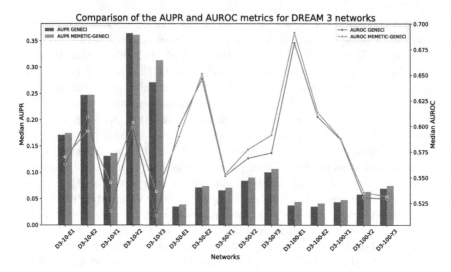

Fig. 3. Comparison of the AUROC and AUPR performance metrics for the GENECI (in blue) and MEMETIC-GENECI (in orange) algorithmic proposals on each of the networks belonging to the third edition of the dream challenges (horizontal axis). For identification, the challenge prefix (D3) is followed by the size of the network (10, 50 or 100) and finally the initial of the organism on which it is based (Y: Yeast, E: E. coli). The bars indicate the medians of the AUPR values and the lines with markers represent the medians of the AUROC values for each network. The AUPR and AUROC values are displayed on separate vertical axes due to their different measurement scales, reserving the left axis for AUPR and the right axis for AUROC. (Color figure online)

4.2   Comparison with GENECI

After configuring the parameters of the memetic algorithm, this section quantifies the improvement achieved by this proposal after adding the additional phase of local search. To this end, the precision results presented in the original GENECI article [33] are compared with those obtained by the best parameter combination seen in the previous section. Specifically, for each network and precision metric, the median of GENECI's executions is compared with the median of the executions of the current proposal. This comparison has been decided to be represented visually for editions 3 and 4 of the DREAM challenges (see Figs. 3 and 4 respectively) and presented quantitatively in Table 3 for the IRMA yeast network.

In Fig. 3, it can be observed that the median accuracies of the solutions from the approach in this work surpass, in most cases, the accuracies provided by the original version of GENECI. Upon closer examination, it is noticed that there is a certain relationship between the size of the networks and the stability of this improvement. That is, for larger networks, the enhancement provided by the additional phase of this approach is more robust and decisive. However, in the case of small networks, more varied differences are observed between the

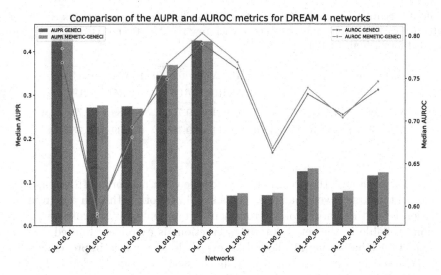

Fig. 4. Comparison of the AUROC and AUPR performance metrics for the GENECI (in blue) and MEMETIC-GENECI (in orange) algorithmic proposals on each of the networks belonging to the fourth edition of the dream challenges (horizontal axis). The nomenclature and interpretation of the graph are identical to those in Fig. 3 (Color figure online).

two algorithms, with ties or even a slight lead of the original version appearing in certain cases. This, in a way, validates the choice of the application domain selected for this proposal which, despite being tested on simulated networks, is intended for inferring real-world networks with significantly larger sizes.

Regarding the instability observed for small-sized networks, it is worth mentioning that these cases have a higher probability of obtaining a poorly distributed sample, as the samples have very few interactions and therefore a good representation is not achieved in any case. Therefore, the instability observed in these cases confirms what was previously mentioned in the parameterization, as even with the introduction of correct interactions, a bad sample can divert the proper evolution of the population. However, thanks to the caution and balance achieved in the parameterization, the impact of these exceptional and indetectable cases a priori is quite moderate on the accuracy of the solutions. It is possible to guide and influence the evolution of individuals without completely damaging their convergence.

In Fig. 4, the precision levels of both proposals for networks from DREAM 4 are compared. In this plot, the connection between the size of the networks and the stability of the improvement provided by the local search phase is once again confirmed. Additionally, in this subset of networks, the correlation between both metrics is observed in greater detail. That is, both metrics seem to simultaneously show the same degree of improvement in most cases. This adds a certain reliability to the proposal of this work.

Table 3. Accuracy values for IRMA networks. In this table, a gene network is contemplated for each pair of columns, where in each row the AUPR and AUROC values are provided for each algorithm.

Technique	IRMA_switch-off		IRMA_switch-on	
	AUROC	AUPR	AUROC	AUPR
Median GENECI	0.8611	0.7865	0.8889	0.75
Median MEMETIC-GENECI	0.8611	0.7865	0.8939	0.7549

Finally, in Table 3, the precision levels for the yeast network of IRMA are presented. In this case, given that it is such a small network with such a high initial precision level, the margin for improvement is minimal. Additionally, the information available in the set of known interactions is extremely limited, around 1 interaction (the minimum allowed). Nevertheless, a subtle improvement has been achieved in the "switch-on" version, maintaining exactly identical values for the "switch-off" instance. The fact that identical values are obtained is due to the small size of the network, causing precision values to be quite staggered.

After analyzing all the sets of networks, it can be checked how the memetic proposal surpasses GENECI in the majority of cases. To provide greater rigor to this comparison, the Wilcoxon test has been calculated, which has provided a p-value of 2.468690e-03 for AUROC and 1.592934e-05 for AUPR. That is, the improvement in the precision of the results is statistically significant.

The ability to achieve statistically significant improvements with such a restricted sample of known interactions (5% of the gold standard) highlights the algorithm's efficacy in integrating and maximizing the informational value of a limited data set. This is especially crucial in the field of computational biology, where the complete and accurate availability of data can be a constant challenge.

It is worthy to note that thanks to the precautions taken during parameterization, this proposal has demonstrated robustness and reliability. During the experimentation, the subset of interactions designated to form the reference in the local search phase was chosen randomly. This random choice has led to the emergence of poorly distributed samples that could disturb the optimization of the population. However, it has been shown that the impact on the deterioration of accuracy has been minimal in these exceptional cases.

Furthermore, it is also important to comment that this proposal has managed to improve results in a set of extensively worked and studied benchmarking networks, whose margin for improvement was initially very limited. The algorithm's ability to find and exploit areas for improvement in these networks indicates its potential to inject the knowledge provided by the expert and maximize its use to discover novel insights in the data.

5   Conclusions and Future Work

This work presents a novel memetic algorithm for the inference of gene regulatory networks (GRNs), that incorporates a local search phase to leverage prior knowledge of gene interactions. This model was applied to a set of networks widely used as benchmarks in the field, which consists of several DREAM challenge networks and the IRMA yeast network. Finally, a 5% of the known interactions of the gold standards were extracted to feed the local search phase of the algorithm, which modifies the individuals to approximate their consensus networks to the known interactions. Results demonstrate a statistically significant improvement in the inference accuracy compared with the previous GENECI model.

The significance of these findings lies in the algorithm's ability to effectively utilize minimal prior knowledge to guide the evolution of gene regulatory network inferences, offering a more precise and adaptable tool for genomic research. This advancement is particularly relevant in the context of computational biology, where the accurate inference of GRNs is crucial for understanding complex biological processes and diseases at the molecular level.

Looking ahead, one promising direction involves dissecting the aggregate terms of the consensus optimization model into multiple objectives, which could enable a more nuanced optimisation process that better captures the complexity of biological networks. In addition, it is essential to evaluate both the original and improved algorithms against a broader academic benchmark. Such extended testing could facilitate more substantial progress towards applications in real-world networks, where the complexities and scale of the data present unique challenges and opportunities for advancing the field of bioinformatics.

Acknowledgments. This work has been partially funded by grant (funded by MCIN/AEI/10.13039/501100011033/) PID2020-112540RB-C41, AETHER-UMA (A smart data holistic approach for context-aware data analytics: semantics and context exploitation). Adrián Segura-Ortiz is supported by Grant FPU21/03837 (Spanish Ministry of Science, Innovation and Universities).

Disclosure of Interests. The authors have no competing interests to declare that are relevant to the content of this article.

References

1. Altay, G., Emmert-Streib, F.: Inferring the conservative causal core of gene regulatory networks. BMC Syst. Biol. **4**(1), 1–13 (2010)
2. Aluru, M., Shrivastava, H., Chockalingam, S.P., Shivakumar, S., Aluru, S.: Engrain: a supervised ensemble learning method for recovery of large-scale gene regulatory networks. Bioinformatics **38**(5), 1312–1319 (2022)
3. Baliarsingh, S.K., Muhammad, K., Bakshi, S.: Sara: a memetic algorithm for high-dimensional biomedical data. Appl. Soft Comput. **101**, 107009 (2021)
4. Cantone, I., Marucci, L., et al.: A yeast synthetic network for in vivo assessment of reverse-engineering and modeling approaches. Cell **137**(1), 172–181 (2009)

5. Correa, L., Borguesan, B., Farfan, C., Inostroza-Ponta, M., Dorn, M.: A memetic algorithm for 3D protein structure prediction problem. IEEE/ACM Trans. Comput. Biol. Bioinf. **15**(3), 690–704 (2016)
6. Escorcia-Rodríguez, J.M., Gaytan-Nuñez, E., et al.: Improving gene regulatory network inference and assessment: the importance of using network structure. Front. Genet. **14**, 1143382 (2023)
7. Faith, J.J., Hayete, B., Thaden, J.T., et al.: Large-scale mapping and validation of escherichia coli transcriptional regulation from a compendium of expression profiles. PLoS Biol. **5**(1), e8 (2007)
8. Finkle, J.D., Wu, J., Bagheri, N.: Windowed granger causal inference strategy improves discovery of gene regulatory networks. Proc. Nat. Acad. Sci. **115**, 2252–2257 (2018)
9. Fujii, C., Kuwahara, H., Yu, G., et al.: Learning gene regulatory networks from gene expression data using weighted consensus. Neurocomputing **220**, 23–33 (2017)
10. Gan, Y., Hu, X., Zou, G., et al.: Inferring gene regulatory networks from single-cell transcriptomic data using bidirectional RNN. Front. Oncol. **12** (2022)
11. García-Nieto, J., Nebro, A.J., Aldana-Montes, J.F.: Inference of gene regulatory networks with multi-objective cellular genetic algorithm. Comput. Biol. Chem. **80**, 409–418 (2019)
12. Ghazikhani, A., Akbarzadeh, T., Monsefi, R.: Genetic regulatory network inference using recurrent neural networks trained by a multi agent system. In: 2011 1st International eConference on Computer and Knowledge Engineering (ICCKE) (2011)
13. Gong, M., Peng, Z., Ma, L., Huang, J.: Global biological network alignment by using efficient memetic algorithm. IEEE/ACM Trans. Comput. Biol. Bioinf. **13**(6), 1117–1129 (2015)
14. Han, P., Gopalakrishnan, C., Yu, H., Wang, E.: Gene regulatory network rewiring in the immune cells associated with cancer. Genes **8**(11), 308 (2017)
15. Hillerton, T., et al.: Fast and accurate gene regulatory network inference by normalized least squares regression. Bioinformatics **38**(8), 2263–2268 (2022)
16. Hurtado, S., Garcia-Nieto, J., Navas-Delgado, I., Nebro, A.J., Aldana-Montes, J.F.: Reconstruction of gene regulatory networks with multi-objective particle swarm optimisers. Appl. Intell. **51**, 1972–1991 (2021)
17. Huynh-Thu, V.A., Irrthum, A., Wehenkel, L., Geurts, P.: Inferring regulatory networks from expression data using tree-based methods. PLoS ONE **5**(9), e12776 (2010)
18. Jiang, H., Turki, T., Zhang, S., Wang, J.T.L.: Reverse engineering gene regulatory networks using graph mining. In: Perner, P. (ed.) MLDM 2018. LNCS (LNAI), vol. 10934, pp. 335–349. Springer, Cham (2018). https://doi.org/10.1007/978-3-319-96136-1_27
19. Khojasteh, H., Khanteymoori, A., Olyaee, M.H.: EnGRNT: inference of gene regulatory networks using ensemble methods and topological feature extraction. Inform. Med. Unlocked **27**, 100773 (2021)
20. Kizaki, N., et al.: The inference method of the gene regulatory network with a majority rule. Nonlinear Theory Appl. IEICE **6**, 226–236 (2015)
21. de Lima Corrêa, L., Dorn, M.: A multi-population memetic algorithm for the 3-D protein structure prediction problem. Swarm Evol. Comput. **55**, 100677 (2020)
22. Liu, L., et al.: Reconstructing gene regulatory networks via memetic algorithm and lasso based on recurrent neural networks. Soft. Comput. **24**, 4205–4221 (2020)
23. Liu, W., et al.: Inferring gene regulatory networks using the improved Markov blanket discovery algorithm. Interdisc. Sci. Comput. Life Sci. 1–14 (2022)

24. Marbach, D., Costello, J.C., Küffner, R., et al.: Wisdom of crowds for robust gene network inference. Nat. Methods **9**(8), 796–804 (2012)
25. Margolin, A.A., Nemenman, I., Basso, K., Wiggins, C., et al.: Aracne: an algorithm for the reconstruction of gene regulatory networks in a mammalian cellular context. BMC Bioinformat. **7**, 1–15 (2006). https://doi.org/10.1186/1471-2105-7-S1-S7
26. Meyer, P., Saez-Rodriguez, J.: Advances in systems biology modeling: 10 years of crowdsourcing dream challenges. Cell Syst. **12**(6), 636–653 (2021)
27. Moerman, T., et al.: GRNBoost2 and Arboreto: efficient and scalable inference of gene regulatory networks. Bioinformatics **35**(12), 2159–2161 (2018)
28. Narasimhan, S., Rengaswamy, R., Vadigepalli, R.: Structural properties of gene regulatory networks: definitions and connections. IEEE/ACM Trans. Comput. Biol. Bioinf. **6**(1), 158–170 (2009)
29. Nazarieh, M., Wiese, A., Will, T., Hamed, M., Helms, V.: Identification of key player genes in gene regulatory networks. BMC Syst. Biol. **10**, 1–12 (2016)
30. Parikshak, N.N., et al.: Systems biology and gene networks in neurodevelopmental and neurodegenerative disorders. Nat. Rev. Genet. **16**(8), 441–458 (2015)
31. Peignier, S., Sorin, B., Calevro, F.: Ensemble learning based gene regulatory network inference. In: 2021 IEEE 33rd International Conference on Tools with Artificial Intelligence (ICTAI), pp. 113–120 (2021)
32. Schmitt, P., et al.: GReNaDIne: a data-driven python library to infer gene regulatory networks from gene expression data. Genes **14**(2), 269 (2023)
33. Segura-Ortiz, A., García-Nieto, J., et al.: GENECI: a novel evolutionary machine learning consensus-based approach for the inference of gene regulatory networks. Comput. Biol. Med. **155**, 106653 (2023)
34. Skok Gibbs, C., et al.: High-performance single-cell gene regulatory network inference at scale: the inferelator 3.0. Bioinformatics **38**(9), 2519–2528 (2022)
35. Watanabe, Y., Seno, S., Takenaka, Y., Matsuda, H.: An estimation method for inference of gene regulatory network using Bayesian network with uniting of partial problems. BMC Genom. **13**, S12 (2012)
36. Wijst, M.G.V.D., Vries, D.H.D., Brugge, H., Westra, H.J., Franke, L.: An integrative approach for building personalized gene regulatory networks for precision medicine. Genome Med. **10**(1), 1–15 (2018)
37. Wu, J., Zhao, X., Lin, Z., Shao, Z.: Large scale gene regulatory network inference with a multi-level strategy. Mol. BioSyst. **12**, 588–597 (2016)
38. Yang, B., Xu, Y.: Reconstructing gene regulation network based on conditional mutual information. In: Proceedings of the 2017 International Conference on Mechanical, Electronic, Control and Automation Engineering (MECAE 2017) (2017)
39. Yasuki, H., Kikuchi, M., Kurokawa, H.: Inferring method of the gene regulatory networks using neural networks adopting a majority rule. In: The 2011 International Joint Conference on Neural Networks (2011)
40. Yin, F., Zhou, J., Xie, W., Zhu, Z.: Inferring sparse genetic regulatory networks based on maximum-entropy probability model and multi-objective memetic algorithm. Memetic Comput. **15**(1), 117–137 (2023)
41. Zarayeneh, N., et al.: Integration of multi-omics data for integrative gene regulatory network inference. Int. J. Data Min. Bioinform. **18**, 223 (2017)
42. Zhao, M., He, W., Tang, J., Zou, Q., Guo, F.: A hybrid deep learning framework for gene regulatory network inference from single-cell transcriptomic data. Briefings Bioinformat. **23**(2), bbab568 (2022)

Human Sex Recognition Based on Dimensionality and Uncertainty of Gait Motion Capture Data

Adam Świtoński[✉][iD] and Henryk Josiński[iD]

Department of Computer Graphics, Vision, and Digital Systems,
Silesian University of Technology, Akademicka 16, 44-100 Gliwice, Poland
{adam.switonski,henryk.josinski}@polsl.pl

Abstract. The paper proposes a method of human sex recognition using individual gait features extracted by measures describing the dimensionality and uncertainty of non-linear dynamical systems. The correlation dimension and sample entropy are computed for time series representing angles of skeletal body joints as well as whole-body orientation and translation. Two aggregation strategies for pose parameters are used – averaging of Euler angles triplets and taking an angle of 3D rotation. In the baseline variant, the distinction between females and males is performed by thresholding the obtained measure values. Moreover, the supervised classification is carried out for the complex gait descriptors characterizing the movements of all bone segments. In the validation experiments, highly precise motion capture measurements containing data of 25 female and 30 male individuals are used. The obtained, at least promising, performance assessed by correct classification rate, the area under the receiver operating characteristic curve, and average precision, is higher than 89%, 96%, and 96%, respectively, and exceeds our expectations. Moreover, the classification accuracy based on a ranking of skeletal joints, as well as whole-body orientation and translation evaluating sex-discriminative traits incorporated in the movements of bone segments, is formed.

Keywords: motion capture · gait analysis · correlation dimension · sample entropy · human sex recognition

1 Introduction

Human sex/gender recognition means females/women and males/men are identified on the basis of their registered behaviors or appearances. It plays an essential role in numerous commercial and non-commercial applications, such as retail and marketing, security and surveillance systems, user personalization, healthcare, gaming industry, smart environments as well as social analyses. This is the reason for diverse and active research studies conducted on this challenge. There are proposed methods, among others, operating on audio data [16], face images

L. Franco et al. (Eds.): ICCS 2024, LNCS 14835, pp. 18–30, 2024.
https://doi.org/10.1007/978-3-031-63772-8_2

[6], keystroke dynamics [27], eye movements [21], handwriting [8] as well as electroencephalogram (EEG) [28], electrocardiogram (ECG) [18], electromyographic (EMG) [7], ground reaction forces [5], inertial (IMU) [17] and motion capture [22] measurements.

In this paper, gait-based sex recognition is taken into account. Thus, females and males are distinguished by the way they walk. In the case of markerless acquisition of human movements performed during gait, it allows to carry on recognition without consciousness of the identified person. There are plenty of approaches proposed for gait-based sex recognition. Primarily, they utilize feature extraction in interpretable and generic variants. In the first one, the meaning of determined features is known. For instance, in [19], eight straightforward discrete variables, such as the angle at touchdown, maximum and minimum peak angles during the stance phase, and the angle at toe-off, are used. In another proposal [12], average values and stride-to-stride standard deviations of stride, swing, stance, and double support time intervals are computed. As regards generic feature extraction, linear dimensionality reduction techniques are mostly chosen. Particularly quite common is Principal Component Analysis (PCA) [14,15,26] maximizing the variance of the resultant features. Moreover, wavelet transform [3] and convolutional neural networks [11] were applied to accomplish the task.

Encouraged by our previous work [22] in which the measures describing gait sequences of bone segment movements having statistically significant differences between populations of females and males are found, we decided to propose the gait-based recognition system and assess its performance. It computes correlation dimension or sample entropy for rotational angles as well as global orientation and translation of the human body. Then, they are used in the classification procedure in two variants. In the first one, the distinction between females and males is performed by separate thresholding the obtained measure values. Moreover, the supervised classification is carried out for the complex gait descriptors characterizing the movements of all bone segments as well as whole-body orientation and translation. Despite the fact that correlation dimension and sample entropy are well-known measures broadly applied in the biosignal and motion data analysis [2,10,24], they were used only in our initial work [22] for the problem of gait-based sex recognition.

2   Related Work

In our previous study [22], we applied correlation dimension, approximate, and sample entropies for the purpose of motion data description. They assess the dimensionality and uncertainty of the processed signal. Primarily, they were computed for rotational data describing angles between adjacent bone segments of the human body. The crucial investigation was related to the comparative analysis of extracted feature values for motion capture sequences representing gait performed by females and males. Descriptive statistics – mean values ± standard deviation and median ± quarter deviations – were determined, as well as non-parametric estimation and statistical hypotheses verification for the noticed differences were carried on.

The correlation dimension obtains greater values by average for females if lower limb movements are taken and smaller if shoulders are analyzed. It is mostly consistent with both entropy measures for which lower limbs behave similarly, but in place of shoulders, head movements achieve greater values for males. It can be interpreted that the system controlling females' hips and knees is more complex, which results in more sophisticated movements, and it is analogous to shoulder and head body segments.

The summary of the work [22] is depicted in Table 1. It contains the p-values of the Mann-Whitney-Wilcoxon test with the null hypothesis that the cumulative distribution functions $F(x)$ (females) and $M(x)$ (males) are the same against the right-tailed and left-tailed alternative ones, meaning $F(x) \geqslant M(x)$ and $F(x) \leqslant M(x)$, respectively. It confirms that the noticed differences are statistically significant – the movements of taken body segments during gait discriminate females and males. This the reason we decided to investigate the performance of gait-based sex recognition using such extracted features. As the approximate and sample entropies are corresponding measures with quite similar discrimination properties obtained, and the first one is considered to be biased statistics, only correlation dimension and sample entropy are taken into account in this study.

Table 1. The p-values of the Mann-Whitney-Wilcoxon test for populations of females and males described by correlation dimension (CD) as well as approximate and sample entropies (AppEnt, SampEnt) calculated for the movements of selected joints [22].

Joint/Segment	Tail	CD	AppEnt	SampEnt
LeftUpLeg	right	$< 0,001$	$< 0,001$	$< 0,001$
RightUpLeg	right	$< 0,001$	$< 0,001$	$< 0,001$
LeftFoot	right	$< 0,001$	$< 0,001$	$< 0,001$
RightFoot	right	$0,007$	$< 0,001$	$< 0,001$
LeftShoulder	left	$< 0,001$	$0,964$	$0,998$
RightShoulder	left	$< 0,001$	$0,999$	$0,891$
Head	left	$0,311$	$< 0,001$	$0,006$

3   Correlation Dimension and Sample Entropy

The system controlling human locomotion can be modeled as a non-linear dynamical system. The movements performed result in data describing poses in consecutive time instants. They are, in fact, observations – the output of the dynamical model. Thus, in the first stage, the reconstruction of the phase space is carried on. For every time instant i a vector $x_i^m = [x_i, x_{i+\tau}, x_{i+2\tau}, ..., x_{i+(m-1)\tau}]$ being the time-delayed measurements is formed. The delay τ is determined by the first local minimum of the mutual information function, and embedding dimension m is obtained using the 'False Nearest Neighbors' approach.

The Grassberger-Procaccia algorithm [9] for estimating correlation dimension is used. It determines a function called correlation sum $CS(r)$ – the fraction of pairs of points (x_i^w, x_j^w) in the phase space whose distances are smaller than r. Then, a logarithmic scale is applied in which the linear approximation of the CS function at the beginning of the range is carried on. Ultimately, the slope coefficient of this approximation stands to be the estimate of the correlation dimension.

The sample entropy [20] is based on all possible patterns containing w consecutive points of the processed time series. The number of similar patterns whose distance is smaller than the assumed radius r is calculated for two subsequent w and $w + 1$ values. Finally, the ratio between them is calculated and transformed by the logarithm function. The similarity radius r is a percentage (20%) of the standard deviation for the entire time series, and w is assumed to be 2.

A more detailed description of the measures taken can be found in [22].

4 Dataset

Highly precise motion capture measurements were used for registration purposes. The acquisition took place in the Human Motion Laboratory (HML) of the Polish-Japanese Academy of Information Technology (PJAIT) (http://www.pja.edu.pl) and was performed by the gold standard Vicon system with spatial accuracy below 1mm and assisted by certified staff. The collected dataset consists of 884 gait sequences of 25 females self-identifying as a woman and 30 males self-identifying as a man.

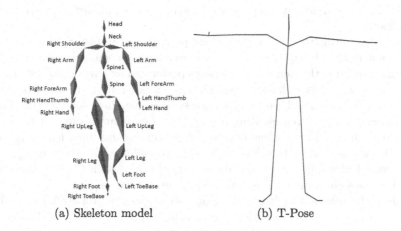

(a) Skeleton model   (b) T-Pose

Fig. 1. Applied skeleton model and T-Pose.

The default Vicon Blade skeleton, containing 22 bone segments as visualized in Fig. 1, was applied. This means that pose space is described by 72 parameters – 22 3D rotations represented by Euler angles triplets as well as whole-body orientation and translation. The parameters specify the human body relative to the reference T-Pose (Fig. 1b).

The gait was performed alongside a straight, five-meter-long route (see Fig. 2) with two interpreted individually paces – the preferred natural and increased ones. The applied frequency of registration was 100 Hz.

It is exactly the same dataset that was used in [22].

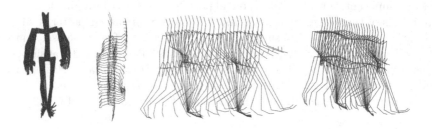

Fig. 2. Example gait instance – front, top, side, and perspective views.

5   Experimental Setup

In the preprocessing stage, the gait main cycle is extracted. It contains two adjacent steps performed by the left and right lower limbs. The extraction algorithm is based on tracking the extremes of distances between ankles as described in [23]. Then, the measures – correlation dimension and sample entropy – are computed for the time series representing every pose parameter and used in the recognition.

Data registration took place over a long period of several years across numerous Vicon updates. Moreover, the exact pose estimation depends on the calibration procedure involving the range of movements performed by participants. It results in varying pose models with the same skeleton, but different meanings of the local coordinates systems and Euler angles triplets. It is visualized for the RightForeArm segment in Fig. 3 representing histograms of correlation dimension values. This is a segment with only one degree of freedom, but for most female recordings, it is incorporated within RX angle while to males – RZ. Thus, recognition taking every Euler angle separately utilizes acquisition-specific features instead of individual ones and causes its performance to be overevaluated.

This is the reason why two aggregation strategies are proposed. In the first one, as we expected, the identical relationships (greater or smaller) between populations of females and males for parameters describing the movements of the same joint or related to whole-body orientation and translation, the average, across these parameters, is computed and taken in further analysis. In the second proposal, before measure computation, rotations are transformed from Euler angle into axis-angle representation and the angle is chosen.

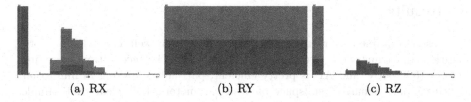

(a) RX   (b) RY   (c) RZ

Fig. 3. Histograms of correlation dimension calculated for Euler angles of RighForeArm segment. The colors blue and red denote female and male samples, respectively. (Color figure online)

Such prepared gait descriptors are utilized in the classification conducted in two variants. In the first one, every segment, as well as whole-body orientation and translation, are classified separately by the thresholding – the females are identified in case of discovered relations greater or smaller. As the optimal threshold value is not obvious and difficult to predict, the performance is assessed by the area under the receiver operating curve (ROC-AUC) and the average precision (AP). It means true (TPR) and false (FPR) positive rates, as well as precision and recall, being relative numbers of actual females predicted as females or males are determined iteratively for successive thresholds and aggregated by the calculated area under obtained in such a way TPR-FPR and precision-recall curves.

In the complex variant, all aggregated measure values for bone segments, as well as whole-body orientation and translation, are classified by the selected supervised techniques. They are namely:

- Naive Bayes with parametric and non-parametric estimation [13],
- k nearest neighbors (kNN) with normalizable Euclidean distance metric [1],
- Random Forest [4] with 100 unconstrained depth trees,
- Multilayer perception (MLP) [29] with a single hidden layer containing a number of units equal to the number of input features.

For this variant, besides ROC-AUC and AP, the CCR ratio is calculated as well. It is the percent of correctly classified samples of the testing set and most intuitively corresponds to the recognition performance.

In the complex variant, a cross-validation algorithm was used to split the data into training and testing parts. Due to correlation dimension and sample entropy may be individual features as well, the custom division into the folds, instead of the default random procedure, was carried out. The folds contain data from single participants only. It means the training set does not have data of the same individuals as the testing one and sex is recognized on the basis of discovered properties for different persons. The number of folds is 55–25 females and 30 males.

In the case of the recognition based on single segments or whole-body orientation or translation (the first variant), there is no training stage – the aggregated/raw measure values directly point to female and male classes. Thus, only the testing set is used, and it contains all samples of the collected dataset.

6   Results

The results obtained for rotations of bone segments, as well as whole-body orientation and translation, are presented in Table. 2. The recognition exceeding 80% of ROC-AUC and AP is pretty efficient, particularly taking in mind that it utilizes only a narrow subspace of pose parameters. It is clear that sample entropy is much more robust in the discrimination of females and males than correlation dimension. The top-ranked bone segments are related to the toe base, neck, spine, foot, and forearm movements. Moreover, the analysis of whole-body translation gives high accuracy of sex prediction. Both aggregation strategies perform similarly – averaging is a bit more efficient for sample entropy and angle of rotation for correlation dimension.

Table 2. Area under receiver operating characteristics curve [%] (RC) and average precision [%] (AP) in the distinction of females and males bone segments movements as well as whole-body orientation and translations during gait based on descriptors containing correlation dimension (CD) and sample entropy (SE) values for average across Euler angles and angle of rotation aggregation strategies.

Segment/global	Average aggregation				Angle of rotation			
	CD		SE		CD		SE	
	RA	AP	RA	AP	RA	AP	RA	AP
Head	52.99	49.58	52.87	52.87	55.83	55.83	52.88	52.88
LeftArm	64.79	64.79	55.16	55.16	65.07	65.07	54.91	54.91
LeftFoot	57.11	57.11	73.29	76.75	57.70	57.70	71.99	75.93
LeftForeArm	53.73	53.73	53.66	56.08	53.73	53.73	53.66	56.08
LeftHand	60.07	60.07	61.91	61.91	61.98	61.98	60.04	60.04
LeftLeg	50.25	50.52	54.78	55.15	50.25	50.52	54.78	55.15
LeftShoulder	55.89	52.22	66.86	73.33	54.99	54.99	66.80	73.39
LeftToeBase	56.30	49.43	86.88	87.18	56.30	49.43	86.88	87.18
LeftUpLeg	52.51	51.56	61.80	63.57	51.59	50.71	58.53	61.00
Neck	53.20	49.84	77.36	77.36	54.88	54.88	78.66	78.66
RightArm	62.62	62.62	56.12	56.12	62.86	62.86	55.80	55.80
RightFoot	52.51	50.47	71.49	71.04	54.66	54.66	70.68	70.39
RightForeArm	53.12	53.12	48.88	51.70	53.12	53.12	48.88	51.70
RightHand	58.44	58.44	57.52	57.52	61.08	61.08	56.58	56.58
RightLeg	53.43	53.43	58.23	57.41	53.43	53.43	58.23	57.41
RightShoulder	51.70	49.81	56.85	64.39	55.60	55.60	55.95	63.19
RightToeBase	60.30	60.30	84.80	85.76	60.30	60.30	84.80	85.76
RightUpLeg	55.27	52.66	66.76	66.39	52.75	51.04	63.54	64.42
Spine	60.74	60.88	78.07	77.98	59.11	58.13	75.64	76.12
Spine1	59.16	59.34	59.36	62.76	58.27	58.57	52.62	57.44
Global rotation	58.64	58.64	66.29	66.29	58.84	58.84	64.65	65.33
Global translation	59.21	60.05	84.67	82.32	-	-	-	-

The results for complete descriptors with measure values of all bone segments as well as whole-body orientation and translation are depicted in Table 3 and Table 4. In the case of the angle of rotation aggregation strategy, which is unworkable for non-rotational pose parameters, averaging is carried out for whole-body translation. The best performance is obtained by sample entropy features, Random Forest classifier and angle of rotation aggregation strategy. It is specified by 89.57%, 96.61%, 96.91% of CCR, ROC-AUC and AP measures, respectively. However, the 5NN classifier with the averaging aggregation strategy is only slightly worse and it has 88.78% CCR. The same general conclusion as previously stated can be drawn – sample entropy is more robust, and both aggregation strategies perform very similarly.

Table 3. Performance [%] of females and males gait classification based on measure extracted for time series of every pose parameter and averaging aggregation strategy.

Classifier	Correlation dimension			Sample entropy		
	CCR	ROC-AUC	AP	CCR	ROC-AUC	AP
NaiveBayes normal	65.65	70.42	69.32	86.17	91.96	92.35
NaiveBayes kernel	65.65	70.32	69.91	86.73	92.64	92.51
1NN	57.71	64.48	58.21	85.49	83.36	74.43
3NN	58.50	63.59	60.04	85.71	91.17	85.93
5NN	57.14	64.66	62.21	88.78	93.34	90.28
RandomForest	65.42	72.87	75.47	89.23	96.38	96.66
MLP	57.82	60.67	62.21	86.28	93.90	95.05

Table 4. Performance [%] of females and males gait classification based on measure extracted for time series of every pose parameter and angle of rotation aggregation strategy.

Classifier	Correlation dimension			Sample entropy		
	CCR	ROC-AUC	AP	CCR	ROC-AUC	AP
NaiveBayes normal	67.23	72.89	75.96	86.96	92.32	93.63
NaiveBayes kernel	68.25	73.17	75.43	87.19	92.92	92.61
1NN	61.34	65.07	58.62	86.17	83.67	74.44
3NN	62.59	66.50	62.51	84.47	91.29	86.12
5NN	62.81	70.18	68.52	87.64	93.10	89.64
RandomForest	67.01	72.56	75.43	89.57	96.61	96.91
MLP	62.36	65.27	68.45	83.22	91.74	93.75

In Table 5, confusion matrices with the number of true and false positives as well as true and false negatives are presented. The percentages of misrecognition of females predicted as males and males as females are very similar.

Table 5. Confusion matrices for females and males gait classification based on sample entropy extracted for time series of every pose parameter with averaging (AVG) and angle of rotation(AR) aggregation strategy.

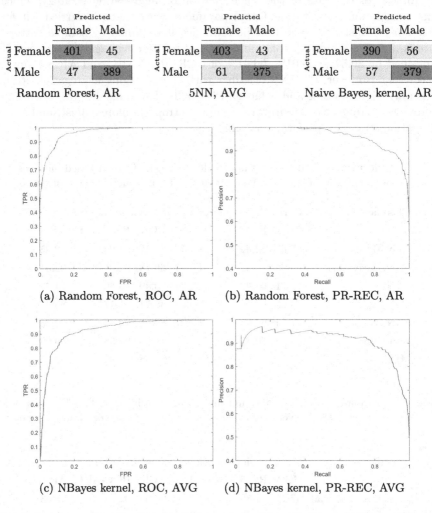

	Predicted	
	Female	Male
Actual Female	401	45
Actual Male	47	389

Random Forest, AR

	Predicted	
	Female	Male
Actual Female	403	43
Actual Male	61	375

5NN, AVG

	Predicted	
	Female	Male
Actual Female	390	56
Actual Male	57	379

Naive Bayes, kernel, AR

(a) Random Forest, ROC, AR   (b) Random Forest, PR-REC, AR

(c) NBayes kernel, ROC, AVG   (d) NBayes kernel, PR-REC, AVG

Fig. 4. Receiver operating characteristics and precision-recall curves for females and males gait distinction based on sample entropy calculated for time series of every pose parameter with averaging (AVG) and angles of rotations (AR) aggregation strategies.

A more detailed analysis of selected classifier workings is presented in Fig. 4, which contains dependencies between TPR and FPR as well as precision and recall measures for two selected cases. If the Random Forest classifier is taken, almost all females are recognized (TPR≈98%), causing only about 30% false detection (FPR). Moreover, for 60% recall, there is a precision greater than 99%.

Table 6. Performance [%] of females and males gait classification based on angle of rotation aggregation strategy and sample entropy extracted for time series of pose parameters except those describing toe base movements (NTB), whole-body translation (NWBT), whole-body orientation and translation (NWB) as well as toe base movements, whole-body orientation and translation (NTBWB).

Classifier	NTB		NWBT		NWB		NTBWB	
	CCR	ROC	CCR	ROC	CCR	ROC	CCR	ROC
NaiveBayes normal	80.39	88.36	86.28	91.31	85.94	91.37	78.00	86.40
NaiveBayes kernel	81.18	88.56	85.26	91.95	86.62	92.03	78.12	86.54
1NN	80.27	77.51	84.24	81.66	83.11	79.84	77.78	73.48
3NN	80.50	85.81	83.11	88.85	83.56	89.02	76.53	82.36
5NN	82.31	89.39	86.73	91.99	85.37	91.30	78.34	86.35
RandomForest	87.30	94.54	87.87	96.11	88.66	96.08	85.15	92.31
MLP	81.97	89.91	84.35	92.70	84.58	93.26	81.63	89.91

As the recognition performance achieved for toe base segments is surprisingly high – their range of movement seems to be insignificant during gait –, some acquisition-specific issues may have an influence on the results. This is the reason we decided to investigate the classification efficiency in the complex variant without taking into account just toe base movements. Additionally, we did the same with whole-body orientation and translation. The results are depicted in Table 6. The accuracy of sex detection is only a bit worse – the percentages of correctly classified gait samples (CCR) are 87.30%, 88.66%, and 85.15% in the case of discarding in the analysis toe base movements, whole-body parameters as well both of them, respectively.

7   Summary and Conclusions

In the paper, the problem of female and male gait distinction is faced. It is based on the interpretable measures describing properties – uncertainty and dimensionality – of nonlinear dynamical systems. Sample entropy and correlation dimension are extracted for time series representing parameters of the human body skeleton model. Two aggregation strategies for the values related to the same segments and whole-body descriptors are proposed. The final recognition is carried out by the thresholding the aggregated measure values and by the selected supervised techniques.

The work is motivated by our results presented in [22] in which statistically significant differences in the correlation dimension and sample entropy values of the time series describing movements of bone segments for the populations of females and males are discovered. However, this is a new research report, which is a natural continuation of the previous one. It goes further and assesses the performance of the recognition – statically significant differences do not mean the efficient classification is feasible and how it should be arranged. Moreover, it verifies the variant in which all Euler angles and translation data are involved.

We state the obtained results to be satisfactory. It is workable to predict females and males on the basis of their gait with almost 90% accuracy (CCR). Moreover, it is feasible to precisely control the balance between properly and improperly recognized females and males depending on expectations according to ROC and precision-recall curves. In addition, discriminative traits in the whole-body and successive bone segment movements are evaluated. Especially lower limb data are highly efficient in sex distinction. The extracted measure values are obviously robust in the faced recognition problem, but they very likely contain individual traits as well. It makes the classification problem to be more difficult and the obtained results even more valuable.

The main limitations of the work are similar to the ones mentioned in [22] and are related to short time series lengths. Moreover, despite the collected dataset being impressive as regards the number of recordings and participants, it has limited diversity. It mainly contains samples of young individuals coming from the same demographic region and which were taken in very similar conditions. An extended investigation with a more diverse dataset depends on recordings availability.

Furthermore, we expect the chaotic properties of the signal may give a valuable description of women's and men's gait discrimination. Due to insufficient time series lengths for computing the Lyapunov exponent, we are going to train a neural network using time courses of known dynamical models as presented in [25] and then apply it for feature extraction of mocap sequences.

Acknowledgements. This publication was supported by the Department of Computer Graphics, Vision and Digital Systems, under the statutory research project (Rau6, 2024), Silesian University of Technology (Gliwice, Poland).

References

1. Aha, D., Kibler, D.: Instance-based learning algorithms. Mach. Learn. **6**, 37–66 (1991)
2. Ahmadi, B., Amirfattahi, R., et al.: Comparison of correlation dimension and fractal dimension in estimating BIS index. Wirel. Sens. Netw. **2**(01), 67–73 (2010)
3. Arai, K., Asmara, R.: Human gait gender classification using 3D discrete wavelet transform feature extraction. Int. J. Adv. Res. Artif. Intell. **3**(2) (2014)
4. Breiman, L.: Random forests. Mach. Learn. **45**(1), 5–32 (2001)
5. Chatain, C., Ramdani, S., Vallier, J.M., Gruet, M.: Recurrence quantification analysis of force signals to assess neuromuscular fatigue in men and women. Biomed. Signal Process. Control **68**, 102593 (2021)
6. Dhomne, A., Kumar, R., Bhan, V.: Gender recognition through face using deep learning. Procedia Comput. Sci. **132**, 2–10 (2018)
7. Di Nardo, F., Mengarelli, A., Maranesi, E., Burattini, L., Fioretti, S.: Gender differences in the myoelectric activity of lower limb muscles in young healthy subjects during walking. Biomed. Signal Process. Control **19**, 14–22 (2015)

8. Gattal, A., Djeddi, C., Bensefia, A., Ennaji, A.: Handwriting based gender classification using COLD and hinge features. In: El Moataz, A., Mammass, D., Mansouri, A., Nouboud, F. (eds.) ICISP 2020. LNCS, vol. 12119, pp. 233–242. Springer, Cham (2020). https://doi.org/10.1007/978-3-030-51935-3_25

9. Grassberger, P., Procaccia, I.: Measuring the strangeness of strange attractors. Physica D **9**(1–2), 189–208 (1983)

10. Harezlak, K., Kasprowski, P.: Application of time-scale decomposition of entropy for eye movement analysis. Entropy **22**(2), 168 (2020)

11. Horst, F., et al.: Explaining automated gender classification of human gait. Gait Posture **81**, 159–160 (2020)

12. Hughes-Oliver, C., Srinivasan, D., Schmitt, D., Queen, R.: Gender and limb differences in temporal gait parameters and gait variability in ankle osteoarthritis. Gait Posture **65**, 228–233 (2018)

13. John, G.H., Langley, P.: Estimating continuous distributions in Bayesian classifiers. In: Eleventh Conference on Uncertainty in Artificial Intelligence, pp. 338–345. Morgan Kaufmann, San Mateo (1995)

14. Kastaniotis, D., Theodorakopoulos, I., Economou, G., Fotopoulos, S.: Gait-based gender recognition using pose information for real time applications. In: 2013 18th International Conference on Digital Signal Processing (DSP), pp. 1–6. IEEE (2013)

15. Kobayashi, Y., Hobara, H., Heldoorn, T.A., Kouchi, M., Mochimaru, M.: Age-independent and age-dependent sex differences in gait pattern determined by principal component analysis. Gait Posture **46**, 11–17 (2016)

16. Kumari, M., Talukdar, N., Ali, I.: A new gender detection algorithm considering the non-stationarity of speech signal. In: 2016 2nd International Conference on Communication Control and Intelligent Systems (CCIS), pp. 141–146. IEEE (2016)

17. Mostafa, A., Barghash, T.O., Assaf, A.A.S., Gomaa, W.: Multi-sensor gait analysis for gender recognition. In: ICINCO, pp. 629–636 (2020)

18. Okin, P.M., Kligfield, P.: Gender-specific criteria and performance of the exercise electrocardiogram. Circulation **92**(5), 1209–1216 (1995)

19. Phinyomark, A., Osis, S.T., Hettinga, B.A., Kobsar, D., Ferber, R.: Gender differences in gait kinematics for patients with knee osteoarthritis. BMC Musculoskelet. Disord. **17**(1), 1–12 (2016)

20. Richman, J.S., Moorman, J.R.: Physiological time-series analysis using approximate entropy and sample entropy. Am. J. Physiol. Heart Circ. Physiol. **278**(6), H2039–H2049 (2000)

21. Sargezeh, B.A., Tavakoli, N., Daliri, M.R.: Gender-based eye movement differences in passive indoor picture viewing: an eye-tracking study. Physiol. Behav. **206**, 43–50 (2019)

22. Świtoński, A., Josiński, H., Polański, A., Wojciechowski, K.: Correlation dimension and entropy in the assessment of sex differences based on human gait data. Front. Hum. Neurosci. **17**, 1233859 (2023)

23. Świtoński, A., Josiński, H., Wojciechowski, K.: Dynamic time warping in classification and selection of motion capture data. Multidimension. Syst. Signal Process. **30**(3), 1437–1468 (2019)

24. Szczesna, A.: Quaternion entropy for analysis of gait data. Entropy **21**(1), 79 (2019)

25. Szczesna, A., Augustyn, D., Harezlak, K., Josinski, H., Switonski, A., Kasprowski, P.: Datasets for learning of unknown characteristics of dynamical systems. Sci. Data **10**(1), 79 (2023)

26. Troje, N.F.: Decomposing biological motion: a framework for analysis and synthesis of human gait patterns. J. Vis. **2**(5), 371–387 (2002)

27. Tsimperidis, I., Arampatzis, A., Karakos, A.: Keystroke dynamics features for gender recognition. Digit. Investig. **24**, 4–10 (2018)
28. Wang, P., Hu, J.: A hybrid model for EEG-based gender recognition. Cogn. Neurodyn. **13**, 541–554 (2019)
29. Witten, I.H., Frank, E., Hall, M.A.: Data Mining: Practical Machine Learning Tools and Techniques, 3rd edn. Morgan Kaufmann Publishers Inc., San Francisco (2011)

A Multi-domain Multi-task Approach for Feature Selection from Bulk RNA Datasets

Karim Salta[1]([✉])[iD], Tomojit Ghosh[2][iD], and Michael Kirby[1][iD]

[1] Colorado State University, Fort Collins, CO 80523, USA
{karim.karimov,michael.kirby}@colostate.edu
[2] University of Tennessee, Chattanooga, TN 37403, USA
tomojit-ghosh@utc.edu

Abstract. In this paper a multi-domain multi-task algorithm for feature selection in bulk RNAseq data is proposed. Two datasets are investigated arising from mouse host immune response to *Salmonella* infection. Data is collected from several strains of collaborative cross mice. Samples from the spleen and liver serve as the two domains. Several machine learning experiments are conducted and the small subset of discriminative across domains features have been extracted in each case. The algorithm proves viable and underlines the benefits of across domain feature selection by extracting new subset of discriminative features which couldn't be extracted only by one-domain approach.

Keywords: Sparse Feature Selection · Multi-Domain Multi-Task Learning · Bulk RNA · VAE · HPC

1 Introduction

In the field of bioinformatics, researchers often use microarray or next-generation sequencing techniques to study the expression levels of genes, with each sample typically having tens of thousands of features. The large number of features often necessitates the use of feature selection algorithms to improve the performance of machine learning tasks such as classification given that many of the observed features may be unrelated to the biological phenomenon of interest. In this way, feature selection algorithms can be used to determine the processes related to a biological mechanism, e.g., the host immune response to infection. Other downstream benefits of feature selection include data visualization and understanding, reduced storage requirements, and faster computations.

Multi-domain feature extraction addresses the problem of leveraging data from disparate sources and is related to the more general problem of multi-domain learning (MDL) [27]. In this paper we address a special case of MDL that involves the classification of data related to the host immune response to infection. The host consists of multiple lines of the Collaborative Cross mouse, the pathogen under consideration is Salmonella. The two domains consist of gene expression data collected from liver and spleen tissues. The proposed machine learning approach has the potential to identify novel biomarkers whose signals are too weak to be captured by analyzing domains individually. The methodology will be demonstrated by selecting potentially important biomarkers that appear to be amplified in strength by the multi-domain data synthesis for characterizing biological processes that exist simultaneously in different tissues.

In a variety of existing methods the designs vary from shallow to deep, with the networks optimized for regression, classification, dimensionality reduction or a combination of multiple tasks, i.e. multi-task learning (MTL) [26]. Most of the feature selection methods fall into the category of MLT methods, with the objective function considering the combination of different goals resulting in better generalization of results. However, the majority of them focus on a single domain feature selection. In this paper we suggest a new MDL method with a multi-task objective (MDL/MTL) function, or, in terms of [28], a multi-domain multi-task (MDMT) method. The MDMT methods have been used widely in Natural Language Processing applications but much less for the analysis of biological data sets.

The feature selection task is frequently performed by the introduction of l_p-norms with $p = 0, 1, 2$, or a combination of norms of weights at intermediate layers, possibly stacked closer to the input. The l_0 is of course appealing but the most expensive to compute [7,8]. However, in [24] it was shown that l_1-norm based methods can be quite competitive when applied to biological data, while not having the complications of dealing with l_0-norm. When applied to biological task, the l_0-norm based methods may require *a priori* knowledge of the size of a subset of biologically significant features, which is certainly not the case for many explorative tasks. Moreover, when the domains are very different the subsets of important features can be very different as well, hence it becomes even harder to approximate the number of selected features. At the same time, l_0-norm based methods haven't been battle-tested across domain as much as l_1-norm based methods, intuitively, the discrete distribution can negatively effect the alignment of disparate domains. Considering all the pros and cons, in this paper we opt to employ ℓ_1-norm based sparsity promotion.

The contributions of this paper include the following: we propose a new sparsity promoting MDMT architecture for feature selection; this approach uses a new masking term that restricts the features that contribute to the cost function; we demonstrate the utility of developed algorithm on gene expression dataset including liver and spleen domains; we conclude that our MDMT approach allows to find new features that are significantly discriminative only across two domains, i.e., are identified when the data is restricted to a single domain. The promise of this approach is an enriched picture of the host immune response that has the potential to lead to a better understanding of the biological process *across tissues*.

The organization of this paper is as follows: In **Related Work**, we present a review of the related articles, situating our study within the broader context of the field and highlighting key contributions from prior research. **Methodology** details the method we employed, outlining the design of a neural network behind our chosen approach and the techniques utilized for design. In **Data**, we describe the data used in our investigation, elaborating on pre-processing steps and labelled groups. **Experiment** delves into the computational experimental setup, discussing the training process, and criteria for rough-tuning. In **Results**, we present the results of our experiments, providing an interpretation and discussion of our findings. Finally, in **Conclusions**, we summarize the highlights of the research and contributions, the implications of our results, and potential avenues for future work in this domain.

2   Related Work

The majority of the feature selection methods study one domain. In what follows we survey a variety of different techniques and algorithms, including linear, non-linear methods and the methods exploiting neural networks. In one of the most influential papers [14], feature selection is cast as a regression task with l_1 regularization of the norm of discriminative vector. This has become a common approach, see also, [15–18]. The Lasso-type methods fail to capture nonlinear interactions between the features. The non-linear methods developed as the kernelized modifications of Lasso method showed decent efficiency when applied to biological data [19–21]. Note that there are also some other methods aiming to sparsify a signal in latent dimensions [22,23].

Deep Feature Selection DFS [7] is one of the first deep neural network algorithms designed specifically for feature selection. DFS employs a one-to-one sparsity layer at the input. The weights on these single connections are penalized with a ℓ_1-norm minimization of norm of weights of this layer in a spirit of [15]; the resulting non-zero weights in the sparsity layer correspond to the selected features. In [3] autoencoders are proposed for feature selection with a sparsity layer used in a fashion similar to DFS. Now the ℓ_1-norm of the weights of sparsity layer is minimized jointly with the reconstruction error. Concrete Autoencoders (CAE) [4] use a concrete selector layer as the first layer in autoencoder setting based on continuous relaxations of concrete random variables suggested in [5]. A supervised CAE method reported in [6] was apparently susceptible to overfitting with limited data. The FsNet paper [6] addressed this problem by introducing small weight-predictor networks. In terms of design [6] is one of the closest to the design developed in this paper, i.e., our method is based on two neural networks: autoencoder and classifier in latent space, but the approach to sparsification is different, and, most importantly, the method in [6] is designed for one domain and the implication of extracted features is quite different.

Most of the feature selection methods can be grouped into three broad categories: filter, wrapper, and embedded methods. In filter methods the features are typically scored, ranked and thresholded with respect to some classification task using different measures such as correlation and mutual information [1]. Filter methods can be very fast, but the quality of extracted features is poor in terms of robustness and adaptability to different datasets. The wrapper methods [2] are universal methods used on top of any learning algorithm based on practical heuristic search of a subset of d features in 2^d space providing the better performance for the underlying algorithm. They are universal and capable of obtaining great results given the large number of samples. For small datasets in high-dimensional space they tend to overfit, and the NP-hardness of the problem makes the computations prohibitively expensive. In embedded methods the feature selection process is typically performed concurrently with some learning algorithm. For example, Iterative Feature Removal (IFR) uses the absolute weights of a sparse SVM model as a criterion for selecting features from a high-dimensional biological data set [25]. Our paper along with the most related works falls into the embedded method category.

3   Methodology

The methods described above only address data residing in one domain. The major question that motivated this study was what are the biological features in datasets sampled from different domains that appear to be related to the host immune response to infection only when studied across domains, naturally leading the consideration of the domain alignment task along with the feature selection. With that said and with a general design of the network in mind we've been looking for the most capable in terms of domain alignment method in application to RNA data. This led us to [30], which is an MDMT method based on a pair of domain specific variational autoencoders (VAE's) [31] generating aligned embeddings for datasets of very different modalities (singe-cell RNA and Chromatin images), and we adapt this method now enhanced with sparsity promoting optimization constraints for feature selection. In order to find a universal representation across tasks, the MTL methods in deep neural networks [29] either improve the architecture of neural networks, or try to find a balance between concurrently trained objectives. This paper benefits from both since our network has shared subnets and at the same time we roughly fine-tune the coefficients used in [30] along with the contribution of sparsity promoting loss function. Utilizing both methods is also justified by the results of generalization of unbalanced optimization methods, e.g. [32–34], indicating that overall they don't outperform the naive approach when all loss functions are weighted with constant scalars.

Our proposed method is based on the network shown in Fig. 1, implemented in PyTorch and trained with the AdamW optimizer. The cost function is a weighted combination of objective functions associated with the reconstruction, classification and sparsification tasks. In our settings we can observe that during the early stages of training the algorithm is learning the shared representations of different domains such that similar samples group together regardless of domain of origin. At the later stages, the sparsification goal is becoming more important with a relatively higher contribution to the total objective and this behaviour continues until the conflicting tasks reach the balance and no further sparsification is possible without a significant loss in classification. We run the training process multiple times. In a spirit of embedded methods, we treat the resulting magnitude of the weights of sparse layer as indicating the importance of different features. However, for the post-processing we employ the frequency of selected features across all runs as it appears to be a more robust metric for feature importance.

It was mentioned before that our design is developed based on the network suggested in [30] with modifications. The classification task is performed in latent space as before, but the inputs of domain-specific VAE's are sparsified by the shared Sparsification Layer (**SL**) and, naturally, the VAE's are trained to reconstruct only these sparsified inputs. Note that in [30], a primary goal is the alignment of the domains in the latent space coupled with the reconstruction of hyper-dimensional RNA data. In contrast to [30], our main goal is the across-domains classification with the sparsification of inputs. The choice of variational modification of autoencoders was dictated by the distribution of latent space provided by this particular modification, allowing further indirect "easy" and relaxed alignment through the shared classifier. Hence, we not only train VAE's solely to reconstruct the sparsified signal, but also we omit the loss function minimizing the KL-divergence across domains from the original design. The resulting network consist of 4 subnets: shared between domains Sparse Layer (**SL**) and **Classifier** subnets, and two domain specific Variational AutoEncoders (**VAE1** and **VAE2**), as depicted in Fig. 1.

The **SL** is a one-to-one mapping: $x \rightarrow \mathbf{W} \odot x$, with the ℓ_1-norm of weights $\|\mathbf{W}\|_1$ penalty used to promote sparsity.

The **VAE1** and **VAE2** subnets are deep fully-connected networks with the following specifications:

- **Encoder**: 2 linear layers with 1024 nodes with batch normalization and Relu-activations, followed by 1 linear mapping to μ and σ living in 128-dimensional space
- **Decoder**: 2 linear layers with 1024 nodes with batch normalization and Relu-activations, followed by 1 layer mapping to input space

Each **Encoder** performs the mapping of the sparsified input to $\mathbf{R}^{128}$: $\mathbf{W} \odot x \rightarrow \mu, \sigma$, while the **Decoder** maps distribution in the latent space into the input space: $\mu + \sigma \rightarrow \tilde{x} \in \mathbf{R}^{34861}$, and the output is further masked by the frozen weights of the sparse layer $\mathbf{W}^* \odot \tilde{x}$ and fed to the MSE loss function of respective VAE.

The **Classifier** is the subnet consisting of 5 linear layers with 1024 nodes with Relu-activations, followed by 1 layer mapping to 2-dimensional space and 1 layer mapping to 1-dimensional space followed by a standard sigmoidal activation function. It is trained to classify embeddings of inputs from both domains in their respective latent spaces.

The overall objective function is a weighted sum of objective functions for different tasks including reconstruction, normalization, classification and sparsi-

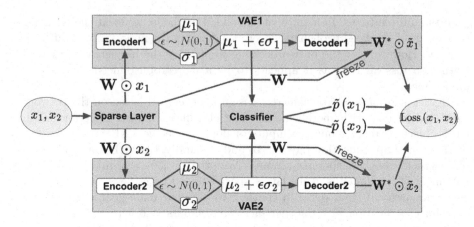

Fig. 1. Network design

fication, namely,

$$
\begin{aligned}
\text{Loss}\,(x_1, x_2) = {} & \alpha \cdot (\text{Loss}_{\text{rec}}\,(x_1) + \text{Loss}_{\text{rec}}\,(x_2)) + \\
& \beta \cdot (\text{Loss}_{\text{var}}\,(x_1) + \text{Loss}_{\text{var}}\,(x_2)) + \\
& \gamma \cdot (\text{Loss}_{\text{class}}\,(x_1) + \text{Loss}_{\text{class}}\,(x_2)) + \\
& \theta \cdot \text{Loss}_{\text{sparse}}
\end{aligned}
\tag{1}
$$

where

$$
\begin{aligned}
\text{Loss}_{\text{rec}}\,(x) &= MSE(\mathbf{W} \odot x, \mathbf{W}^* \odot \tilde{x}) \\
\text{Loss}_{var}(x) &= D_{KL}\left(N\left(\mu, \sigma^2 I\right), N(0, I)\right) \\
\text{Loss}_{\text{class}}\,(x) &= \log \text{Loss}\,(p(x), \tilde{p}(x)) \\
\text{Loss}_{\text{sparse}} &= \|\mathbf{W}\|_1
\end{aligned}
\tag{2}
$$

The block-scheme of one training process is given by Algorithm refalg:training. After the rough-tuning of the hyper-parameters from [30] along with the sparsity contribution and parameters for optimizers, we set the following hyper-parameters:

- $\alpha, \beta, \gamma, \theta = 10,\ 10^{-4},\ 1,\ 10^{-4}$
- LR's for AdamW optimizer for VAE's, SL and Classifier $= 10^{-4}$
- all other parameters for AdamW optimizer are set to Pytorch default

4   Data

The data consists of mice bulk RNA sequences extracted from two different tissues: spleen and liver, used as two different domains for the purposes of this paper. The mice that were exposed to Salmonella infection were monitored and

---

Algorithm 1: Training

Input	:	$\theta, \alpha, \beta, \gamma$ weight of loss functions
		$argsOpt1, argsOpt2$ parameters of Adam optimizers for VAE's
		$argsOptCls$ parameters of Adam optimizers for Classifier
		$argsOptSL$ parameters of Adam optimizers for SL

Output : **W** - weights of SL

Data:

   $[x_1, x_2]$ list of pairs of equally seized batches sampled from both datasets
   20 000 elements, i.e epochs of training

Initialize : SL, VAE1, VAE2, Classifier (initialize subnets)

   OptSL = AdamW($argsSL$)
   Opt1 = AdamW($argsOpt1$)
   Opt2 = AdamW($argsOpt2$)
   OptCls = AdamW($argsCls$)

1 **for** $x_1, x_2 \in [x_1, x_2]$ **do**
2     $sp(x_1), sp(x_2) = \mathbf{W} \odot x_1, \mathbf{W} \odot x_2$
3     $\tilde{x}_1, \mu_1 + \epsilon\sigma_1 = \text{VAE1}(sp(x_1))$
4     $\tilde{x}_2, \mu_2 + \epsilon\sigma_2 = \text{VAE2}(sp(x_2))$
5     $\tilde{p}(x_1) = \text{Classifier}(\mu_1 + \epsilon\sigma_1)$
6     $\tilde{p}(x_2) = \text{Classifier}(\mu_2 + \epsilon\sigma_2)$
7   Calculate Loss as in (1)
8   Backpropagate Loss
9   Step all optimizers

---

categorized by health status as tolerant, resistant, susceptible, or delayed susceptible, the latter two being related to strains unifying the mice who died within 1 or 3 weeks respectively. In all our experiments we combine these latter two groups into one susceptible group. With this new labelling we have 31 and 9 tolerant samples, 27 and 7 resistant samples and 90 and 53 susceptible samples for spleen and liver domains respectively for all infected mice. Also, the data includes control samples representing the mice who had never been exposed to infection labeled as "never infected". This group accounts for 104 samples, with 93 samples from spleen and 11 samples from liver. Phenotypes of these samples are determined based on their genetic strains. Initial bulk RNA dataset was TMM-normalized, the outliers and duplicates have been detected and dropped out. Finally, the domain-specific data, i.e. combined RNA data for samples from spleen and combined RNA data for samples from liver, have been z-scored for each domain separately and filtered for common across tissues genes in all feature selection algorithms resulting in data samples consisting of 34,861 genes, i.e., the dimension of the input space.

5 Experiment

We consider three distinct types of experiment. In the first type the goal is to extract a small subset of features that discriminate among the phenotypes susceptible and tolerant. The second type extracts features discriminating among the phenotypes susceptible and resistant. The third type extracts the features discriminating between infected and never infected mice. The exact Python code with the training models and post-processing utilities is available at [35]. With a use of Ray package [36] the experiment was run on 16 V100 GPU's in a multiprocessing mode for 10 different random samplings of 85% of data, with 90 different weights initializations for each, summing up to 900 runs. You can see the typical evolution of training process including 60,000 epochs for all three experiments in Figs. 2, 3, 4.

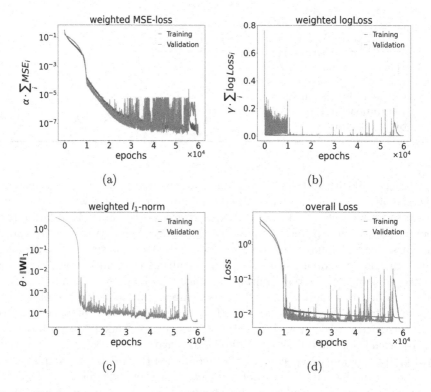

Fig. 2. Weighted components of overall losses for phenotypes tolerant versus susceptible across domain experiment: (a) - reconstruction errors, (b) - classification errors, (c) - sparsity loss, (d) - overall loss.

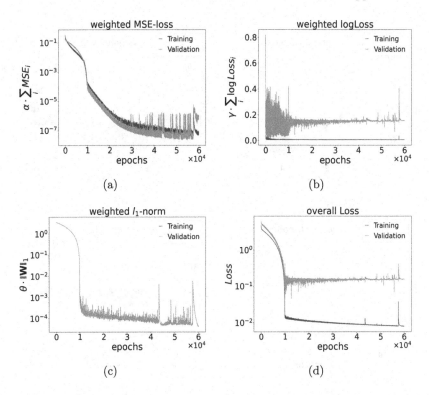

Fig. 3. Weighted components of overall losses for phenotypes resistant versus susceptible across domain experiments: (a) reconstruction error, (b) classification error, (c) sparsity loss, (d) total loss.

The number of epochs was fixed at 20,000 based on the indication of the flattening of the sparsity curves as long as it doesn't effect the accuracy of classification and the reconstruction loss is relatively small. Again, the reconstruction is not in the primary focus of this method, i.e., it wasn't the major task to consider when deciding on the number of epochs, especially, since the required for the across-domain classification alignment in the latent space was typically achieved even after 20,000 epochs, see the Fig. 5. The visualization of loss indicates that further sparsification slightly decreases the classification accuracy for two out of three experiments, by limiting the number of epochs we also prevent this long-run negative effect. The PCA images for all figures represent the PCA of combined representations of both domains in their bottleneck layers, and indicate a proper clustering and separability across domains, i.e., good alignment in latent space.

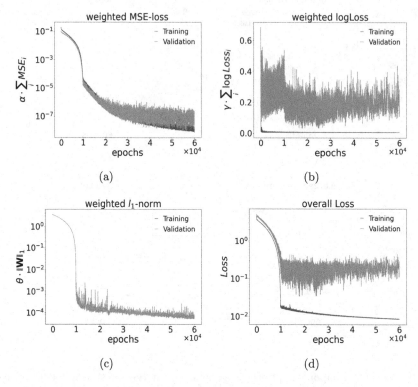

Fig. 4. Weighted components of overall losses for infected mice versus never infected mice across domain experiment: (a) reconstruction error, (b) classification error, (c) sparsity loss, (d) total loss.

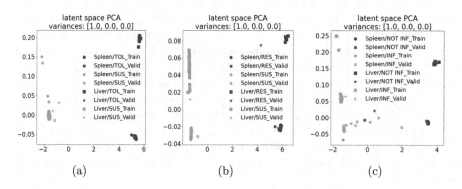

Fig. 5. Latent space PCA's for all three across domain experiments: (a) phenotypes tolerant (TOL) versus susceptible (SUS), (b) phenotypes resistant (RES) versus susceptible (SUS), (c) infected (INF) mice versus never infected mice (NOT INF).

6 Results

The post-processing of the sparsity layer weights was conducted in a same way for all 3 experiments. Firstly, all the weights across 900 runs have been aggregated, normalized and the Elbow Method was applied to find a threshold, and later for each run the weights below the threshold were set to zero. At the next step we calculated the frequencies of features appearing in subsets of features with non-zero weights across all runs. The resulting distributions of frequencies are shown in Fig. 6, along with resulting number of features selected in two consecutive steps by Elbow Method.

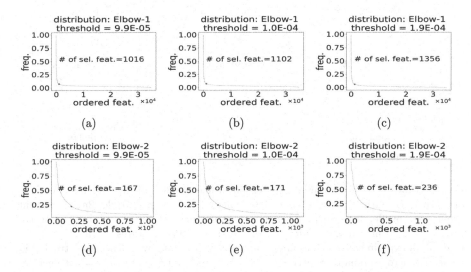

Fig. 6. Results from the across domain (MDMT) experiments. The features selected are ordered by frequency for 3 experiments: panels (a) and (d) correspond to the phenotypes tolerant versus susceptible; panels (b) and (e) correspond to resistant versus susceptible; panels (c) and (f) correspond to infected versus never infected.

Importantly, in addition to the cross domain learning, we also selected features for each domain separately for all three experiments. This allows us to evaluate the distinct characteristics of single domain and multi-domain alignment for feature extraction. In Fig. 7 panels (a), (b), and (c), we can see the distribution of features for all three across domain experiments grouped by overlapping with features selected in one-domain experiments. "Both" features is a subset of these features that also appear in both the spleen and liver domains. "None" refers to the features that are the features that only appear in across domain results. "Spleen" are the features appearing in both across domains and spleen domain results but not in the liver domain results. Finally, "Liver" features are the features appearing in both across domains and liver domain results but not in the spleen domain results. We can see that for all the experiments

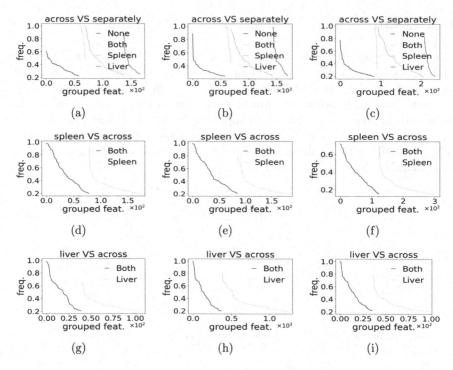

Fig. 7. Grouped by overlapping distribution of features for all experiments. Columns are associated with experiments, i.e. 1st - tolerant versus susceptible, 2nd - resistant versus susceptible, 3rd - infected versus never infected. (a),(b),(c) distributions of features extracted in across domains experiments; (d),(e),(f) features extracted in separate spleen experiment, features are grouped by overlapping with features extracted in respective across domains experiment: "Both" - overlapping, "Spleen" - not overlapping; (g),(h),(i) features extracted in separate liver experiment, features are grouped by overlapping with features extracted in respective across domains experiment: "Both" - overlapping, "Liver" - not overlapping.

apart from the features captured from domain-specific experiments some new highly-weighted features were captured in across domain experiments (the blue line denoting "None"). These features that are only present in the across domain experiment reflect the new information being captured by the proposed method. These correspond to biomarkers that we suspect have a potentially unique role in the host response to infection.

Figures 7, panels (d),(g),(e),(h),(f), and (i) show the distributions of features extracted in one domain experiments grouped by overlapping with respective across domains results. Even though the majority of the most discriminative features extracted in separate experiments are also extracted in across domain experiments, some highly-weighted features extracted from one domain experiments are apparently not captured in respective across domains experiments. These results might indicate that the improvement in robustness is needed, but at the same time they may contain biological insights about the across domain importance of some features that couldn't be obtained from studying only one domain for Salmonella infection or about the difference in manifestations of infection in different tissues. Based on the results from additional experiments with other datasets most likely the latter is true.

Returning to Fig. 5, we see the results of the embedding of the sparsified input in a two-dimensional latent space using the features found in the across domain architecture for all three experiments. We observe that the tolerant and resistant versus susceptible experiments give excellent classification on test data in the latent space. In contrast, infected versus control mice are not as easily discriminated.

7   Conclusions

This paper proposes a novel architecture for multi-domain, multi-task feature selection. This area of research has a relatively small literature, possibly because of the complexity associated with simultaneously exploring multiple incommensurate measurement domains. The proposed approach leverages prior art in multi-domain learning while adding a masked feature selection approach that serves to identify biologically relevant aspects of the host immune response to infection. We demonstrate that the demands of the MDMT problem formulation can be successfully addressed with the proposed architecture. Further, the application of the approach leads to the discovery of novel biomarkers whose signals appear to be amplified by the multi-domain approach; indeed, a fraction of these biomarkers do not appear in either single experiment. Hence, this approach holds the promise of generating new biological insights that might go undetected using single domain methodologies. Additionally, we observed the MDMT features provide excellent classification results between susceptible and tolerant or resistant phenotypes.

There are several possible modifications of our method. In this paper the optimization problem descends in the direction of the gradient of weighted objectives. There is growing evidence that isolating descent directions to improve individual cost functions may lead to improved solutions. Additionally, alternative data reduction and reconstruction mappings could be explored reflecting recent developments in deep neural networks including graph convolutional neural networks or transformers. This preliminary work is focused on algorithm development; we propose to explore the biological ramifications of the biomarkers in future work.

We believe further development in these directions could lead to even more efficient methods and provide biologists with a new perspective on understanding the evolution of infections in tissues.

Acknowledgements. We would like to thank Helene Andrews-Polymenis and David Threadgill for providing the data for this investigation.

References

1. Blum, A.L., Langley, P.: Selection of relevant features and examples in machine learning. Artif. Intell. **97**(1), 245–271 (1997)
2. John, G.H., Kohavi, R., Pfleger, K.: Irrelevant features and the subset selection problem. In: Proceedings of 11th International Conference on Machine Learning, pp. 121–129 (1994)
3. Han, K., Wang, Y., Zhang, C., Li, Ch., Xu, C.: AutoEncoder Inspired Unsupervised Feature Selection, arXiv, arXiv:1710.08310 (2017). https://arxiv.org/abs/1710.08310
4. Fatih Balin, M., Abid, A., Zou, J.Y.: Concrete autoencoders: differentiable feature selection and reconstruction. In: ICML (2019)
5. Maddison, C. J., Mnih, A., Teh, Y. W.: The Concrete Distribution: A Continuous Relaxation of Discrete Random Variables, arXiv, arXiv:1611.00712 (2016). https://arxiv.org/abs/1611.00712
6. Singh, D., et al.: Fsnet: feature selection network on high-dimensional biological data. arXiv, arXiv:2001.08322 (2020). https://arxiv.org/abs/2001.08322
7. Li, Y., Chen, C., Wasserman, W.: Deep feature selection: theory and application to identify enhancers and promoters. J. Comput. Biol. **23**(5), 322–336 (2016)
8. Feng J., Simon, N.: Sparse-Input Neural Networks for High-dimensional Nonparametric Regression and Classification, arXiv, arXiv:1711.07592 (2017). https://arxiv.org/abs/1711.07592
9. Hinton, G., Salakhutdinov, R.: Reducing the dimensionality of data with neural networks. Science **313**, 504–507 (2006)
10. Salah, R., Pascal, V., Xavier, M., Xavier, G, Yoshua, B.: Contractive auto-encoders: explicit invariance during feature extraction. In: Proceedings of the 28th International Conference on Machine Learning, ICML 2011, New York, pp. 833–840 (2011)
11. Hinton, G., Osindero, S., Teh, Y.: A fast learning algorithm for deep belief nets. Neural Comput. **18**, 1527–1554 (2006)
12. Yamada, Y., Lindenbaum, O., Negahban, S., Kluger, Y.: Feature selection using stochastic gates. In: International Conference on Machine Learning, PMLR, vol. 119, pp. 10648–10659 (2020)
13. Lemhadri, I., Ruan, F., Abraham, L., Tibshirani, R.: Lassonet: a neural network with feature sparsity. J. Mach. Learn. Res. **22**(127), 1–29 (2021)
14. Tibshirani, R.: Regression shrinkage and selection via the lasso. J. Roy. Stat. Soc. B **58**, 267–288 (1996)
15. Zou, H., Hastie, T.: Regularization and variable selection via the elastic net. J. R. Stat. Soc. Series B Stat. Methodol. **67**(2), 301–320 (2005)
16. Tibshirani, R., Saunders, M., Rosset, S., Zhu, J., Knight, K.: Sparsity and smoothness via the fused lasso. J. Roy. Stat. Soc. B (Stat. Methodol.) **67**(1), 91–108 (2005)
17. Zou, H.: The adaptive lasso and its oracle properties. J. Amer. Stat. Assoc. **101**(476), 1418–1429 (2006)
18. Meinshausen, N.: Relaxed lasso. Comput. Stat. Data Anal. **52**(1), 374–393 (2007)
19. Yamada, M., Jitkrittum, W., Sigal, L., Xing, E.P., Sugiyama, M.: High-dimensional feature selection by feature-wise kernelized lasso. Neural Comput. **26**(1), 185–207 (2014)
20. Liu, H., Wasserman, L., Lafferty, J.D.: Nonparametric regression and classification with joint sparsity constraints. In: Proceedings of Advances in Neural Information Processing Systems, pp. 969–976 (2009)

21. Shevade, S.K., Keerthi, S.S.: A simple and efficient algorithm for gene selection using sparse logistic regression. Bioinformatics **19**(17), 2246–2253 (2003)
22. Chan, A.B., Vasconcelos, N., Lanckriet, G.R.G.: Direct convex relaxations of sparse SVM. In: International Conference on Machine Learning (2007)
23. Gurram, P., Kwon, H.: Optimal sparse kernel learning in the empirical kernel feature space for hyperspectral classification. IEEE J. Sel. Top. Appl. Earth Obs. Remote Sens. **7**(4), 1217–1226 (2014)
24. Ghosh, T., Karimov, K. ,Kirby, M.: Sparse linear centroid-encoder: a biomarker selection tool for high dimensional biological data. In: IEEE International Conference on Bioinformatics and Biomedicine (BIBM), pp. 3012–3019 (2023)
25. O'Hara, S., Wang, K., Slayden, R.A. et all.: Iterative feature removal yields highly discriminative pathways. BMC Genom. **14**(1), 1–15 (2013)
26. Caruana, R.: Multitask learning. Mach. Learn. **28**(1), 41–75 (1997)
27. Joshi, M., Cohen, W.W., Dredze M., Rosé, C.P.: Multi-domain learning: when do domains matter? In: Proceedings of the 2012 Joint Conference on Empirical Methods in Natural Language Processing and Computational Natural Language Learning, pp. 1302–1312 (2012)
28. Yang, Y., Hospedales, T.M.: A unified perspective on multi-domain and multi-task learning. arXiv, arXiv:1412.7489 (2014). https://arxiv.org/abs/1412.7489
29. Ruder, S., An Overview of Multi-Task Learning in Deep Neural Networks. arXiv, arXiv:1706.05098 (2017). https://arxiv.org/abs/1706.05098
30. Yang, K.D., Belyaeva, A., Venkatachalapathy, S., et al.: Multi-domain translation between single-cell imaging and sequencing data using autoencoders. Nat. Commun. **12**(31) (2021). https://doi.org/10.1038/s41467-020-20249-2
31. Kingma, D.P., Welling, M.: Auto-Encoding Variational Bayes. arXiv, arXiv:1312.6114 (2013). https://arxiv.org/abs/1312.6114
32. Chen, Z., Badrinarayanan, V., Lee, C.Y., Rabinovich, A.: GradNorm: gradient normalization for adaptive loss balancing in deep multitask networks. In: International Conference on Machine Learning (2017)
33. Sener, O., Koltun, V.: Multi-task learning as multi-objective optimization. In: Neural Information Processing Systems (2018)
34. Guo, M., Haque, A., Huang, D.-A., Yeung, S., Fei-Fei, L.: Dynamic task prioritization for multitask learning. In: Ferrari, V., Hebert, M., Sminchisescu, C., Weiss, Y. (eds.) ECCV 2018. LNCS, vol. 11220, pp. 282–299. Springer, Cham (2018). https://doi.org/10.1007/978-3-030-01270-0_17
35. https://github.com/kkarimov/iccs2024
36. https://github.com/ray-project/ray

Neural Dynamics in Parkinson's Disease: Integrating Machine Learning and Stochastic Modelling with Connectomic Data

Hina Shaheen[1]([⊠])(iD) and Roderick Melnik[2](iD)

[1] Faculty of Science, University of Manitoba, Winnipeg, MB R3T 2N2, Canada
Hina.Shaheen@umanitoba.ca
[2] MS2Discovery Interdisciplinary Research Institute, Wilfrid Laurier University,
Waterloo, ON N2L 3C5, Canada
rmelnik@wlu.ca

Abstract. Parkinson's disease (PD) is a neurological disorder defined by the gradual loss of dopaminergic neurons in the substantia nigra pars compacta, which causes both motor and non-motor symptoms. Understanding the neuronal processes that underlie PD is critical for creating successful therapies. This work presents a novel strategy that combines machine learning (ML) and stochastic modelling with connectomic data to understand better the complicated brain pathways involved in PD pathogenesis. We use modern computational methods to study large-scale neural networks to identify neuronal activity patterns related to PD development. We aim to define the subtle structural and functional connection changes in PD brains by combining connectomic with stochastic noises. Stochastic modelling approaches reflect brain dynamics' intrinsic variability and unpredictability, shedding light on the origin and spread of pathogenic events in PD. We created a hybrid modelling formalism and a novel co-simulation approach to identify the effect of stochastic noises on the cortex-BG-thalamus (CBGTH) brain network model in a large-scale brain connectome. We use Human Connectome Project (HCP) data to elucidate a stochastic influence on the brain network model. Furthermore, we choose areas of the parameter space that reflect both healthy and Parkinsonian states and the impact of deep brain stimulation (DBS) on the subthalamic nucleus and thalamus. We infer that thalamus activity increases with stochastic disturbances, even in the presence of DBS. We predicted that lowering the effect of stochastic noises would increase the healthy state of the brain. This work aims to unravel PD's complicated neuronal activity dynamics, opening up new options for therapeutic intervention and tailored therapy.

Keywords: Brain networks · Machine learning · Laplacian operator · Neural dynamics · Wiener process · Neurodegenerative disorders

Stochastic modelling of brain networks.

© The Author(s), under exclusive license to Springer Nature Switzerland AG 2024
L. Franco et al. (Eds.): ICCS 2024, LNCS 14835, pp. 46–60, 2024.
https://doi.org/10.1007/978-3-031-63772-8_4

1   Introduction

Parkinson's disease (PD) stands as one of the most prevalent neurodegenerative disorders, characterized by the progressive loss of dopaminergic neurons in the substantia nigra pars compacta, leading to debilitating motor symptoms such as tremors, rigidity, and bradykinesia [1,2]. Despite significant advancements in therapeutic approaches, including pharmacological interventions and DBS, our understanding of the complex interplay between neuronal dynamics, disease progression, and treatment outcomes remains incomplete [3,4].

Recent years have witnessed a paradigm shift in neuroscientific research, driven by the convergence of computational methodologies, artificial intelligence (AI) techniques, including ML tools, and advancements in neural engineering [5]. Among these approaches, ML holds promise in deciphering intricate patterns within vast datasets, offering insights into disease mechanisms and personalized treatment strategies. Concurrently, DBS has emerged as a potent therapeutic modality, modulating aberrant neuronal circuits to alleviate motor symptoms in PD patients [6]. The discipline of ML, which is a subdivision of AI, has experienced rapid growth and has recently impacted medical fields like neurosurgery [5]. A literature review focusing on the application of ML in DBS has not yet been published despite the field's growing interest in the area.

In parallel, stochastic modelling has gained traction to capture the inherent randomness and complexity of neuronal activity [7], shedding light on the dynamic nature of neurological disorders such as PD [8]. By integrating these diverse methodologies, researchers aim to unravel the underlying mechanisms governing neuronal dysfunction in PD, thereby paving the way for more effective interventions and improved patient outcomes. Furthermore, recent studies employing multiscale mathematical modelling have highlighted the efficacy of nonlinear reaction-diffusion equations in discerning neuropathological conditions [9]. Notably, connectomic data has revealed the extensive impact of DBS across various cortical and subcortical regions [10]. Discrete brain network models operating in a spatio-temporal domain elucidate the dynamics of model parameters, thereby simulating large-scale brain activity [3,10,11].

In essence, neurons constitute the fundamental units of our nervous system, with the basal ganglia (BG) comprising three critical nuclei: the subthalamic nucleus (STN), the globus pallidus internus (GPi), and the globus pallidus externus (GPe) [2]. Neurons utilize neurotransmitters for intercellular communication and employ action potentials to transmit signals within the cell upon receiving external stimuli (I_{app}). Notably, using a reduced number of neurons, such as 10 neurons per nucleus, yields similar outcomes to those obtained with 100 neurons. Thus, each nucleus in our study comprises 10 cells [2].

In the present study, we adopted a novel co-simulation approach utilizing a modified Rubin-Terman model for subcortical brain regions surrounding the basal ganglia across the entire cerebral hemisphere from our previous study [2]. This approach incorporates stochastic noise, explicitly incorporating a Wiener process, to capture additional variability and complexity in brain dynamics [12,13]. Therefore, we integrate a discrete brain network model for each cortical region, incorporating stochastic noise at the macroscopic scale to better align with experimental data on neuron firing characteristics. Following the strategy outlined in [2], we explore critical aspects of the model dynamics, including the influence of stochastic noise on healthy and diseased states. Our findings demonstrate that the eigendecomposition of the Laplace operator, incorporating stochastic noise, can predict the collective dynamics of human brain activity at the macroscopic scale [2]. These findings suggest that the disruption of multivariate connection-wise functional connectivity patterns holds promise for discriminating PD patients based on cognitive status, supporting previous observations of altered functional connectivity associated with cognitive impairment in PD. Our research uncovers significant findings regarding the influence of stochastic noise on brain dynamics. Specifically, we observed that in the presence of stochastic noise, the activity of the thalamus reaches a critical threshold, contrasting with scenarios lacking noise. Furthermore, our analysis revealed that stochastic noise amplifies the membrane potential of the thalamus, potentially exacerbating brain disease states. This effect of stochastic noise is pronounced, leading to burst oscillations in the membrane potential across all selected regions, even in the presence of DBS. Our study highlights the brain's resilience as it endeavours to maintain a healthy state for a prolonged period following DBS despite stochastic noise.

The rest of the paper is organized as follows. In Sect. 2, we describe our model in its different components: (i) a discrete and (ii) a stochastic discrete brain network model of the CBGTH. Section 3 presents numerical results based on the developed stochastic discrete brain network model for the cortex-thalamus-basal-ganglia systems. The computational results were obtained using codes developed in C-language and SHARCNET supercomputer facilities, and the simulation results were visualized in MATLAB. Implications of these results and their importance are discussed in Sect. 4. Finally, we conclude our findings and outline future directions in Sect. 5.

2   Methods

This section highlights the discrete and stochastic brain network model of CBGTH. In this section, we present (a) the discrete model of the CBGTH network mediated by Laplacian terms and (b) the stochastic brain network model of the CBGTH system, giving particular attention to stochastic noises. We evaluated the behaviour of stochastic noises in the brain regions such as Gpe, GPi, STN and thalamus (TH) and firing patterns under healthy and pathological states to validate the features of the CBGTH model. We then use data to examine the firing rates of the coupled neurons on each node in the brain network. Finally, the effects of noise in the presence of DBS on STN and thalamus are evaluated.

2.1   Discrete Brain Network Model of CBGTH

The network comprises nodes delineated within the brain connectome, often corresponding to established brain atlas regions. We aim to construct a model capable of capturing temporal voltage variations across different nodal points.

The brain connectome is represented as a weighted network $\mathcal{G}$ consisting of V nodes and E edges, derived from diffusion tensor imaging (DTI) and tractography techniques [2], as adopted from the HCP dataset. The edges of this network symbolize axonal bundles within white-matter tracts. To generate a network approximation of the diffusion terms, we utilize a weighted graph Laplacian, where the weights of the weighted adjacency matrix $\mathbf{W}$ are determined by the ratio of the mean fibre number n_{ij} to the mean squared length l_{ij}^2 connecting nodes i and j, expressed as:

$$W_{ij} = \frac{n_{ij}}{l_{ij}^2}, \quad i = 1, \ldots, V. \tag{1}$$

These weights align with the inverse length-squared dependency observed in the canonical discretization of the continuous Laplace (diffusion) operator [2]. Additionally, we define the diagonal weighted degree matrix as:

$$D_{ii} = \sum_{j=1}^{V} W_{ij}, \quad i, j = 1, \ldots, V. \tag{2}$$

Furthermore, the graph Laplacian $\mathbf{L}$ with (i, j)-entry is defined as:

$$L_{ij} = \rho(D_{ij} - W_{ij}), \quad i, j = 1, \ldots, V, \tag{3}$$

where ρ represents the diffusion coefficient.

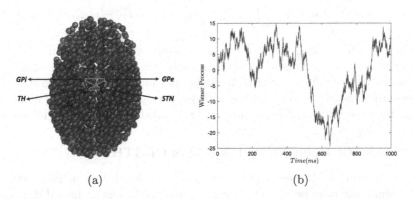

(a)   (b)

Fig. 1. (Color online) (a) Discrete brain network connectome in a healthy condition (left) (axial view from bottom). The four nodes are STN, GPe, GPi, and TH, and we replaced the spiking node "cortex" with the whole brain connectome (b) Stochastic noises applied to STN, GPe, GPi, and TH (Color figure online)

The adjacency matrix for simulations is derived from diffusion tensor magnetic resonance images obtained from 418 healthy HCP subjects sourced from the Budapest Reference Connectome v3.0 [2]. Figure 1(a) showcases a network composed of $V = 4$ nodes and $E = 6$ edges representing brain regions like the putamen, globus pallidus, and thalamus. Each node is assumed to occupy a surface area of $1.5\,\mathrm{cm}^2$. Each node linked with STN, GPi, GPe, and TH carries the voltage v^{sn}, v^{gi}, v^{ge}, and v^{th}, respectively. The network equations for the continuous model take the form of a system of first-order ordinary differential equations as follows:

$$\frac{dv^{sn}}{dt} = -d_{v^{sn}} \sum_{k=1}^{V} L_{1k}v_k + \frac{1}{c_m}\left(-I_{Na}^{sn} - I_K^{sn} - I_L^{sn} - I_T^{sn} - I_{Ca}^{sn} - I_{ahp}^{sn} - I_{ge \to sn} + I_{snapp} \right),$$
$$(4)$$

$$\frac{dv^{gi}}{dt} = -d_{v^{gi}} \sum_{k=1}^{V} L_{2k}v_k + \frac{1}{c_m}\left(-I_{Na}^{gi} - I_K^{gi} - I_L^{gi} - I_T^{gi} - \tag{5}$$
$$I_{Ca}^{gi} - I_{ahp}^{gi} - I_{sn \to gi} - I_{ge \to gi} + I_{giapp} \right),$$

$$\frac{dv^{ge}}{dt} = -d_{v^{ge}} \sum_{k=1}^{V} L_{3k}v_k + \frac{1}{c_m}\left(-I_{Na}^{ge} - I_K^{ge} - I_L^{ge} - I_T^{ge} - \tag{6}$$
$$I_{Ca}^{ge} - I_{ahp}^{ge} - I_{sn \to ge} - I_{ge \to ge} + I_{geapp} \right),$$

$$\frac{dv^{th}}{dt} = -d_{v^{th}} \sum_{k=1}^{V} L_{4k}v_k + \frac{1}{c_m}\left(-I_{Na}^{th} - I_K^{th} - I_L^{th} - I_T^{th} - I_{gi \to th} + I_{smc} \right), \tag{7}$$

with non-negative initial conditions for all variables v^{sn}, v^{gi}, v^{ge}, and v^{th}. Additionally, $d_{v^{sn}}, d_{v^{gi}}, d_{v^{ge}}$, and $d_{v^{th}}$ represent the diffusion terms corresponding to each node. The weights in the weighted adjacency matrix represent the spread of transneuronal degeneration from one node to its neighbours. Next, we introduce stochastic noise into the discrete brain network model to observe its influence.

2.2   Stochastic Brain Network Model of CBGTH

The integration of ML techniques with stochastic modelling in brain studies holds significant promise for advancing our understanding of neural dynamics and function [14]. In this section, we develop a discrete brain network model incorporating the addition of stochastic noise. The noise levels are crucial for ensuring the proper functioning of signals within the nervous system [15]. Studies have suggested that in computational models of neurodegenerative conditions such as PD, increased external noise levels are necessary for optimal function, reflecting the aging process and reduced plasticity [16]. Consequently, noise stimulation could be an alternative therapeutic approach for alleviating PD symptoms [15]. Therefore, based on the model presented in Sect. 2.1, we have added the noise terms as follows:

$$\frac{dv^{sn}}{dt} = -d_{v^{sn}} \sum_{k=1}^{V} L_{1k} v_k + \frac{1}{c_m} \bigg(-I_{Na}^{sn} - I_K^{sn} - I_L^{sn} - I_T^{sn} - I_{Ca}^{sn} - \tag{8}$$

$$I_{ahp}^{sn} - I_{ge \to sn} + I_{snapp} \bigg) + \sigma_1 \cdot dW_1(t),$$

$$\frac{dv^{gi}}{dt} = -d_{v^{gi}} \sum_{k=1}^{V} L_{2k} v_k + \frac{1}{c_m} \bigg(-I_{Na}^{gi} - I_K^{gi} - I_L^{gi} - I_T^{gi} - I_{Ca}^{gi} - \tag{9}$$

$$I_{ahp}^{gi} - I_{sn \to gi} - I_{ge \to gi} + I_{giapp} \bigg) + \sigma_2 \cdot dW_2(t),$$

$$\frac{dv^{ge}}{dt} = -d_{v^{ge}} \sum_{k=1}^{V} L_{3k} v_k + \frac{1}{c_m} \bigg(-I_{Na}^{ge} - I_K^{ge} - I_L^{ge} - I_T^{ge} - I_{Ca}^{ge} - \tag{10}$$

$$I_{ahp}^{ge} - I_{sn \to ge} - I_{ge \to ge} + I_{geapp} \bigg) + \sigma_3 \cdot dW_3(t),$$

$$\frac{dv^{th}}{dt} = -d_{v^{th}} \sum_{k=1}^{V} L_{4k} v_k + \frac{1}{c_m} \bigg(-I_{Na}^{th} - I_K^{th} - I_L^{th} - I_T^{th} - I_{gi \to th} + I_{smc} \bigg) + \sigma_4 \cdot dW_4(t),$$

$$\tag{11}$$

where $dW_i(t)$ represents the increment of the Wiener process $W_i(t)$ and σ_i are the scaling factors (representing the intensity of the noise) for each equation. When numerically integrating these stochastic differential equations, we generated increments of the Wiener process at each time step dt to represent the stochastic component using the Euler-Maruyama method. Incorporating noise into the CBGTH system within a discrete brain network model provides valuable insights into how the brain functions [16]. Figure 1(b) showcases a Wiener process or stochastic noises added into the CBGTH system. Since noise is present throughout various neural processes, from perceiving sensory signals to generating motor responses, it profoundly affects neuronal dynamics. Therefore, understanding the impact of noise is crucial for comprehending the brain's behaviour [17]. The significance of this impact will be explored further in the following Sect. 3. Moreover, the DBS current is added to the spatio-temporal model to the membrane potential equations of STN as follows:

$$\frac{dv^{sn}}{dt} = -d_{v^{sn}} \sum_{k=1}^{V} L_{1k} v_k + \frac{1}{c_m} \bigg(-I_{Na}^{sn} - I_K^{sn} - I_L^{sn} - I_T^{sn} - I_{Ca}^{sn} - \tag{12}$$

$$I_{ahp}^{sn} - I_{ge \to sn} + I_{snapp} + I_{DBS} \bigg) + \sigma_1 \cdot dW_1(t),$$

where $c_m = 1\mu F/cm^2$ and I_{DBS} is adopted from [2]. According to Eq. (12), the DBS electrode has been applied to the STN node in the discrete brain network connectome. The relevant parameters are given in Table 1 (the other relevant parameters are adopted from [2], (pd is a parameter, and $pd = 0$ indicates that the network is in healthy states, while $pd = 1$ shows that the network is in Parkinsonian states). The $--$ represents no connection to neurons.

Table 1. Parameter set for the CBGTH network [2].

	STN neuron	GPe/GPi neuron	TH neuron
I_{Ca}	$2(c^2)(v-140)$	$0.15(s_\infty(v))^2(v-120)$	$--$
I_{ahp}	$20(v+80)\,(w/(w+15))$	$10(v+80)(w/(w+10))$	$--$
$I_{ge\to sn}$	$0.5S_{ge\to sn}(v+85)$	$--$	$--$
$I_{ge\to ge}$	$--$	$0.5Sge \to ge(v+85)$	$--$
$I_{ge\to gi}$	$--$	$0.5S_{ge\to gi}(v+85)$	$--$
$I_{sn\to ge}$	$--$	$0.15S_{sn\to ge}v$	$--$
$I_{sn\to gi}$	$--$	$0.15S_{sn\to gi}v$	$--$
$I_{gi\to th}$	$--$	$--$	$0.112S_{gi\to th}(v+85)$
I_{snapp}	$33 - 10pd$	$--$	$--$
I_{giapp}		$22 - 6pd$	$--$
I_{geapp}		$21 - 13pd + (-1.5)$	$--$

3   Results

In this section, we will investigate how stochastic noise impacts the CBGTH system in healthy and PD brain states by integrating ML and stochastic modelling with connectomic data.

Importantly, noise introduces stochastic fluctuations into the brain network, affecting the timing and reliability of neural signal transmission [18,19]. In the context of the basal ganglia-thalamocortical circuit, where precise timing is crucial for motor control and cognitive processes, the impact of noise may lead to alterations in information processing and integration [18]. Moreover, neural noise, originating from various sources such as sensory input, cellular processes, and electrical activity, significantly influences the functioning of the nervous system. While it can hinder information processing, it also contributes to brain function by shaping functional networks, enhancing synchronization, and impacting task performance [19]. The brain's dynamics, characterized by subject-specific parameters and diverse outputs, make it a noisy dynamical system. Recent research indicates that noninvasive brain stimulation can alter the signal-noise relationship, but the precise relationship between noise amplitude and the global effects of local stimulation remains uncertain [18].

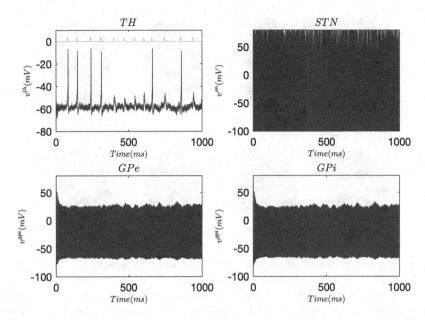

Fig. 2. (Color online) The effect of stochastic noise on the membrane voltages of the discrete brain network's TH, STN, GPe and GPi neurons in a healthy state. The red pulse trains in the top right panel denote SMC signal. (Color figure online)

The impact of stochastic noise on the selected regions, namely the TH, STN, GPi, and GPe, in the healthy brain is illustrated in Fig. 2. It is observed that burst oscillations occur across all brain regions, particularly in the thalamus and subthalamic nucleus, where the oscillations persistently burst. Consequently, this heightened neural activity exacerbates the healthy state of the brain. As a result, it impairs membrane potential and disrupts the normal functioning of neurons. These findings underscore the significant adverse effects of stochastic noise on the brain and its constituent regions, potentially leading to the development or exacerbation of brain injury [20, 21].

In the Parkinsonian state, the membrane voltages of key neuronal populations, including STN, GPi, GPe, and TH neurons within a discrete brain network, exhibit dynamic fluctuations over time, as depicted in Fig. 3. The initial equilibrium has been set to $-65\,\mathrm{mV}$; these neurons display varying voltage concentrations due to stochastic effects and diffusion processes. The color scale of voltage concentrations in Fig. 3 is plotted using MATLAB jet colormap. The recorded voltages at specific time points, such as $t = 354.56\,\mathrm{ms}$, $356.8\,\mathrm{ms}$, $356.23\,\mathrm{ms}$, $370\,\mathrm{ms}$, $374.67\,\mathrm{ms}$, $383.2\,\mathrm{ms}$, $385.05\,\mathrm{ms}$, and $385.57\,\mathrm{ms}$, reveal temporal changes in neuronal activity. Notably, certain neuronal populations exhibit elevated voltages relative to others at different time points, as indicated by the color nodes. For instance, at $t = 356.23\,\mathrm{ms}$, the voltage of TH neurons surpasses that of other neurons. In contrast, at $t = 356.8\,\mathrm{ms}$, the GPe and TH neurons exhibit higher voltage concentrations within the CBTH circuitry in the Parkinsonian

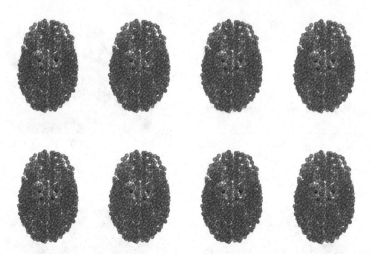

Fig. 3. (Color online) The effect of stochastic noise on the membrane voltage distributions of TH (top right), STN (top left), GPe (bottom right) and GPi (bottom left) neurons in the brain over time in Parkinson's state (axial views from below). Top panel (left to right): $t = 354.56\,\text{ms}$, $356.8\,\text{ms}$, $356.23\,\text{ms}$, $370\,\text{ms}$, and for the bottom panel (left to right): $t = 374.67\,\text{ms}$, $383.2\,\text{ms}$, $385.05\,\text{ms}$, $385.57\,\text{ms}$. (Color figure online)

state. These voltage dynamics underscore the intricate interplay of stochastic noise and diffusion processes in shaping neuronal activity patterns associated with PD [2].

In Fig. 3, stochastic noise is crucial in modulating the membrane voltage distributions of key neuronal populations implicated in PD. Over time, stochastic fluctuations in membrane potentials within these neural networks can exacerbate pathological activity patterns in the Parkinsonian state. In the TH neurons involved in dopamine production, stochastic noise may contribute to the dysregulation of dopamine levels characteristic of Parkinson's. Similarly, in the STN, known for its involvement in motor control, stochastic noise might amplify aberrant firing patterns associated with movement dysfunction. Meanwhile, within the GPe and GPi, integral components of the basal ganglia circuitry, stochastic fluctuations could disrupt the delicate balance of inhibitory signalling, further exacerbating motor symptoms [14,16]. These stochastic influences underscore the complexity of Parkinson's pathophysiology and highlight the importance of understanding noise modulation within neural circuits for developing effective therapeutic interventions [20].

Next, the DBS has been applied to STN neurons in the PD state of the brain. The application of DBS to STN neurons in the Parkinsonian state of the brain often results in a temporary restoration of healthy neural activity within the basal ganglia circuitry, leading to symptom alleviation in PD patients. As depicted in Fig. 4, we applied the DBS in a spatio-temporal domain for a smaller amount of time. It is interesting to know that in the presence of a diffusion operator, neurons maintained a healthy state for a sufficient time after the DBS had

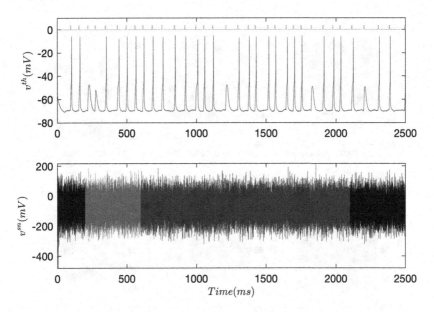

Fig. 4. (Color online) Effect of stochastic noise on membrane voltages of the TH (top) and the STN (bottom) neurons of discrete brain network in the Parkinson's state (black color). The effects of open-loop DBS on the STN neurons are presented in red color (bottom). However, the blue color shows a healthy state after the DBS is applied to the PD state. The red pulse trains in the top panel denote the SMC signal. (Color figure online)

been applied. We see the healthy state of STN neurons in blue color, as shown in Fig. 4. However, despite the therapeutic benefits of DBS, the long-term maintenance of a healthy state remains challenging. This is evident in the observed disturbances in thalamic activity characterized by bursts of oscillations and fluctuating membrane potentials, as shown in Fig. 4. Even with the presence of DBS, stochastic noise and diffusion processes continue to exert adverse effects on neural activity within the brain. Stochastic noise, arising from random fluctuations in ion channels and synaptic activity, can disrupt the finely tuned balance of excitation and inhibition within neural networks [21,22]. Additionally, diffusion processes, which govern the spread of neurotransmitters and other signalling molecules, can lead to spatial and temporal variations in neuronal activity.

In Fig. 4, the persistence of disturbances in thalamic activity despite DBS suggests that stochastic noise and diffusion processes may interact with the therapeutic intervention, leading to unintended consequences. These adverse effects underscore the complexity of neural dynamics in PD and highlight the need for further research to better understand and mitigate the impact of stochastic noise and diffusion on DBS efficacy [23,24]. Additionally, advancements in DBS technology and optimization of stimulation parameters may help minimize these adverse effects and improve long-term therapeutic outcomes for Parkinson's patients.

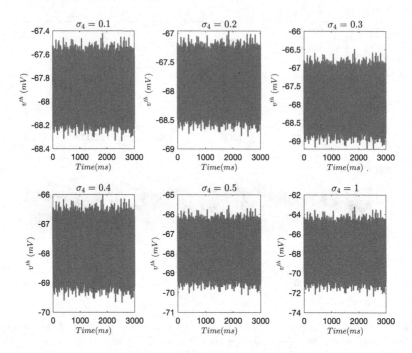

Fig. 5. Color online) The effect of stochastic noise on pathological activity patterns within the thalamus. (Color figure online)

Stochastic noise can significantly impact thalamic activity, disrupting its normal functioning. As depicted in Fig. 5, the thalamus may exhibit erratic fluctuations in membrane potentials and firing patterns in the presence of stochastic noise. This can lead to disturbances in sensory processing, motor control, and cognitive functions that rely on thalamic signalling. Moreover, Figs. 2, 3, 4 and 5 were plotted using $\sigma_1 = 0.1, \sigma_2 = 0.4, \sigma_3 = 0.4, \sigma_4 = 0.5$. As seen in Fig. 5, we observed that stochastic noise tends to drive the membrane potential of thalamic neurons towards the PD state. The fluctuations in the membrane potential exhibit low-frequency oscillations. These oscillations appear to have a regular pattern but are modulated by the stochastic noise added to the system. The frequency and amplitude of these oscillations may vary depending on the system's parameters and the noise level. The stochastic noise introduced in the system causes the membrane potential to fluctuate randomly around a mean value. As the noise level (σ_4) increases, the amplitude of the fluctuations also increases. This suggests that noise can significantly influence the dynamics of the system. Incorporating stochastic noise in the thalamic membrane potential exacerbates the pathological activity patterns associated with PD. Conversely, when the noise is absent, the membrane potentials tend towards a healthier state, particularly in the presence of DBS. These observations underscore the critical role of stochastic noise in modulating thalamic activity, thereby influencing the balance between pathological and healthy states in neurological disorders such as PD.

4   Discussion

In the current work, we used resting-state functional connectomes and machine-learning approaches such as parallel computing in classifying brain connectomes in healthy and PD states, with and without the stochastic process. To solve the network model computationally, we utilized the Euler-Maruyama method with a time-step $dt = 0.001$, with consistent results across various time-step values. Throughout all simulations, we used brain connectome data sourced from https://braingraph.org, with no significant changes observed over time.

Additionally, obtaining precise data regarding cortical-BGTH tractography proves challenging due to various limitations in structural MRI data, as highlighted by Meier et al. [8]. Petersen et al. have recently introduced an advanced axonal pathway atlas for the human brain, integrating findings from histological studies, imaging data, and expert insights [25,26]. Earlier work optimized connection probabilities and weights among BG regions to align with empirical fMRI data on an individual basis [2]. However, many studies resort to normative connectome atlases due to the complexities associated with acquiring and interpreting patient-specific diffusion-weighted imaging data. Yet, the potential benefits of patient-specific connection data remain uncertain.

In this work, we modified the Rubin and Terman model to better align with experimental evidence on neuron firing characteristics [2, 27]. Our study focuses on a network mathematical model that enables experimentalists to quickly evaluate membrane potentials throughout the cortex's four central nuclei and BGTH regions. Importantly, significant connection changes were detected in the PD brain, especially in stochastic noise, consistent with previous findings [14,19,20]. Unlike prior studies focusing on group differences, our study examined the discriminatory potential of resting-state functional connectivity at the individual level. This provides evidence that connectivity patterns with stochastic noise can distinguish PD patients with cognitive impairment from those without. Notably, our study is the sole one to show this capacity. In PD patients, functional connectivity reductions were observed across significant brain regions, with a disproportionate involvement of occipital-temporal and occipital-frontal connections compared to healthy controls. These findings contribute to the understanding of PD-associated cognitive impairment, corroborating previous neuroimaging modalities' observations.

Furthermore, this study underscores the potential of resting-state functional connectivity measures for individual-level discrimination in PD, providing valuable insights into the disease's pathophysiology. Our examination of stochastic noise's influence on various brain regions, including the STN, GPi, GPe, and thalamus, enhances our understanding of the complexities of PD. Moreover, noise can modulate the excitability of neurons within the brain network, influencing their firing patterns and synchronization properties. This modulation could change the overall network dynamics, affecting the balance between inhibitory and excitatory signals and potentially leading to dysregulated activity associated with neurological disorders such as PD. This discovery not only calls into question existing therapy options but also demonstrates the potential of our hybrid modelling approach for identifying subtle elements of brain dysfunction. Overall, investigating the influence of noise on the CBTH system using a discrete brain network paradigm gives valuable information on the system's resilience, flexibility, and susceptibility to dysfunction. It provides insights into the processes underpinning neurological diseases and may aid in developing therapeutic approaches to restore normal network function.

5 Conclusions

In conclusion, this study presents a novel approach utilizing a fusion of ML, stochastic modelling, and connectomic data to delve into the intricate neural pathways implicated in Parkinson's disease (PD) pathogenesis. By harnessing modern computational methodologies, we've endeavoured to decode the nuanced changes in structure and function within the PD-afflicted brain. Our findings shed light on the subtle alterations in neuronal activity patterns associated with PD progression, illuminating potential targets for therapeutic intervention. The hybrid modelling framework and innovative co-simulation technique developed in this research offer a deeper understanding of the impact of stochastic disturbances on the CBGTH network within the context of large-scale brain connectivity maps derived from the HCP. Notably, our analysis reveals that even in the presence of DBS, stochastic influences can lead to heightened activity in the thalamus, a key node in PD pathology.

In the future, we aim to analyze high temporal and spatially resolved cerebral data sources from functional near-infrared spectroscopy and EEG, PET, and MRI/fMRI data from healthy patients with neurodegenerative conditions such as PD. Also, the effect of stochastic noises in other regions, such as GPe, GPi and the whole cortex, will be analyzed. This work lays the groundwork for novel therapeutic strategies tailored to individual patients by elucidating the complex dynamics of neuronal activity underlying PD. The integration of ML, stochastic modelling, and connectomic data holds promise for advancing our understanding of PD pathophysiology and accelerating the development of personalized treatment approaches. Ultimately, the goal is to translate these insights into tangible clinical benefits, offering hope to those affected by this debilitating neurological disorder.

References

1. Salaramoli, S., Joshaghani, H.R., Hosseini, M., Hashemy, S.I.: Therapeutic effects of selenium on alpha-synuclein accumulation in substantia Nigra pars compacta in a rat model of Parkinson's disease: behavioral and biochemical outcomes. Biol. Trace Element Res. 1–11 (2023)
2. Shaheen, H., Pal, S., Melnik, R.: Multiscale co-simulation of deep brain stimulation with brain networks in neurodegenerative disorders. Brain Multiphys. **3**, 100058 (2022)

3. Shaheen, H., Melnik, R., The Alzheimer's Disease Neuroimaging Initiative: Bayesian inference and role of astrocytes in amyloid-beta dynamics with modelling of Alzheimer's disease using clinical data. arXiv Preprint arXiv:2306.12520 (2023)

4. Johnson, K.A., Okun, M.S., Scangos, K.W., Mayberg, H.S., de Hemptinne, C.: Deep brain stimulation for refractory major depressive disorder: a comprehensive review. Mol. Psychiatry 1–13 (2024)

5. Peralta, M., Jannin, P., Baxter, J.S.: Machine learning in deep brain stimulation: a systematic review. Artif. Intell. Med. **122**, 10219 (2021)

6. Tai, A.M., et al.: Machine learning and big data: implications for disease modelling and therapeutic discovery in psychiatry. Artif. Intell. Med. **99**, 101704 (2019)

7. Thieu, T.K.T., Melnik, R.: Coupled effects of channels and synaptic dynamics in stochastic modelling of healthy and Parkinson's-disease-affected brains. AIMS Bioeng. **9**(2), 213–238 (2022)

8. Oliveira, A.M., Coelho, L., Carvalho, E., Ferreira-Pinto, M.J., Vaz, R., Aguiar, P.: Machine learning for adaptive deep brain stimulation in Parkinson's disease: closing the loop. J. Neurol. **270**(11), 5313–5326 (2023)

9. Meier, J.M., et al.: Virtual deep brain stimulation: multiscale co-simulation of a spiking basal ganglia model and a whole-brain mean-field model with the virtual brain. Exp. Neurol. 114111 (2022)

10. Peng, G.C., et al.: Multiscale modeling meets machine learning: what can we learn? Arch. Comput. Methods Eng. **28**(3), 1017–1037 (2021)

11. Seguin, C., Sporns, O., Zalesky, A.: Brain network communication: concepts, models and applications. Nat. Rev. Neurosci. **24**(9), 557–74 (2023)

12. Novelli, L., Friston, K., Razi, A.: Spectral dynamic causal modeling: a didactic introduction and its relationship with functional connectivity. Network Neurosci. 1–25 (2024)

13. Vashistha, R., et al.: ParaPET: noninvasive deep learning method for direct parametric brain PET reconstruction using histoimages. EJNMMI Res. **14**(1), 10 (2024)

14. Woźniak, S., Pantazi, A., Bohnstingl, T., Eleftheriou, E.: Deep learning incorporating biologically inspired neural dynamics and in-memory computing. Nat. Mach. Intell. **2**(6), 325–336 (2020)

15. Shi, P., Li, J., Zhang, W., Li, M., Han, D.: Characteristic frequency detection of steady-state visual evoked potentials based on filter bank second-order underdamped tristable stochastic resonance. Biomed. Signal Process. Control **84**, 104817 (2023)

16. Liu, C., Wang, J., Deng, B., Li, H., Fietkiewicz, C., Loparo, K.A.: Noise-induced improvement of the Parkinsonian state: a computational study. IEEE Trans. Cybern. **49**(10), 3655–3664 (2018)

17. Charalambous, E., Djebbara, Z.: On natural attunement: shared rhythms between the brain and the environment. Neurosci. Biobehav. Rev. **155**, 105438 (2023)

18. Shaheen, H., Melnik, R.: Deep brain stimulation with a computational model for the cortex-thalamus-basal-ganglia system and network dynamics of neurological disorders. Comput. Math. Methods **2022**, 8998150 (2022)

19. Zheng, Y., et al.: Noise improves the association between effects of local stimulation and structural degree of brain networks. PLoS Comput. Biol. **19**(5), e1010866 (2023)

20. Touboul, J.D., Piette, C., Venance, L., Ermentrout, G.B.: Noise-induced synchronization and antiresonance in interacting excitable systems: applications to deep brain stimulation in Parkinson's disease. Phys. Rev. X **10**(1), 011073 (2020)

21. Staffaroni, A.M., et al.: A longitudinal characterization of perfusion in the aging brain and associations with cognition and neural structure. Hum. Brain Mapp. **40**(12), 3522–3533 (2019)

22. Liang, J., Yang, Z., Zhou, C.: Excitation-inhibition balance, neural criticality, and activities in neuronal circuits. Neuroscientist 10738584231221766 (2024)

23. Seguin, C., Jedynak, M., David, O., Mansour, S., Sporns, O., Zalesky, A.: Communication dynamics in the human connectome shape the cortex-wide propagation of direct electrical stimulation. Neuron **111**(9), 1391–1401 (2023)

24. Carron, R., Chaillet, A., Filipchuk, A., Pasillas-Lépine, W., Hammond, C.: Closing the loop of deep brain stimulation. Front. Syst. Neurosci. **7**, 112 (2013)

25. Abós, A., et al.: Discriminating cognitive status in Parkinson's disease through functional connectomics and machine learning. Sci. Rep. **7**(1), 45347 (2017)

26. Petersen, M.V., et al.: Holographic reconstruction of axonal pathways in the human brain. Neuron **104**(6), 1056–1064 (2019)

27. Rubin, J.E., Terman, D.: High frequency stimulation of the subthalamic nucleus eliminates pathological thalamic rhythmicity in a computational model. J. Comput. Neurosci. **16**(3), 211–235 (2004)

Investigation of Energy-Efficient AI Model Architectures and Compression Techniques for "Green" Fetal Brain Segmentation

Szymon Mazurek[1,2]✉ ⓘ, Monika Pytlarz[1] ⓘ, Sylwia Malec[1] ⓘ,
and Alessandro Crimi[1,2] ⓘ

[1] Sano Centre for Computational Personalized Medicine,
Nawojki 11, 30-072 Cracow, Poland
{s.mazurek,a.crimi}@sanoscience.org
[2] AGH University of Krakow, Adam Mickiewicz Avenue 30, 30-059 Cracow, Poland
https://www.sano.science, https://www.agh.edu.pl

Abstract. Artificial intelligence has contributed to advancements across various industries. However, the rapid growth of artificial intelligence technologies also raises concerns about their environmental impact, due to associated carbon footprints to train computational models. Fetal brain segmentation in medical imaging is challenging due to the small size of the fetal brain and the limited image quality of fast 2D sequences. Deep neural networks are a promising method to overcome this challenge. In this context, the construction of larger models requires extensive data and computing power, leading to high energy consumption. Our study aims to explore model architectures and compression techniques that promote energy efficiency by optimizing the trade-off between accuracy and energy consumption through various strategies such as lightweight network design, architecture search, and optimized distributed training tools. We have identified several effective strategies including optimization of data loading, modern optimizers, distributed training strategy implementation, and reduced floating point operations precision usage with light model architectures while tuning parameters according to available computer resources. Our findings demonstrate that these methods lead to satisfactory model performance with the low energy consumption during deep neural network training for medical image segmentation.

Keywords: medical imaging · segmentation · green learning · fetal brain · sustainable AI

1 Introduction

1.1 Fetal Brain Segmentation

Magnetic resonance imaging (MRI) is a popular non-invasive method for evaluating the development of the central nervous system of the fetus during pregnancy. In recent years, neuroimaging has become popular for studying the fetal

L. Franco et al. (Eds.): ICCS 2024, LNCS 14835, pp. 61–74, 2024.
https://doi.org/10.1007/978-3-031-63772-8_5

brain. However, manual segmentation of the structures of the fetal brain is time-consuming and subject to variability between observers. Therefore, researchers have used artificial intelligence to automate and standardize this process [5]. Fetal brain segmentation faces challenges due to the small size of the fetal brain, extra tissues that need to be distinguished, motion artifacts, and limited image quality from fast 2D sequences. Additionally, only narrow datasets that vary in image acquisition parameters are publicly available, which complicates the training of the Deep Learning (DL) algorithm. In terms of DL methods, various approaches have been tested, including unsupervised training, atlas fusion, deformation, and parametric methods. Many methods incorporate preliminary steps, such as fetal brain location and region of interest (ROI) cropping. Super-resolution reconstruction algorithms have enabled 3D segmentation, but standardization of these techniques is lacking. The majority of techniques make use of convolutional neural networks (CNNs), especially the U-Net architecture. Reconstruction algorithms and transfer learning methods are widely used to address data issues. Due to their ability to minimize partial volume effects, 3D segmentation techniques are becoming more and more common; but their performance depends on image quality [5].

1.2 "Green" Deep Learning

Over the past decade, significant advances have been made in artificial intelligence and machine learning (ML), driven largely by the accessibility of large datasets and the rise of DL. Deep neural networks (DNNs) play a key role in DL, and a prominent focus while developing a new DNN architecture is on achieving state-of-the-art (SOTA) results. However, the progress of the current SOTA is often achieved by increasing the complexity of the model, leading to a 300,000-fold increase in computational load in a span of six years [24]. Although these networks deliver impressive results, improving their capabilities requires substantial data and computational resources, resulting in high energy usage with notable economic and environmental implications. Consequently, there is a critical need for energy-efficient DL to address concerns related to finances, ecology, and practical usability [15]. Moreover, these models are extremely data-hungry, making their direct application to tasks, such as fetal brain segmentation, difficult. As a result, the exploration of their lightweight counterparts becomes a reasonable step towards addressing climate change; they will also facilitate the application of SOTA medical image analysis techniques in scenarios where energy is limited or costly, such as in developing nations and on portable devices relying on batteries. Moreover, they will encourage the adoption of improved training methods to replace the current conventional approach, which typically involves extensive searching for optimal hyperparameters and a trial-and-error process.

When looking for energy usage optimizations for DL, one could divide the areas of interest into the following categories: infrastructure and software, architectural design, efficient data use, optimization of training, and inference. The foundation for effective running of DL models is a reliable software and hardware system, including a careful selection of libraries and the use of graphics processing units, tensor processing units, or neuromorphic computing [15,16]. Another crucial component in minimizing computational effort is the selection of the right architecture. Modeling efficiency can be achieved through the use of compact neural networks and various automation and assembly techniques [16,32]. Even lightweight and optimized architectures often require a large amount of data to achieve peak performance. To reduce the data-related cost of training, data augmentation and active learning can be used [16,32]. Various techniques have been suggested to reduce the expense of training itself, such as different initialization techniques, normalization, progressive training, and mixed precision training [16,32]. After successful training, the energy efficiency of inference should be considered. Common compression methods contain quantization, pruning, low-rank factorization, deployment sharing, and knowledge distillation [16,32]. The need for reducing energy usage in DL applications is recognized, also in the medical domain. Parsa et. al. [17] proposed an interesting approach that decomposes the diagnosis process into subtasks evaluated by separate networks of varying complexity, achieving remarkable energy savings. Yu et. al. [33] tested numerous machine learning techniques in clinical prediction tasks, exploring different approaches to reduce energy requirements for their training. Sathish et.al. [23] propose a model quantization technique in medical imaging tasks, achieving major energy usage reductions while performing inference on CPUs. The code is available in https://github.com/szmazurek/efficient_segmentation.

1.3   Contribution

We aim to evaluate the environmental impact of deep learning by looking at energy consumption during model training. Existing solutions concentrate only on one category, such as architecture design [30] or hardware acceleration [31]. In contrast, our study intends to thoroughly examine the impact of different optimization techniques, starting from architecture selection, efficient data usage, and training acceleration, leading to the evaluation of the model's environmental effect. We investigated a variety of energy-efficient techniques to apply to the segmentation of the fetal brain in MRI, to develop a model with the best ratio of Dice score to energy consumption. We draw several observations and recommendations for creating energy-aware DL algorithms from the obtained results. We hope that our findings will help guide future research in the domain, helping researchers and practitioners navigate the landscape of available techniques for energy-efficient medical DL.

2   Methods

2.1   Setup and Hardware

All experiments were carried out using Python 3.11.5 with Pytorch 2.0.1 [18] and Lightning 2.0.3 libraries [7] with CUDA 11.7. Data loading and processing were performed with the Monai 1.2.0 [3] library. Energy usage was tracked using Codecarbon 2.3.1 and logged with Weights and Biases 0.15.4. We use an HPC environment containing 4 Nvidia A100 GPUs with 40 GB of RAM memory, 120 cores of 2 AMD EPYC 7742 64-core processors, and up to 500 GB of RAM memory.

2.2   Used Datasets

For our project, we incorporated the Openneuro dataset, a library of 1241 manually traced fetal fMRI images of 207 fetuses [22]. To comply with BIDS standards, the authors merged 3D volumes (raw and mask) into a 4D time series file [27]. The masks were drawn in a single volume from a period of fetal stillness. Within pre-processing, we cleaned the data from files missing matching masks .nii or raw .nii, extracted 3D volumes from functional MRI (fMRI) times series, and sliced 3D volumes into 2D pairs of slices raw vs. mask. Furthermore, we used a second dataset of fMRI, T2-weighted, and diffusion-weighted MRI scans [6]. Then we merged these datasets, doing the subject-wise division. In total, we obtained 40945 2D slices, which were used for the experiments. Data were split by patient into training, validation, and test subsets. For the test 10% patients were randomly chosen. Another 10% was allocated from the remaining patients for validation. The rest was used during training.

2.3   Energy Usage and Performance Measure

For tracking the relationship between performance and energy consumption, we measured the total energy consumed by the hardware used in training (CPUs, GPUs, and RAM memory) during the evaluation process. We chose Dice/kJ (kilojoules) as the metric describing this relationship.

2.4   Experimental Design

We designed the experiments as follows: first, we aimed to establish a baseline performance. We chose U-Net [21] as our baseline due to its ubiquitous usage in medical image segmentation tasks. This model was optimized using the Adam optimizer with a learning rate of 0.001. DDP in the default version was chosen for the communication between GPUs. Floating point operations were reduced to mixed 16-bit bfloat precision. We decided to use an early stopping algorithm to prevent overfitting and stop training when no improvement in validation loss is achieved in 15 consecutive epochs. The best parameters of the model are saved and used later in the inference. The batch size was set to 128 images per GPU. Training with a larger batch size would be possible, however, using a too-large value can cause training instability and require tuning of other parameters that would offset the large gradient values effects.

Immediately after establishing the baseline, we switched to Attention-Squeeze-Unet, as it has shown a relatively small reduction in test Dice score compared to U-Net, while drastically reducing training time, allowing us to conduct upcoming experiments faster. The rest of the configuration remained unchanged. We then proceeded to evaluate the techniques described in the later part of this section incrementally. That is, for a chosen technique, we ran the experiments and measured the performance. If it improved the Dice/kJ score, it was incorporated into the setup and the next technique was evaluated. It is important to note that these experiments were inspired by the MICCAI 2023 E2MIP [6], therefore with the experiments, we aimed to align with its evaluation criteria. Due to this fact, we chose the hardware setup used by the challenge authors. Additionally, we assumed that the algorithm we propose will be trained and evaluated on unknown data. We therefore avoid taking any solutions that require tuning specific to the used data distribution.

2.5   Evaluated Techniques

Data Caching and Loader Configuration. During neural network training, the speed with which the data can be provided to the model is often a bottleneck. Training on large datasets involves a lot of I/O operations, as the data needs to be loaded from the memory. Also, the pre-processing applied to the data on the fly slows down the process. This can be alleviated by using data caching in memory, especially if computational resources allow it. We evaluated the potential solution to this problem, Monai's CacheDataset abstraction. It first preloads the data into RAM and applies deterministic transformations, resulting in gains in model throughput at the cost of increased memory consumption. Furthermore, we examined the impact of the configuration of data-loading utilities offered by popular machine-learning frameworks such as Pytorch:

– Number of workers determines the number of concurrent processes involved in accessing the data.
– Data prefetching is an operation of loading and pre-processing the data by each worker into a buffer, from where the data are immediately accessed when the model requests for it.
– Workers persistence is a term referring to the handling of parallel dataloading processes. When using it, worker processes are not destroyed upon the completion of an epoch, hence there is no re-spawn overhead.
– Memory pinning enables the allocation of a predetermined memory subspace from which the transfer of data to GPU is increased.

Hyperparameter Search Techniques. We adopted the automatic learning rate tuning offered by the Lightning API to select the initial learning rate [7]. This was the only chosen automatic tuning method, as the full architecture and hyperparameter search are compute-costly procedures.

Data Augmentation. Medical images can take advantage of number data augmentation techniques for natural image analysis such as geometric transformations (rotations, horizontal reflections, cropping, shifting), the addition of random noise, or gamma correction [13]. Medical image datasets should not have been augmented via transformations influencing the color such as the modification of the saturation or the hue of the natural image; therefore, we applied simple geometric transforms, rotations, and flips, to reduce overfitting.

Efficient Model Architectures. Architectural design and the overall size are the key determinants of model resource requirements. We explored the landscape of lightweight DL models. We chose architectures to evaluate based on their number of parameters and existing code implementations. Finally, we choose the following: MobileNetV3-small [12], MicroNet [4], EfficientNet [25], Squeeze-Unet [1] and Attention-Squeeze-Unet [19]. Several additional models were directly taken from or inspired by implementation in [29]: SQNet, LinkNet, SegNet, ENet, ERFNet, EDANet, ESPNetv2, FSSNet, ESNet, CGNet, DABNet, ContextNet and FPENet.

Quantization for Training and Inference. Quantization is a technique for reducing model size that converts model weights from high-precision floating point to low-precision floating point or integer representations, such as 16-bit or 8-bit. By converting the weights of a model from a high-precision to a lower-precision representation, the model size and inference speed can be increased without sacrificing too much precision. Additionally, quantization improves the efficacy of a model by reducing memory bandwidth requirements and increasing cache utilization [2]. However, quantization can introduce new challenges and trade-offs between accuracy and model size, especially when using low-precision integer formats such as INT8.

Loss Functions. The choice of the loss function can help the model both to achieve better final performance and to increase convergence speed. In the experiments, we evaluated the most popular loss functions used in training neural networks for segmentation, such as Dice, Binary Cross Entropy (BCE), and Matthews Correlation Coefficient (MCC).

Pruning. Pruning is a known technique to reduce the model size by removing a subset of parameters. This is intended to increase throughput and lower the computation requirements. Usually, this reduction comes with the cost of reduced performance; therefore, the procedure usually involves iterations of applying pruning followed by fine-tuning the model to regain the performance. Pruning can be classified as unstructured or structured. Unstructured pruning aims to remove chosen weights without altering the network structure. The objective of structured pruning is to remove a group of parameters, thus reducing the size of neural networks. It also involves establishing the importance of the parameters to prune and the relationships between them. This is achieved using various algorithms, including the method evaluated in this study, a Dependency Graph (DepGraph). This method explicitly models the dependency between layers and comprehensively group coupled parameters for pruning, showing great results on benchmark tasks [8].

Gradient Averaging. Gradient averaging stabilizes and speeds up training in distributed settings, when training CNNs with batch gradient descent. From the available methods, stochastic weight averaging (SWA) was chosen for evaluation [14]. This method tracks the learned parameters for every epoch as the training nears the end and replaces the final ones with the average of them. Research has consistently shown that performance gains were observed in various problems solved using neural networks with nearly no additional computational costs.

Choosing Optimizer and Its Configuration. To perform optimization in neural networks, various algorithms were proposed. The choice of the optimizer and its hyperparameters has a significant impact on the convergence of the model, training time, and generalizability. We oriented ourselves towards adaptive optimization algorithms, which can tune the learning rates per parameter during training based on gradients. The initial choice of learning rate less influences them, as it is only used mostly as an upper limit value for the aforementioned choice of the one specific for a given parameter. Therefore, we sought to examine the Novograd optimizer, a relatively novel method for adaptive optimization [10]. This algorithm allows for adaptive parameter updates and reduces the memory footprint of the Adam optimizer by half. It was also shown to be more robust to the choice of learning rates, therefore being the go-to choice when the data are unknown or when the cost of hyperparameter search is to be avoided. It can also be paired with AMSGrad [20], a solution to problems with convergence of Adam optimizer where learning rate updates were too aggressive, the closer the optima. We also evaluated the influence of the exponential decay rate for the first and second gradient moments estimates (β_1 and β_2).

PowerSGD. PowerSDG is a low-rank gradient compressor based on power iteration that compresses gradients quickly, aggregates them using all-reduce, and achieves test performance comparable to stochastic gradient descent (SDG) [28].

Training Parallelization. Distributed data parallel (DDP) is a method for training parallelization of DL models based on copying the model to each computing device, performing the forward pass on a subset of training batches, and accumulating the results in the main process to update the model parameters. We evaluated its performance as well as the available configuration options offered by Lightning API implementation, such as static

We also evaluated Bagua, a set of distributed training algorithms [9]. The authors have demonstrated that these algorithms benefit the training speed compared to other strategies, including Pytorch DDP implementation. The solutions offered were incorporated via the lightning-bagua library, a plugin that allows us to use these algorithms with the Pytorch Lightning framework [7]. The algorithms offered by Bagua are related to distributed communication (bytegrad, asynchronous model average, improved all-reduce, low-precision decentralized SGD) and optimization algorithm (QAdam [26]).

3   Results and Discussion

Data caching was the first efficiency technique explored as the solutions were evaluated in an HPC cluster environment with significant available compute resources. Storing training and validation data in RAM after first loading from the disk led to shortening the time of epochs by nearly 400% - a significant improvement that offsets the initial cost of loading cache within a few epochs.

The use of the Novograd optimizer versus the Adam originally used has shown improvements in training speed.

We also experimented with setting different parameters β_1 and β_2. This led to slower training, drastically increasing the energy cost. These parameters could probably be tuned via an extensive hyperparameter search; however, energy constraints do not allow this in this setting.

The learning rate was chosen by an automatic search included in Pytorch Lightning. We decided to incorporate this step despite the benefits of using Novograd optimizer, which led to 2–3% improvements in final performance. However, increasing the search time above 100 steps did not improve performance while increasing execution time. We also decided to incorporate learning rate scheduling and performing the learning rate decay by a factor of 10 when no improvement in validation loss was observed for the past 5 epochs, up to a minimum value of 10^{-6}. This also led to a slight improvement in the final performance, increasing the test Dice score by another 1–2%.

The next step was to choose the number of workers to load the data. We decided to check 4, 6, and 8 workers for the data loader to avoid unnecessary profiling runs. These experiments were done with every worker pre-fetching 2 batches of data. Six workers provided the best performance to energy consumed ratio. Going beyond a certain number of workers seems to decrease performance due to communication overhead, even if a suitable number of processor cores are available. Finally, we also decided to use pinned memory and persistent workers, as the memory resources allow for. This can also increase the throughput by removing the need to spawn worker processes every epoch and allocating memory for CPU to GPU data transfer.

Next, we add augmentation of the training images via rotation and random flipping. It led to improved performance at virtually no cost, increasing the test Dice score by nearly 5%. Up to this point, the experiments were conducted using a Dice loss. We evaluated the usefulness of MCC and BCE losses. They led to worse performance without any reduction in energy usage.

To further improve communication with DDP, we disabled the search for unused parameters with every training step and set the computational graph to static. These changes allowed to increase the model's throughput and reduce training time. Using a bucket view for gradient reduction between devices led to decreased performance without energy benefits.

For comparison with DDP, experiments were conducted with algorithms offered by Bagua. The speedup in our case was not significant. Compared to a properly configured basic DDP strategy, it has shown slightly lower performance and higher energy consumption.

Next, we evaluated the usage of PowerSGD. We have observed significant performance drops, as the compression is lossy and its parameters need to be properly tuned to avoid these drops while maintaining the benefits of faster communication. Similar results were observed when we evaluated the SWA. Research has shown the benefits of increased performance with no performance loss; however, once again the length of the training had to be known to properly tune the algorithm's hyperparameters. We did not observe any performance benefits in our tests.

We also examined the effects of pruning. Iterative pruning of the least important parameters evaluated using DeGraph was performed. During tests, we observed drastic drops in model performance. We, therefore, concluded that pruning on models that are already relatively small (less than 10M parameters) has to be done with caution, as the capacity of the model to learn the patterns in the data can turn out to be too low after applying the technique.

3.1 Final Model Choice

As the pipeline and methods used were configured, we evaluated all the models listed. On the basis of these findings, we chose Attention-Squeeze-Unet as the final. The model has reached a high Dice score on the test data set relative to the baseline Unet. It has also shown the best energy per epoch ratio. To justify our choice, we present graphs of the performance of the model relative to energy consumption in Figs. 2 to 4. We also report the segmentation to highlight qualitatively how noisy were the data, as depicted in Fig. 1, where the overlay between an original MRI slice and a segmented results are showing, pointing out how cumbersome the data were and how challenging would be to obtain higher Dice score (Fig. 3).

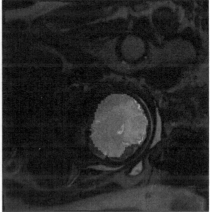

Fig. 1. Example DWI (left) and T2W (right) slices with segmentation mask obtained from final Attention-Squeeze-Unet trained with the optimized pipeline. Samples come from the E2MIP testing samples shared in the initial challenge announcement as an example.

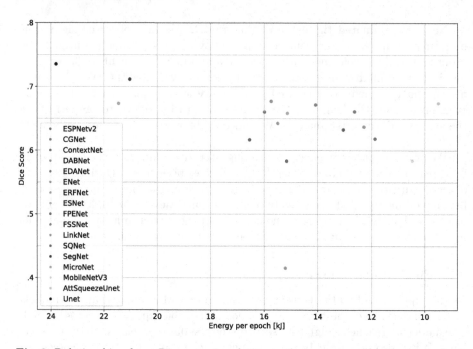

Fig. 2. Relationship of test Dice score in relation to the energy consumed per epoch of training.

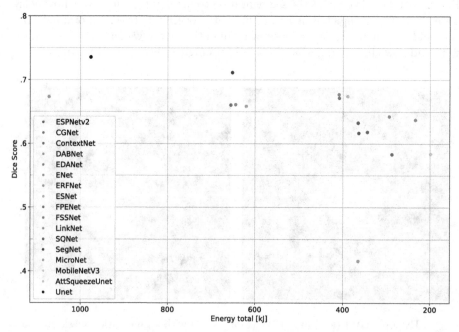

Fig. 3. Relationship of test Dice score in relation to the energy consumed during training.

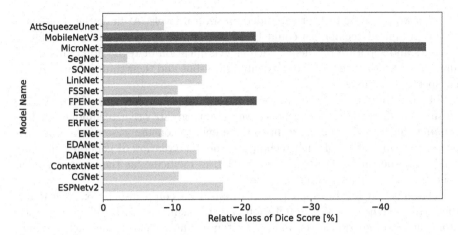

Fig. 4. Percentage of lost test Dice score for a given model tested relative to baseline value obtained by Unet.

3.2 Conclusions and Recommendations

We have shown that proper use of available methods can lead to satisfactory model performance while maintaining low energy consumption when training DNNs for medical image segmentation. In this study, we focused mostly on the training part of the process, since inference uses negligible amounts of energy when done only once. If the model is to be deployed and perform inference many times, then methods that were deemed too costly to optimize, such as pruning and quantization, can be considered, as the gains made for long-term operations can be substantial.

Recommendations for Energy-Efficient Fetal Brain Segmentation:

- Optimization of data loading - configuration of the data loading pipeline should be considered every time. Caching provides the most significant speedups but should be used with caution only in environments with no memory constraints.
- Choice of optimizer - modern optimizers reach the performance on par with the established ones while allowing for memory footprint reduction. Adaptive methods also seem to be robust to the hyperparameter choice, leading to lower energy used for their configuration.
- Optimal distributed strategy and reducing floating point operation precision can offer significant throughput improvements without loss in performance.
- Model architecture - using already existing or creating custom architecture that uses a small number of parameters leads to faster training, smaller compute requirements, and potentially still maintains satisfying levels of performance.
- Usage of methods requiring parameter tuning should be considered when computing resources allow for their tuning, otherwise they may result in suboptimal performance.

This work can be further expanded on several levels. At first, the robustness of the presented techniques could be examined by applying them to different datasets and separate problems, such as classification or regression. This would allow us to see the generality of presented solutions and their tuning requirements for specific use cases.

In this study, due to the nature of the examined problem, we focused more on the training process. However, as we note, there are other techniques such as pruning that can dramatically impact the energy consumption of the inference process. Examination of such techniques and the creation of general guidelines would be especially relevant for applied DL in medicine, as the created models operate mostly in the inference mode.

The effect of hardware-specific customization could also be examined further. HPC environments are notoriously heterogeneous, with different accelerators, node connection links, and filesystems. Thus, when using such systems, it would be beneficial to establish an approach for estimating optimal training configuration before starting full-scale experiments. Additionally, modern GPU accelerators have varying efficiency/energy curves, which means that there exists a "sweet spot" of clock frequency for a given GPU model that results in the best FLOPS to used energy ratio [11]. Furthermore, using custom ASIC platforms like TPU could lead to further energy consumption reductions, especially in the inference phase.

In summary, by enhancing the efficiency of machine learning algorithms through optimization techniques, the computational demands can be significantly reduced. This not only can accelerate model training but also minimizes the carbon footprint associated with the vast computational resources required. Taking into consideration this goal, we investigated and recommended some optimization ideas to reduce energy consumption and carbon emissions. We would like to stress the importance of sustainable computational approaches, and we would like to invite the scientific community to focus further on caring about carbon footprints.

Acknowledgments. This publication is supported by the European Union's Horizon 2020 research and innovation programme under grant agreement Sano No 857533. This publication is supported by Sano project carried out within the International Research Agendas programme of the Foundation for Polish Science, co-financed by the European Union under the European Regional Development Fund. This research was supported in part by the PLGrid infrastructure on the Athena computer cluster.

References

1. Beheshti, N., Johnsson, L.: Squeeze U-Net: a memory and energy efficient image segmentation network. In: Proceedings of the IEEE/CVF Conference on Computer Vision and Pattern Recognition Workshops, pp. 364–365 (2020)
2. Cai, Z., He, X., Sun, J., Vasconcelos, N.: Deep learning with low precision by half-wave gaussian quantization. In: Proceedings of the IEEE Conference on Computer Vision and Pattern Recognition, pp. 5918–5926 (2017)

3. Cardoso, M.J., et al.: MONAI: an open-source framework for deep learning in healthcare (2022). https://doi.org/10.48550/ARXIV.2211.02701. https://arxiv.org/abs/2211.02701

4. Chen, S.Y., Chen, G.S., Jing, W.P.: A miniaturized semantic segmentation method for remote sensing image. arXiv preprint arXiv:1810.11603 (2018)

5. Ciceri, T., Squarcina, L., Giubergia, A., Bertoldo, A., Brambilla, P., Peruzzo, D.: Review on deep learning fetal brain segmentation from magnetic resonance images. Artif. Intell. Med. **143**, 102608 (2023). https://doi.org/10.1016/j.artmed.2023.102608

6. Faghihpirayesh, R.: E2MIP Challenge, MICCAI 2023 (2023). https://github.com/Faghihpirayesh/E2MIP_Challenge_FetalBrainSegmentation

7. Falcon, W.: The PyTorch Lightning team: PyTorch Lightning (2019). https://doi.org/10.5281/zenodo.3828935

8. Fang, G., Ma, X., Song, M., Mi, M.B., Wang, X.: DepGraph: towards any structural pruning. In: Proceedings of the IEEE/CVF Conference on Computer Vision and Pattern Recognition, pp. 16091–16101 (2023)

9. Gan, S., et al.: BAGUA: scaling up distributed learning with system relaxations (2021). https://arxiv.org/abs/2107.01499

10. Ginsburg, B., et al.: Stochastic gradient methods with layer-wise adaptive moments for training of deep networks (2019)

11. Hodak, M., Gorkovenko, M., Dholakia, A.: Towards power efficiency in deep learning on data center hardware. In: 2019 IEEE International Conference on Big Data (Big Data), pp. 1814–1820 (2019). https://doi.org/10.1109/BigData47090.2019.9005632

12. Howard, A., et al.: Searching for MobileNetV3. In: Proceedings of the IEEE/CVF International Conference on Computer Vision, pp. 1314–1324 (2019)

13. Hussain, Z., Gimenez, F., Yi, D., Rubin, D.: Differential data augmentation techniques for medical imaging classification tasks. In: AMIA Annual Symposium Proceedings, vol. 2017, p. 979 (2017)

14. Izmailov, P., Podoprikhin, D., Garipov, T., Vetrov, D.P., Wilson, A.G.: Averaging weights leads to wider optima and better generalization (2018)

15. Mehlin, V., Schacht, S., Lanquillon, C.: Towards energy-efficient deep learning: an overview of energy-efficient approaches along the deep learning lifecycle. arXiv preprint arXiv:2303.01980 (2023)

16. Menghani, G.: Efficient deep learning: a survey on making deep learning models smaller, faster, and better. ACM Comput. Surv. **55**(12) (2023)

17. Parsa, M., Panda, P., Sen, S., Roy, K.: Staged inference using conditional deep learning for energy efficient real-time smart diagnosis. In: 2017 39th Annual International Conference of the IEEE Engineering in Medicine and Biology Society (EMBC), pp. 78–81 (2017). https://doi.org/10.1109/EMBC.2017.8036767

18. Paszke, A., et al.: PyTorch: an imperative style, high-performance deep learning library (2019)

19. Pennisi, A., Bloisi, D.D., Suriani, V., Nardi, D., Facchiano, A., Giampetruzzi, A.R.: Skin lesion area segmentation using attention squeeze U-Net for embedded devices. J. Digit. Imaging **35**(5), 1217–1230 (2022)

20. Phuong, T.T., Phong, L.T.: On the convergence proof of AMSGrad and a new version (2019). http://arxiv.org/abs/1904.03590

21. Ronneberger, O., Fischer, P., Brox, T.: U-Net: convolutional networks for biomedical image segmentation. In: Navab, N., Hornegger, J., Wells, W.M., Frangi, A.F. (eds.) MICCAI 2015. LNCS, vol. 9351, pp. 234–241. Springer, Cham (2015). https://doi.org/10.1007/978-3-319-24574-4_28

22. Rutherford, S., et al.: Automated brain masking of fetal functional MRI with open data. Neuroinformatics **20**(1), 173–185 (2021). https://doi.org/10.1007/s12021-021-09528-5

23. Sathish, R., Khare, S., Sheet, D.: Verifiable and energy efficient medical image analysis with quantised self-attentive deep neural networks. In: Albarqouni, S., et al. (eds.) Distributed, Collaborative, and Federated Learning, and Affordable AI and Healthcare for Resource Diverse Global Health, pp. 178–189. Springer, Cham (2022). https://doi.org/10.1007/978-3-031-18523-6_17

24. Schwartz, R., Dodge, J., Smith, N.A., Etzioni, O.: Green AI. Commun. ACM **63**(12), 54–63 (2020). https://doi.org/10.1145/3381831

25. Tan, M., Le, Q.: EfficientNet: rethinking model scaling for convolutional neural networks. In: International Conference on Machine Learning, pp. 6105–6114 (2019)

26. Tang, H., et al.: 1-bit Adam: communication efficient large-scale training with Adam's convergence speed (2021). https://arxiv.org/abs/2102.02888

27. Turk, E., et al.: Functional connectome of the fetal brain. J. Neurosci. **39**(49), 9716–9724 (2019)

28. Vogels, T., Karimireddy, S.P., Jaggi, M.: PowerSGD: practical low-rank gradient compression for distributed optimization. In: Advances in Neural Information Processing Systems, vol. 32 (2019)

29. Wang, Y.: Efficient-segmentation-networks PyTorch implementation (2019). https://github.com/xiaoyufenfei/Efficient-Segmentation-Networks

30. Wu, D., et al.: LightNet: a novel lightweight convolutional network for brain tumor segmentation in healthcare. IEEE J. Biomed. Health Inform. (2023). https://doi.org/10.1109/JBHI.2023.3297227

31. Xiong, S., et al.: MRI-based brain tumor segmentation using FPGA-accelerated neural network. BMC Bioinformatics **22**(1), 421 (2021)

32. Xu, J., et al.: A survey on green deep learning. arXiv preprint arXiv:2111.05193, November 2021

33. Yu, J.R., et al.: Energy efficiency of inference algorithms for clinical laboratory data sets: green artificial intelligence study. J. Med. Internet Res. **24**(1), e28036 (2022). https://doi.org/10.2196/28036

Negation Detection in Medical Texts

Maria Chiara Martinis[1,2]([✉]) [iD], Chiara Zucco[1,2] [iD], and Mario Cannataro[1,2] [iD]

[1] Data Analytics Research Center, University "Magna Græcia",
88100 Catanzaro, Italy
{martinis,chiara.zucco,cannataro}@unicz.it
[2] Department of Medical and Surgical Sciences, University "Magna Græcia",
88100 Catanzaro, Italy

Abstract. Negation detection refers to the automatic identification of linguistic expression that convey negation within a textual content. In medical and biomedical context, the negation detection plays a pivotal role in understanding clinical documentation and extracting meaningful insights. In this paper, we survey 16 articles published from 2005 to 2023 and focusing on negation detection within medical domain. Our evaluation framework encompass both methodological aspects and application-oriented considerations. Specifically, we discuss the used approaches, the employed methodology, the specific tasks addressed, the target language of textual analysis, and the evaluation metrics used. On the application front, for each reviewed study, we delineate the medical domains under investigation (e.g., cardiology, oncology), the types of data analyzed, and the availability of datasets. The majority of reviewed works are conducted in English, with a prevalence of machine learning and deep learning approaches, and classic classification evaluation metrics. Application domains exhibit heterogeneity, with a slight predominance in oncology, and diverse data sources including EHRs, abstracts, scientific papers, and web-derived information (e.g., Wikipedia or blog entries). Throughout this review, we will identify limitations and gaps in this research area, as well as examine the benefits it could bring to the scientific community and the methods currently employed.

Keywords: NLP · Negation detection · medical domain

1 Introduction

Negation detection (ND) is a critical element in the field of Natural Language Processing (NLP) and has significant implications for a wide range of applications. This process involves recognizing instances in the text where a statement is negated or its meaning is reversed, proving crucial for understanding conveyed information in Morante et al. [10].

Its importance lies in the fact that negation can completely change the interpretation of a statement, often adding complexity to text comprehension. For example, distinguishing between "The patient is not experiencing a fever" and

L. Franco et al. (Eds.): ICCS 2024, LNCS 14835, pp. 75–87, 2024.
https://doi.org/10.1007/978-3-031-63772-8_6

"The patient has a fever" has substantial implications in decision-making in the healthcare sector.

Negation detection requires in-depth text analysis, including lexical indicators, syntactic patterns, and contextual clues.

Recent developments in the field of Natural Language Processing (NLP), particularly the use of machine learning models such as deep neural networks and transformer-based architectures, have sparked growing interest, promising new perspectives to overcome these challenges.

The goal of this article is to provide an overview of the state of the art in "negation detection," specifically identifying the latest methodologies and emerging technologies, emphasizing the specific linguistic challenges influencing this research field.

We will analyze the practical applications of negation detection in key sectors, such as medicine, highlighting how accurate negation identification can significantly improve result accuracy and relevance.

Through this investigation, we aim to offer an in-depth overview of current frontiers in negation detection research, identify ongoing challenges, and propose possible directions for future developments.

One of the early methods in the literature, explained by Chapman et al. in [1], introduces the negation identification method called NegEx. This approach relies on a set of predefined rules to detect the presence of negations in texts.

In this study, the authors verify, using their algorithm with preprocessed sentences, whether the conclusions and diseases indexed from discharge summaries are negated or affirmed by the dictating physician.

NegEx, despite its usefulness in identifying negations in very complex sentences, has significant limitations. Incorrectly assigning negations to analyzed sentences, particularly within patient treatment diagnoses, could lead to an inaccurate diagnosis analysis.

For this reason, Mehrabi et al. [9], developed a new negation algorithm called DEEPEN, specifically designed to address the false positives of NegEx. The system was developed and tested using electronic health record (EHR) data from Indiana University (IU) and subsequently evaluated on the Mayo Clinic dataset to assess its generalizability.

By incorporating the dependencies of NegEx, DEEPEN can reduce the incorrect negations that NegEx assigns.

The present work aims to review 16 articles on negation detection within medical domain published from 2005 to 2023 to provide an overview of existing methodologies related to the identification of "negation' across different medical-related written texts.

Our evaluation framework encompasses methodological aspects as well as application-oriented considerations. More in details, for each reviewed work, we analyze the proposed approaches, the methods employed, the specific tasks addressed (i.e. negation recognition and/or negation scope resolution), the target language of textual analysis, and evaluation metrics utilized. Regarding applications, we examine the medical domains under investigation (e.g., cardiology, oncology), the source of data analyzed, and dataset availability.

The rest of the paper is organized as follows: Sect. 2 provides a brief overview of the main approaches to negation detection; Sect. 3 describes the main approaches to negation detection. Finally, Sect. 4 concludes the document.

2 Basic Definition and Open Problems

Negation detection and negation scope negation resolution (NSR) are two fundamental methods in the field of Natural Language Processing (NLP). Both of these concepts are useful for analyzing and interpreting natural language accurately, providing tools for various applications in the field of Natural Language Processing.

Negation detection refers to identifying, within a text, expressions indicating the negation of a specific statement or concept.

The capability is fundamental for correctly understanding the meaning of the text and interpreting the information accurately as in Morante et al. [10]. On the other hand, scope negation resolution involves identifying the expression or concept negated within a sentence or a broader context.

In Dalloux et al. [2], the parts of the text affected or involved by a specific negation are identified. The determination described is crucial for accurately interpreting the meaning of a sentence and may vary depending on the linguistic context and the structure of the sentence itself.

The detection of negation in the field of Natural Language Processing (NLP) is of fundamental importance, especially when dealing with medical texts. The presence of negations can significantly impact understanding of the meaning of individual sentences or entire texts within the medical context.

In addressing this complexity, both basic and advanced approaches, as well as specific considerations for negation detection in a scientific context, are outlined. An important challenge in negation detection is represented by the linguistic diversity found in scientific texts. Languages used in scientific contexts often employ technical and specialized languages with complex syntactic structures, making the accurate identification of negations challenging.

The use of negated expressions through elaborate syntactic constructs requires an advanced approach to ensure precise detection.

Beyond the difficulties faced in identifying negation in scientific documents, a critical aspect involves resolving scope negation. This term pertains to recognizing or expanding the negation within a singular sentence or a more extensive context. A precise interpretation hinges on a thorough understanding of how negation impacts the message's structure.

Ultimately, this process leads to a heightened understanding of how negation operates in the message, pinpointing the scope in which it is employed.

The detection of negation in Natural Language Processing (NLP) can present several challenges, including:

1. Linguistic Diversity:
   Languages show variations in grammatical structure and how negation is expressed, posing a challenge to establishing universal rules or models;

2. Technical Language:
   When dealing with scientific or technical texts, the presence of specialized terminology and intricate syntactic structures can impede the precise identification of negation;

3. Ambiguity:
   Negation might be conveyed in an unclear or implicit manner, demanding an understanding of the contextual nuances for precise detection;

4. Availability of Datasets for Specific Languages or Domains:
   The presence of datasets suitable for training models may be limited, particularly for languages that are less common or specific professional domains.

To address these challenges, various types of approaches are highlighted, such as the utilization of rules-based techniques, machine learning, and deep learning to provide more sophisticated solutions. Among the rules-based approaches, the use of linguistic rules emerges as a fundamental starting point. By defining keywords associated with negation, such as "non," "neither," and "none," it's possible to structure an identification mechanism that handles cases where negation is expressed through complex syntactic constructs [19]. In parallel, lexical analysis proves useful in creating a list of negative terms, employing specific techniques to detect them in the text. The application of Part-of-Speech (POS) tagging allows for the examination of the grammatical structure of sentences, identifying typical patterns associated with negation. Advanced approaches go further, adopting techniques from Machine Learning and Deep learning. Machine learning involves training a classifier based on a dataset annotated with negative information. Models like recurrent neural networks (RNN) or transformer neural networks (BERT) demonstrate high performance in NLP tasks.

Embedding word vectors, representing words as vectors, leverages semantic similarity to identify words associated with negation. Specific considerations for scientific articles include attention to technical terminology, handling the complex structures often present in scientific publications, and collecting or creating specific datasets for model training and evaluation.

Adapting the model to the language and scientific context, along with the use of appropriate metrics such as precision, recall, and F1-score, contributes to ensuring an accurate evaluation of the system's performance. When experimenting with different approaches, the goal is to find the most suitable solution for the specific context of scientific analysis.

3   Main Approaches to Negation Detection

In this section, we analyze the methodologies employed to detect negation, with a particular focus on the medical context.

The analysis involves various strategies, such as rule-based approaches (Sect. 3.1), machine learning models (Sect. 3.2), and deep learning techniques (Sect. 3.3), to address challenges, as shown in Table 1.

Table 1. Example of Approaches and Methodologies for Negation Detection in Different Languages.

Authors	Approaches	Year
Elkin et al. [3]	Rules-based	2005
Zamaraeva et al. [17]	Rules-based	2018
Hammami [6]	Rules-based	2021
Mutalik et al. [12]	Machine learning	2001
Chapman et al. [1]	Machine learning	2001
Huang et al. [7]	Machine Learning-Ruled based	2007
Vincze et al. [16]	Machine Learning	2008
Savova et al. [14]	Machine learning	2010
Funkner et al. [5]	Machine Learning	2014
Mukherjee et al. [11]	Machine Learning	2017
Sun et al. [15]	Machine Learning	2021
Mehrabi [9]	Deep Learning	2015
Dalloux [2]	Deep Learning	2019
Daan de Jong [8]	Deep Learning	2021
Van Es et al. [4]	Deep Learning	2023

3.1   Rule-Based Approaches

Rule-based approaches rely on predefined rules and linguistic models to recognize negation expressions in this context. The following articles belong to this category. Hammami et al. [6] introduce a novel rule-based approach, employing natural language processing (NLP) techniques for classifying Italian pathological reports based on the ICD-OM coding scheme.

This approach facilitates the identification and categorization of morphological content in Italian pathological reports by analyzing negation patterns. The method involves selecting two negation categories during training, assigning identified negation terms to these categories, and merging negation terms with negated terms into single bigrams based on their positions. The final classification algorithm achieved a noteworthy micro-F1 score of 98.14% across 9594 pathological reports in the test dataset.

The study proposed by Elkin et al. [3], aims to compare the effectiveness of an automated system in assigning negation to clinical concepts within compositional expressions with manually assigned negation. The dataset comprises 41 clinical documents (medical evaluations) analyzed through the Mayo Vocabulary Server Parsing Engine, utilizing SNOMED-CT^{TM} for conceptual coverage.

Validation of the identification of concepts and textual cues related to negation is conducted through a review by a medical terminologist. The results of recall were 97.2% and a specificity of 98.8% in negation assignment, indicating high precision compared to manual evaluation.

The Zamaraeva et al. in [17] focus on optimizing feature extraction in pathology reports, with particular attention to precise negation detection. The authors propose a targeted approach to accurately recognize and delineate the context of negation within medical texts.

This improvement in negation detection contributes to a more accurate extraction of relevant information from pathological reports, enhancing the overall quality of the extracted features. The document introduces new methodologies for negation detection and discusses the results obtained, highlighting the potential impact on medical practice.

3.2   Machine Learning Approaches

Mukherjee et al. [11] propose a generic parser called NegAIT (Negation Assessment and Inspection Tool) is being developed and thoroughly examined to annotate the presence of negations in a text. This parser takes text as input and identifies occurrences of morphological negations, negative phrases, and double negations.

It is implemented in an integrated Java-Scala environment, using a rule-based approach through the Open Domain Informer (ODIN) event extraction framework. The steps to recognize negation include text tokenization, sentence splitting, word stemming and analysis with the Stanford parser in [8].

Manually crafted rules combining regular expressions and lexicons are applied to identify various types of negation. Due to the initial lexicon generating many false positives, additional negative words from other lexicons have been incorporated.

The work in [5], focuses on the development of a negation detection module for the analysis of unstructured clinical documents in the Russian language, employing a machine learning-based approach. The module has been trained and tested using anonymized electronic health records of patients with acute coronary syndrome. The notable effectiveness of the module is particularly evident in predicting the potential need for surgical interventions for patients affected by acute coronary syndrome.

The adopted methodology involves annotating clinical documents for specific pathologies, normalizing annotated texts, and utilizing a gradient-boosting classifier for training and optimization. Relevant outcomes emerge in refining predictive models related to surgical interventions, based on text characteristics, while the impact on outcome prediction models is less pronounced.

This negation detection module represents a crucial component within a broader application for clinical document analysis, enhancing prediction accuracy, and contributing significantly to advanced practices in the clinical context.

Sun et al. [15] discusses the participation of the MedAI system in SemEval-2021 Task 10, focusing on domain adaptation for negation detection. The authors introduce a new method called "Negation-aware Pre-training," designed to enhance the model's ability to detect negations in source-free contexts.

The study explores the challenges associated with negation detection in the medical context and outlines the methodology employed by the MedAI system. This approach may involve utilizing pre-training techniques specifically crafted to handle negations and adapting the model to the unique characteristics of the medical domain. The document also presents experimental results, highlighting the effectiveness of the proposed approach in improving negation detection performance in medical text, even without access to labeled source data.

Naldi et al. [13], present an evaluation method based on a test set to assess the effectiveness of a sentiment analysis tool in identifying negations within a medical sentence. They examined a basic test set containing over thirty manually labeled sentences. However, they observed that widely used sentiment analysis packages continue to exhibit inefficiency in handling negative sentences, primarily due to their test set-based nature. This approach forces the algorithm to deal with highly critical sentences, whose polarity may be challenging for the algorithm itself to comprehend.

A potential solution could involve adopting machine learning-based approaches. However, this requires a substantial amount of labeled data for training, which makes challenging the training of such approaches.

The negation annotator in cTAKES utilizes the NegEx algorithm in [1], based on models, to identify words and phrases indicative of negation near mentions of named entities. Similarly, the status annotator employs a similar approach to identify relevant words and phrases indicating the status of a named entity.

For negation detection, each discovered named entity is associated with one of the dictionary's semantic types and includes attributes such as the associated text span ('span'), the associated terminology/ontology code ('concept'), the negation of the named entity ('negation'), and the status associated with the named entity ('status').

The status value is set to "possible" for future events, while allergies to a specific drug are managed by setting the negation attribute of that drug to "is negated." Non-patient experiences are marked as "family history of" if applicable.

cTAKES is distributed with highly performant machine learning modules and models and operates on the Apache Unstructured Information Management Architecture (UIMA) URL http://incubator.apache.org/uima/. Vincze et al. [16], analyze a corpus annotated for negations, speculations, and their linguistic extensions, emphasizing that detecting signals of negation and uncertainty is more straightforward in clinical documents due to the abundance of keywords.

The study highlights the importance of detecting negations and scopes in the biomedical context, with over 10% of sentences containing modifiers that significantly influence semantic meaning. The BioScope corpus is designed to facilitate the development of automatic systems for detecting negations and scopes, addressing challenges related to keyword identification and linguistic scope determination.

Mutalik et al. in [12], test the hypothesis that the detection of negated concepts in dictated medical documents can be effectively achieved using tools designed for the analysis of formal languages. They developed a program called Negfinder, incorporating a lexical scanner and a parser based on a subset of context-free grammars.

This was analyzed on a diverse set of 40 medical documents to recognize negation patterns in the text. The parser's performance was evaluated on two test sets: one visually inspected for false positives and false negatives, and another independently examined by both a human observer and Negfinder, ultimately comparing the results and achieving a sensitivity of 95.7%.

Jamai et al. in [7], a novel hybrid method is outlined, which integrates the use of regular expression matching with grammatical analysis. This approach aims to overcome the previously mentioned constraint in automatically detecting negations within clinical radiology reports. Jamai et al. in [14] aim to develop and evaluate an open-source natural language processing system designed to extract information from free-text electronic health records. The system, named cTAKES (Clinical Text Analysis and Knowledge Extraction System), is available as open source at http://www.ohnlp.org and extensively leverages existing open-source technologies, including the Unstructured Information Management Architecture framework and the OpenNLP natural language processing toolkit. cTAKES has been specifically trained for the clinical domain and demonstrates a high precision of 94.9%.

3.3   Deep Learning Approaches

Daan de Jong in [8], presents a strategy based on the use of neural networks to resolve the scope of predicted negation cues in two phases. The main objective is to address the challenge of accurately identifying the scope of words or phrases indicating negation.

The first phase focuses on predicting negation cues, while the second phase is dedicated to resolving the associated scope of these cues. In the initial phase, a neural network is trained to predict negation cues in the text. Subsequently, in the second phase, another neural network is employed to resolve the scope of these predicted negation cues. By utilizing a two-step approach, the aim is to enhance accuracy in identifying the meaning or scope of negations within the context of the text.

In this study [2], Dalloux et al. analyze a natural language processing (NLP) technique, specifically "negation detection", in French biomedical texts. To conduct this analysis, the authors have first identified the words that express speculation and negation. Subsequently, they identified their contexts, namely, the tokens within the sentences that are influenced by the presence of negation or speculation. This approach was tested in two French datasets, annotated with negation and speculation signals along with their respective scopes. The approach examined utilizes CRF and BiLSTM, achieving precision results of 97.21%.

Van Es et al. in [4] provide a detailed explanation of the implementation of both approaches, discussing the results obtained through a thorough and comparative evaluation. Metrics such as precision, recall, and F1-score are analyzed to assess the performance of the methods in different conditions and clinical contexts. The authors also examine cases where one approach might be preferable over the other, considering the specific challenges present in Dutch clinical texts.

The ultimate goal of the paper is to offer a clear guide on choosing between rule-based and machine learning methods for negation detection in Dutch clinical texts, taking into account the effectiveness and practicality of each approach in specific clinical contexts.

Mehrabi et al. in [9] developed a new negation algorithm called DEEPEN, specifically designed to address the false positives of NegEx. The system was developed and tested using electronic health record (EHR) data from Indiana University (IU) and subsequently evaluated on the Mayo Clinic dataset to assess its generalizability. By incorporating the dependencies of NegEx, DEEPEN can reduce the incorrect negations that NegEx assigns.

4 Discussions and Results

By delving into the challenges and methodologies analyzed in the preceding sections, our objective is to highlight the role of negation detection in ensuring the accurate comprehension of information in medical texts, significantly enhancing the quality of analyses conducted.

As stated before, our evaluation framework encompasses two main aspects: methodological aspects as well as application-oriented insights. More in details, we evaluated the 16 reviewed works methodological aspects based on:

- approach types (rule-based, ML-based, DL-based);
- proposed methodology;
- the specific tasks addressed i.e. Negation Detection or Negation Scope Resolution (NSR);
- the target language od the analysis,
- the evaluation metrics.

Table 2 summarize the findings. When considering the application-oriented aspects, the focus is posed on the application areas such as medical specialties, the source of the analyzed data, and dataset availability.

Among the issues that emerged, significant challenges are identified stemming from the substantial lack of datasets specifically compiled in a suitable technical language to correctly identify negation expressions.

This can make the training process of negation detection models difficult, limiting the ability to adequately extract the linguistic nuances present in medical texts.

In Table 3 a detailed representation of the application scopes addressed in the articles and the availability of the training/evaluation dataset is provided. The discussed analysis will be important for identifying emerging trends and

Table 2. The table shows information about the proposed methodology, the specific task, the target language, and the metrics used to compare the 16 works under review.

Article	Methods	Task	Language	Metrics
Elkin et al. [3]	Rules	ND	English	Precision
Zamarava et al. [17]	Rules	ND	English	Not Available
Hammami et al. [6]	Rules	ND	Italian	F1 score
Mutalik et al. [12]	Regular expressions	ND	English	Not Available
Chapman et al. [1]	Regular expression	ND	English	Specificity
Huang et al. [7]	Hybrid approaches, regular expression matching with grammatic parsing	ND	English	Specificity
Vincnze et al. [16]	Annotation dataset	ND-NSR	English	Not Available
Savova et al. [14]	Machine Learning	ND	English	Accuracy and F1 score
Funkner et al. [5]	Multiclass classification	ND	Russian	F1 score
Mukherjee et al. [11]	Binary classification	ND	English	Precision and Recall
Sun et al. [15]	Self-supervised Model	ND	English	F1 score
Meharabi [9]	Graph-based and transition-based	ND	English	Precision and Recall
Dalloux [2]	BiLSTM	ND-NSR	French	Precision
Daan de Jong [8]	BiLSTM	NSR	English	F1 score
Van Es et al. [4]	BiLSTM	ND	German	Not Available

assessing the influence of various methodologies in the field of negation detection. Furthermore, the ambiguity of the data and the approaches used could add further complexity to these challenges, requiring a more careful and detailed analysis.

Our study has emphasized the importance of negation detection in ensuring accurate comprehension of information in medical texts, thus enhancing the quality of analyses conducted.

Table 3. The table shows the application domain, source and availability of the dataset

Article	Application area	Source	Dataset Availability
Elkin et al. [3]	Internal Medicine	Electronic Health Records	Not Available
Zamarava et al. [17]	Oncology	Pathological reports	Available
Hammami et al. [6]	Oncology	Pathological reports of oncology	Not Available
Mutalik et al. [12]	Surgery	medical documents	Not Available
Chapman et al. [1]	Not Available	Electronic Health Records	Not Available
Huang et al. [7]	Radiology	Clinical Radiology Reports	Not Available
Vincze et al. [16]	Not Available	Medical reports	Available
Savova et al. [14]	Not Available	Clinical reports	MAYO
Funkner et al. [5]	Cardiology	Electronic Medical Reports related to acute coronary syndrome	Not Available
Mukherjee et al. [11]	Not Available	6 corpora: patient blogs, Cochrane reviews, PubMed abstracts, clinical trial texts, and English Wikipedia articles for different medical topics	Not Available
Sun et al. [15]	Not Available	Clinical notes	i2b2 2010 Challenge Dataset, available under data use agreements
Mehrabi [9]	Not Available	Electronic Health Records	MAYO clinical
Dalloux [2]	Cardiology, Urology, Oncology	French biomedical documents	Available
Daan de Jong [8]	Radiology and Biological research	Radiology clinical reports, full papers, and scientific abstracts in the biological domain	Bioscope corpus
Van Es et al. [4]	Not Available	Clinical corpus	Erasmus Dutch Clinical Corpus

One of the issues highlighted in this paper is the limited availability of datasets specifically designed for negation detection, especially in the medical field. This shortage represents a barrier to the advancement of precise models for this technique. As a future work, this could be addressed by creating specific data sets for negation detection with a particular focus on medical domains.

Given this, future efforts could focus not only on generating new datasets but also on adapting and expanding existing linguistic resources. This initiative would facilitate the development of advanced existing models, leading to further growth in research in this field.

Acknowledgement. This work was funded by the Next Generation EU - Italian NRRP, Mission 4, Component 2, Investment 1.5, call for the creation and strengthening of 'Innovation Ecosystems', building 'Territorial R&D Leaders' (Directorial Decree n. 2021/3277) - project Tech4You - Technologies for climate change adaptation and quality of life improvement, n. ECS0000009. This work reflects only the authors' views and opinions, neither the Ministry for University and Research nor the European Commission can be considered responsible for them.

References

1. Chapman, W.W., Bridewell, W., Hanbury, P., Cooper, G.F., Buchanan, B.G.: A simple algorithm for identifying negated findings and diseases in discharge summaries. J. Biomed. Inform. **34**(5), 301–310 (2001)

2. Dalloux, C., Claveau, V., Grabar, N.: Speculation and negation detection in French biomedical corpora. In: Mitkov, R., Angelova, G. (eds.) Proceedings of the International Conference on Recent Advances in Natural Language Processing (RANLP 2019), pp. 223–232. INCOMA Ltd., Varna, Bulgaria (2019). https://doi.org/10. 26615/978-954-452-056-4_026, https://aclanthology.org/R19-1026

3. Elkin, P.L., et al.: A controlled trial of automated classification of negation from clinical notes. BMC Med. Inform. Decis. Mak. **5**, 1–7 (2005)

4. van Es, B., et al.: Negation detection in Dutch clinical texts: an evaluation of rule-based and machine learning methods. BMC Bioinform. **24**(1), 10 (2023)

5. Funkner, A., Balabaeva, K., Kovalchuk, S.: Negation detection for clinical text mining in Russian. Stud. Health Technol. Inform. **270**, 342–346 (2020). https:// doi.org/10.3233/SHTI200179

6. Hammami, L., et al.: Automated classification of cancer morphology from Italian pathology reports using natural language processing techniques: a rule-based approach. J. Biomed. Inform. **116**, 103712 (2021). https://doi.org/10.1016/j.jbi.2021. 103712, https://www.sciencedirect.com/science/article/pii/S1532046421000411

7. Huang, Y., Lowe, H.J.: A novel hybrid approach to automated negation detection in clinical radiology reports. J. Am. Med. Inform. Assoc. **14**(3), 304–311 (2007). https://doi.org/10.1197/jamia.M2284

8. de Jong, D.: Scope resolution of predicted negation cues: a two-step neural network-based approach. CoRR abs/2109.07264 (2021). https://arxiv.org/abs/2109.07264

9. Mehrabi, S., et al.: DEEPEN: a negation detection system for clinical text incorporating dependency relation into NegEx. J. Biomed. Inform. **54**, 213–219 (2015)

10. Morante, R., Blanco, E.: Recent advances in processing negation. Nat. Lang. Eng. **27**(2), 121–130 (2021). https://doi.org/10.1017/S1351324920000534

11. Mukherjee, P., et al.: NegAit: a new parser for medical text simplification using morphological, sentential and double negation. J. Biomed. Inform. **69**, 55–62 (2017). https://doi.org/10.1016/j.jbi.2017.03.014, https://www.sciencedirect.com/science/article/pii/S1532046417300631

12. Mutalik, P.G., Deshpande, A., Nadkarni, P.M.: Use of general-purpose negation detection to augment concept indexing of medical documents: a quantitative study using the UMLs. J. Am. Med. Inform. Assoc. **8**(6), 598–609 (2001)

13. Naldi, M., Petroni, S.: A testset-based method to analyse the negation-detection performance of lexicon-based sentiment analysis tools. Computers **12**(1) (2023). https://doi.org/10.3390/computers12010018, https://www.mdpi.com/2073-431X/12/1/18

14. Savova, G.K., et al.: Mayo clinical text analysis and knowledge extraction system (cTAKES): architecture, component evaluation and applications. J. Am. Med. Inform. Assoc. **17**(5), 507–513 (2010). https://doi.org/10.1136/jamia.2009.001560

15. Sun, J., Zhang, Q., Wang, Y., Zhang, L.: MedAI at SemEval-2021 task 10: negation-aware pre-training for source-free negation detection domain adaptation. In: Palmer, A., Schneider, N., Schluter, N., Emerson, G., Herbelot, A., Zhu, X. (eds.) Proceedings of the 15th International Workshop on Semantic Evaluation (SemEval-2021), pp. 1283–1288. Association for Computational Linguistics, Online (2021). https://doi.org/10.18653/v1/2021.semeval-1.183, https://aclanthology.org/2021.semeval-1.183

16. Vincze, V., Szarvas, G., Farkas, R., Móra, G., Csirik, J.: The bioscope corpus: biomedical texts annotated for uncertainty, negation and their scopes. BMC Bioinform. **9**(11), 1–9 (2008)

17. Zamaraeva, O., Howell, K., Rhine, A.: Improving feature extraction for pathology reports with precise negation scope detection. In: Bender, E.M., Derczynski, L., Isabelle, P. (eds.) Proceedings of the 27th International Conference on Computational Linguistics, pp. 3564–3575. Association for Computational Linguistics, Santa Fe, New Mexico, USA (2018). https://aclanthology.org/C18-1302

18. Frederic: Categorization and construction of rule based systems. Commun. Comput. Inf. Sci. **459** (2014). https://doi.org/10.1007/978-3-319-11071-4-18

19. Liu, H., Gegov, A., Stahl, F.: Categorization and construction of rule based systems. In: Mladenov, V., Jayne, C., Iliadis, L. (eds.) EANN 2014. CCIS, vol. 459, pp. 183–194. Springer, Cham (2014). https://doi.org/10.1007/978-3-319-11071-4_18

EnsembleFS: an R Toolkit and a Web-Based Tool for a Filter Ensemble Feature Selection of Molecular Omics Data

Aneta Polewko-Klim[1]([✉]) [ID], Paweł Grablis[1], and Witold Rudnicki[1,2,3] [ID]

[1] Faculty of Computer Science, University of Bialystok, K. Ciołkowskiego 1M,
15-245 Białystok, Poland
anetapol@uwb.edu.pl
[2] Computational Center, University of Bialystok, K. Ciołkowskiego 1M,
15-245 Białystok, Poland
[3] Interdisciplinary Centre for Mathematical and Computational Modelling,
University of Warsaw, Kupiecka 32, 03-046 Warsaw, Poland

Abstract. The development of more complex biomarker selection protocols based on the machine learning (ML) approach, with additional processing of information from biological databases (DB), is important for the accelerated development of molecular diagnostics and therapy.

In this study, we present *EnsembleFS* user-friendly R toolkit (R package and Shiny web application) for heterogeneous ensemble feature selection (EFS) of molecular omics data that also supports users in the analysis and interpretation of the most relevant biomarkers. *EnsembleFS* is based on five feature filters (FF), namely, U-test, minimum redundancy maximum relevance (MRMR), Monte Carlo feature selection (MCFS), and multidimensional feature selection (MDFS) in 1D and 2D versions. It uses supervised ML methods to evaluate the quality of the set of selected features and retrieves the biological characteristics of biomarkers online from the nine DB, such as Gene Ontology, WikiPathways, and Human Protein Atlas. The functional modules to identify potential candidate biomarkers, evaluation, comparison, analysis, and visualization of model results make *EnsembleFS* a useful tool for selection, random forest (RF) binary classification, and comprehensive biomarker analysis.

Keywords: ensemble feature selection · machine learning · omics

1 Introduction

The molecular omics data are generally unbalanced and high-dimensional with a low sample size, and have complex correlation structures. Although multiple bioinformatic tools have been developed to analyze omics data, the practical process of selecting, evaluating, and analyzing crucial biomarkers from these data is a significant challenge for researchers.

Various feature selection (FS) methods implemented in the R and Python packages are usually used to construct the computational pipeline for the discovery of biomarkers from omics data. Researchers often use open-source software

L. Franco et al. (Eds.): ICCS 2024, LNCS 14835, pp. 88–96, 2024.
https://doi.org/10.1007/978-3-031-63772-8_7

for biomarker identification available in public repositories, such as GitHub, as well as non-commercial automated software, such as, for example, the Omic-Selector [18]. However, the use of these tools usually requires a certain level of experience in programming and statistics, knowledge of ML methods, and specific hardware resources. Moreover, these ready-made FS procedures are usually designed and optimized for specific types of omics data [3] and particular research tasks. Only a handful of tools are specialized in selecting biomarker candidates from omics datasets for supervised ML methods. In the literature, we found only a few tools that partially address this issue, such as the Feature-Select software in MATLAB [13], the OmicSelector tools based on deep learning [18], the standalone program BioDiscML [11], MRMD3.0 Python tool [8], and the mixOmics R package [17]. FeatureSelect uses three classes of FS methods and then applies optimization algorithms to find the optimal feature subset and create predictive models. MRMD3.0 uses seven FF, three wrapper methods, and seven embedding methods to search for the best features for classifiers. BioDiscML provides multiple multivariate data analyses and uses wrappers to find the optimal combination of features to predict outcomes. The MixOmics offers multivariate FS methods for exploring and integrating biological data sets. The software libraries mentioned above focus on FS and ML techniques that can find the minimal and optimal combination of features for predicting models or multiomics data integration tasks. These tools do not have functionalities that allow the user to retrieve biological information about biomarkers from the database and use methods that are susceptible to noise and instability.

Here, we propose a comprehensive tool for biomarker discovery in quantitative omics data that allows users to: (i) select the top biomarker candidates using either an ensemble of FS methods or any individual FS method, (ii) build and evaluate predictive models using the RF classifier, (iii) evaluate the quality of the feature set, (iv) benchmark model results for various FS methods, (v) retrieve biological information about the top biomarkers from the nine biological DB, e.g. molecular function, cellular component, and biological process from the Gene Ontology (GO) [2], signalling pathways from the Kyoto Encyclopedia of Genes and Genomes (KEGG) [15], and disease phenotypes from the Human Phenotype Ontology (HP) [10].

2   Methods

2.1   Feature Selection and Classification Algorithms

The *EnsembleFS* uses the U-test, MRMR [4], MCFS [5], MDFS-1D and MDFS-2D methods [14] to remove irrelevant variables. These feature filters are not related to the classifier and have better generalization properties than wrappers and embedded methods [9].

The MDFS measures the decrease in the information entropy of the decision variable due to the knowledge of k-dimensional tuples of variables and measures the influence of each variable in the tuple [14]. This FF performs an exhaustive search of all possible k-tuples and assign to each variable a maximal information gain due to a given variable that was achieved in any of the k-tuples that included this variable. The 2D version of this algorithm (MDFS-2D) can capture synergistic interactions between feature pairs and the decision variable.

The MRMR method is based on mutual information (MI) as a measure of the relevancy and redundancy of features, where the feature redundancy is an aggregate MI measure between each pair of features in the selected feature subset, and relevance to a class variable is an aggregate MI measure between each feature with respect to the class variable.

The MCFS method is based on a Monte Carlo approach to select informative features. This algorithm is capable of incorporating interdependencies between features. The MCFS offers several cutoff methods (e.g. critical angle, k-means, and permutations) for discerning informative and non-informative features.

The random forest algorithm [1] was used to construct predictive models. This algorithm works well in data sets with a small number of objects, has few tuneable parameters that do not relate directly to the data, is very rarely faulty, and usually gives results that are often the best or very close to the best results achievable by any classification algorithm [6].

Algorithm 1: EFS$(l, f, S = \{P_1, \ldots, P_k\})$ the ensemble FS algorithm with RF classifier (1).

---

input : Feature filters f_j, $j = 1, \ldots, m$

Dataset $S = \{(y, X)\}$ with n entries of p features $V = \{v_1, \ldots, v_p\}$

belonging to one of two classes, randomly split into k partitions P_i

output: Combined set of informative features F

Ranked informative feature set F_j, $j = 1, \ldots, m$

Performance estimation metric E_j, $j = 1, \ldots, m$

Feature selection stability measure A_j, $j = 1, \ldots, m$

repeat r **times**

    foreach S_i **do**

        Generate the training set $S_{\backslash i}(V) \leftarrow S(V) \setminus P_i(\text{V})$

        foreach f_j **do**

            Perform feature selection on the training set $W_i \leftarrow f(S_{\backslash i}(V))$

            Collect the ranked informative feature set $W_i = \{v_1, \ldots, v_d\}$

            Remove highly correlated features with W_i

            Build the model on the training set $L_i \leftarrow l(S_{\backslash i}(U_i))$ using top N features U_i with W_i

            Performance estimation E_i: use the model L_i on a test set P_i

        end

    end

end

$E_j \leftarrow \frac{1}{r \cdot k} \Sigma E_i$, $i = 1, \ldots, r \cdot k$

Assess the FS stability A_j of $r \cdot k$ feature sets U_i

Collect the feature set F_j from $r \cdot k$ sets U_i by using the majority voting strategy, for each of m feature filters

Collect combined feature set $F = \bigcup_{j=1}^{m} F_j$ or $F = \bigcap_{j=1}^{m} F_j$

2.2   Ensemble Feature Selection

The process of selecting relevant features from the original dataset and the model-building procedure executed in the *EnsembleFS* is shown in Algorithm 1. The area under the receiver operator curve (AUC), the accuracy (ACC) and the Matthews correlation coefficient (MCC) are used to assess the performance of the model, and the Lustgarten stability measure (ASM) was used [12] to assess the stability of selection.

3   *EnsembleFS* an R Toolkit

EnsembleFS is based on carefully chosen statistical and ML methods recommended for biomedical data and uses feature filters based on alternative approaches: statistical, information theory, and methods sensitive to interactions between variables. It allows the user to select and rank relevant biomarkers from quantitative omics data using an ensemble of various FS algorithms (U-test, MRMR, MCFS, MDFS-1D, and MDFS-2D). The user can modify the list of FS methods used and the values of their parameters. The quality of the feature set can be verified by applying the RF classification algorithm within a stratified k-fold cross-validation (CV) or alternatively within size-k random sampling, repeated n times. Redundant and correlated features can be removed from extracted feature subsets. The stability of the feature sets returned by *EnsembleFS* is measured using ASM, while the performance of the predictive models is estimated using AUC, ACC, and MCC metrics. *EnsembleFS* allows to set the values of selected parameters for ML models, such as the top N number of features that the classifier will use. It also allows the user to compare the results of the predictive models obtained using the features returned by different FS methods. The results of the ML models are visualized in the form of interactive tables and plots. The final feature set for each filter is obtained by majority voting on the CV results. The combined set of biomarkers for biological analysis is obtained, depending on the user choice, as the intersection or union of the results of all the filters used.

  EnsembleFS accepts data that include different biomarker identifiers (ID), such as Ensembl gene ID, NCBI Entrez gene ID, and Uniprot IDs. It should be underlined that *EnsembleFS* allows the generation of final report files that include the results of ML models and crucial information on the top genes from nine biological DBs.

3.1   Web Application

EnsembleFS web application (app) has an interactive web interface for data analysis and visualization. This tool consists of a module for selecting relevant biomarkers and a module for collecting biological information on genes. The functionalities of these modules are available via *Feature Selection* tab and *Gene Information* tab, respectively. The general functional specification of these basic software modules is presented in Fig. 1. *EnsembleFS* web app is available online at https://uco.uwb.edu.pl/apps/EnsembleFS (webserver demo). It is open source, free software under an MIT license. *EnsembleFS* web app architecture, the source code, workflow, tutorial and the exemplary report of feature selection and modelling results are described in detail in the *Home* tab, *Help* tab, and project home page https://github.com/biocsuwb/EnsembleFS.

3.2   R Package

EnsembleFS R package includes software to select, collect, analyze and interpret the top biomarkers with omics data. Compared to the *EnsembleFS* web app, the R package includes the *ensembleFS()* dynamic function that allows the user to easily add any other FS method to the default list of five basic FF (*methods* input parameter). Users can manually set all the hyperparameters of the model. The size of input data is limited only by the computer's performance. Source code, examples, implementation details, and documentation are available on GitHub (https://github.com/biocsuwb/EnsembleFS-package).

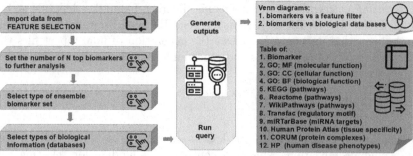

Fig. 1. Main functionality modules of *EnsembleFS* web app: A) Feature Selection tab, B) Gene information tab. Cuboids represent the interaction between *EnsembleFS* and the user, and the octagons represent *EnsembleFS* processes. For notes, see text.

4 Use Case

To demonstrate selected capabilities of *EnsembleFS* web app in a real case study, we used RNA-seq data from the TCGA-LUAD (https://www.cancer.gov/tcga) program [7]. The description of the data set, the preprocessing procedure, and the example results of feature selection and classification of tumor vs normal tissue are included in [16] and the Help tab → Example sub-tab. For testing purposes, we used only 574 samples and 2000 differentially expressed genes (DEGs) with the highest difference in gene expression level. To find the most relevant DEGs for the classification of tissue types, we performed the ensemble FS. Default parameters were used for all FS methods. The 0.3 random sampling, with 10 iterations, was selected for model validation. We conducted the quantitative analysis of the most informative DEGs and compared the performance of the prediction models for each FS method with the top N features. Our analysis shows that the five FS methods identified 1608 unique DEGs in total. MDFS-2D filter identified the highest number of relevant features (1024 DEGs) in ten subsets of features. The best predictive model (ACC = 0.996 ± 0.016) was obtained with the top 30 features selected by the MRMR method. The best overall predictive results were achieved by the RF classifier with at least the top 75 features for all filters (Fig. 2 (left panel)). In this regard, the DEGs returned by individual FS algorithms for N = 75 were chosen for further biological analysis. It should be noted that the final DEG sets selected by different filters for N = 75 were quite divergent (Fig. 2(right panel)). Although 136 DEGs were selected in total, none of the DEGs was identified by all FS methods.

In the next step, we submitted queries to the nine biological DBs and collected information on the 51 genes found. Molecular function categories for 22 genes were indicated. The fifty-nine GO biological processes were examined for 23 genes. Nineteen significant cellular components were found for 12 genes. Seven

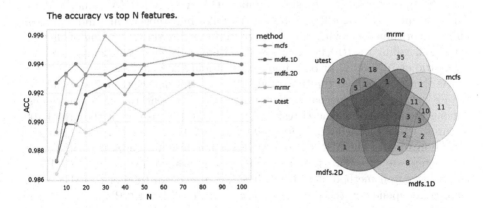

Fig. 2. Left panel: the average values for the ACC between 10 feature subsets as a function of N top features for all filters for LUAD data. Right panel: the final number of the most relevant biomarkers with FS methods for N = 75. See notation in text.

Table 1. Execution times (hh:mm:ss) for a single iteration of the FS algorithm and RF classification for the TCGA-LUAD dataset with 574 samples and p biomarkers. The execution time of information searches in nine biological DB for the m-number of most relevant biomarkers with the EFS method (union of top features with five FS methods). Default parameters were used for each FS method. The 3-fold CV and 0.3 random sampling (RS) were used for model validation (V). The calculations were performed on an Intel Core i5-12400 CPU using 32 GB RAM. For notes, see text.

p	m	V	U-test	MDFS-1D	MDFS-2D	MRMR	MCFS	Ensemble	DB query
100	80	CV	00:00:05	00:00:04	00:00:04	00:00:04	00:00:13	00:00:31	00:05:44
	82	RS	00:00:02	00:00:01	00:00:02	00:00:02	00:00:05	00:00:11	00:05:26
200	140	CV	00:00:09	00:00:09	00:00:08	00:00:05	00:00:20	00:00:52	00:09:51
	149	RS	00:00:03	00:00:03	00:00:03	00:00:02	00:00:07	00:00:19	00:09:38
1000	205	CV	00:02:18	00:02:16	00:02:19	00:00:07	00:01:14	00:08:17	00:14:23
	265	RS	00:00:49	00:00:47	00:00:49	00:00:02	00:00:27	00:02:57	00:13:59

metabolic and signalling pathways were selected from the KEGG and 14 pathways from the WikiPathways. Higher tissue-specific expression was observed in five genes. Twenty-six protein complex-coding genes were found. Higher tissue-specific expression was observed in 5 genes. And finally, 20 disease-related phenotypes were detected. Our analysis revealed a series of genes (SLC25A10, TGFBR1, and SFTPC, etc.) related to lung cancer.

5   Computational Aspects

To test the speed efficiency of the *EnsembleFS* web app, we reviewed its performance for various data sizes. Table 1 presents example execution times of the FS and RF algorithm for previously described TCGA-LUAD RNA sequence data. The one run of the algorithm involved the following steps: calling individual or all FS algorithms, removing correlated features, estimating a ranking of the features, and calling the RF classification algorithm for random sampling (train-test split ratio of 70%–30%) or one time for the 3-fold CV method.

As shown in Table 1, the time of algorithm performance strongly depends on the FS method and hyperparameter tuning when the feature number increases. Among the applied FF, the U-test and MDFS are the fastest for the initial 100 features, while the MRMR is for 1000 features. It should be added that the execution time of the MRMR and MCFS algorithms increases if their default parameters are changed, that is, a number of relevant features for the MRMR and other cutoff methods for the MCFS. The execution time of the MDFS-2D algorithm depends on the processor's architecture (CPU or GPU).

6 Summary

We developed the *EnsembleFS* R toolkit (R package and web app) for individual FS or ensemble FS of high-dimensional molecular data and automatic collection of information on the most relevant genes from the nine well-known biological databases. In this work, we present the selected capabilities and the advantages of the *EnsembleFS* web application. Our results show that *EnsembleFS* is an excellent tool for selection, binary classification, and comprehensive analysis of biomarkers. We provide *EnsembleFS* as a freely accessible web server for users (30 MB limit on the total amount of data). For much larger data, we recommend using the *EnsembleFS* R package. In the current version *EnsembleFS* uses five feature filters and the random forest classifier. In the future, we plan to increase the number of feature selection and classification methods.

References

1. Breiman, L.: Random forests. Mach. Learn. **45**, 5–32 (2001)
2. Gene Ontology Consortium: The gene ontology resource: enriching a gold mine. Nucleic Acids Res. **49**(D1), D325–D334 (2021)
3. Determan, C.: Optimal algorithm for metabolomics classification and feature selection varies by dataset. Int. J. Biol. **7**(1) (2015)
4. Ding, C., Peng, H.: Minimum redundancy feature selection from microarray gene expression data. J. Comput. Biol. Bioinform. **3**(2), 185–205 (2005)
5. Draminski, M., Koronacki, J.: rmcfs: an R package for monte Carlo feature selection and interdependency discovery. J. Stat. Softw. **85**(12), 1–28 (2018)
6. Fernandez-Delgado, M., et al.: Do we need hundreds of classifiers to solve real world classification problems? J. Mach. Learn. Technol. **15**, 3133–3181 (2014)
7. Hammerman, P., et al.: Comprehensive genomic characterization of squamous cell lung cancers. Nature **489**, 519–525 (2012)
8. He, S., et al.: MRMD3. 0: a Python tool and webserver for dimensionality reduction and data visualization via an ensemble strategy. J. Mol. Biol., 168116 (2023)
9. Jović, A., et al.: A review of feature selection methods with applications. In: 38th International Convention on Information and Communication Technology, Electronics and Microelectronics (MIPRO), pp. 1200–1205 (2015)
10. Köhler, S., et al.: The human phenotype ontology in 2021. Nucleic Acids Res. **49**(D1), D1207–D1217 (2021)
11. Leclercq, M., et al.: Large-scale automatic feature selection for biomarker discovery in high-dimensional omics data. Front. Genet. **10**, 452 (2019)
12. Lustgarten, J., et al.: Measuring stability of feature selection in biomedical datasets. In: AMIA Annual Symposium Proceedings, pp. 406–410 (2009)
13. Masoudi-Sobhanzadeh, Y., et al.: FeatureSelect: a software for feature selection based on machine learning approaches. BMC Bioinformatics **20**(170) (2019)
14. Mnich, K., Rudnicki, W.R.: All-relevant feature selection using multidimensional filters with exhaustive search. Inf. Sci. **524**, 277–297 (2020)
15. Okuda, S., et al.: KEGG atlas mapping for global analysis of metabolic pathways. Nucleic Acids Res. **36**, W423-6 (2008)

16. Polewko-Klim, A., Rudnicki, W.R.: Analysis of ensemble feature selection for correlated high-dimensional RNA-Seq cancer data. In: Krzhizhanovskaya, V.V., et al. (eds.) ICCS 2020. LNCS, vol. 12139, pp. 525–538. Springer, Cham (2020). https://doi.org/10.1007/978-3-030-50420-5_39

17. Rohart, F., et al.: mixOmics: an R package for 'omics feature selection and multiple data integration. PLoS Comput. Biol. **13**(11), e1005752 (2017)

18. Stawiski, K., et al.: OmicSelector: automatic feature selection and deep learning modeling for omic experiments. bioRxiv (2022). https://doi.org/10.1101/2022.06.01.494299

A Method for Inferring Candidate Disease-Disease Associations

Pietro Cinaglia[1,3](✉)[iD] and Marianna Milano[2,3][iD]

[1] Department of Health Sciences, Magna Graecia University, 88100 Catanzaro, Italy
[2] Department of Experimental and Clinical Medicine, Magna Graecia University, 88100 Catanzaro, Italy
[3] Data Analytics Research Center, Magna Graecia University, 88100 Catanzaro, Italy
{cinaglia,m.milano}@unicz.it

Abstract. The analysis of Disease-Disease Associations (DDA) and Gene-Disease Associations (GDA) is a relevant task in bioinformatics. These are analysed to investigate the interactions between sets of diseases and genes as well as their similarity, e.g., to improve the phases of diagnosis, prognosis and treatment in medicine. Generally, the extraction of information of interest from large-scale data, usually heterogeneous and unstructured, is performed via time-consuming processes. Therefore, several computational approaches have been focused on their prediction through data integration and machine learning techniques.

This paper presents a solution for Inferring DDA (*IDDA*) by integrating curated biomedical ontologies and medical dictionaries. It is able to extract a set of DDA using an in-house score based on the GDA. A preliminary step based on data enrichment retrieves the information about gene and disease, and it integrates these with a set of curated biological data ontologies and dictionaries. Specifically, *IDDA* extracts DDAs based on an in-house score, which uses GDAs for its evaluations. In a preliminary step, it performs data enrichment to retrieve concepts both for diseases and genes, by integrating several curated biomedical ontologies and medical dictionaries.

Keywords: bioinformatics · gene-disease · disease-disease · ontologies · data integration

1 Background

In recent years, a large amount of genomic and biological data is analysed in clinical research trials to evaluate novel treatments, to correlate human diseases with genomics data, as well as for knowledge extraction [3,5,7,8]. For instance, the genes involved in a disease are analysed for knowledge extraction, to understand its key factors (e.g., molecular basis and biological mechanisms), as well as to evaluate treatments and diagnosis. Furthermore, disease profiling also uses -omics data (e.g., genomics, transcriptomics, metabolomics) for evaluating susceptibility, progression and manifestation.

L. Franco et al. (Eds.): ICCS 2024, LNCS 14835, pp. 97–104, 2024.
https://doi.org/10.1007/978-3-031-63772-8_8

Data integration allows cataloguing and analysing heterogeneous and unstructured information from different types and models, such as transcription factor binding sites, protein interactions, Gene-Disease Associations (GDAs), drug-target associations, medical ontologies and dictionaries, as well as literature repositories [4]. To give a non-exhaustive, genes associated with similar disorders show a higher likelihood of interaction, and diseases with common genes could share similar origins or mechanisms, by extension [13]. Similarly, Disease-Disease Associations (DDAs) represents relationships among diseases, and are useful to investigate diagnosis, prognosis, and treatments.

Experimental methods for GDA are expensive and time-consuming [9], therefore, several computational methods were developed to infer GDA. Generally, these identify concepts from medical literature, as well as by integrating protein interactions, functional annotation of signalling pathways, gene expression, medical vocabulary, disease concepts, and other biomedical data source.

In this scenario, ontologies play a crucial role in obtaining an interdisciplinary view from large and heterogeneous sources [6].

An ontology consists of a formal representation of relationships and properties existing among a set of concepts [2].

Usually, network-based scoring methods are applied to infer DDAs, establishing relationships between two or more diseases, based on biological assumptions; a non-exhaustive example may be: if two known disease gene sets are associated with related diseases, they should be close to each other in the protein or gene network.

DOSE [20] is a well-known tool for scoring similarities between diseases. It uses Disease Ontology (DO) [15] to associate each disease with an identifier, in order to compute semantic similarity between correlated concepts; genetic information and diseases not mapped by DO are disregarded. *DOSE* can apply both Jiang [14] and Wang [19] scores for inference.

In this paper, we present *IDDA*, a solution for Inferring Disease-Disease Associations (IDDA) by integrating curated biomedical ontologies and medical dictionaries.

2   Materials and Methods

In this section, we describe the methodology applied by our solution for integrating and processing the following sets of data: ClinVar [16], MedGen [11], DO, Gene Ontology (GO) [1], and DisGeNet [17].

IDDA integrates the mentioned datasets, to produce its own dataset which enables the enrichment of information related to genes and diseases.

2.1   Datasets

In this section, we propose a description for each dataset of curated biomedical ontologies and dictionaries used by *IDDA*.

ClinVar provides an archive for human medically relevant variants and phenotypes. The phenotypic descriptions available in ClinVar are based on the information maintained by MedGen.

MedGen is a catalogue of human disorders and phenotypes with a genetic component, released by the National Center for Biotechnology Information.

Human disorders are also catalogued in DO, which consists of a set of terms linked hierarchically by using interrelated subtypes.

GO describes the fundamental characteristics of genes and their products in a species-independent manner.

A set of curated GDAs is available in DisGeNET using the UMLS Concept Unique Identifier (UMLS-CUI) and Entrez gene unique integers (GeneID) to identify the disease and the gene, respectively.

2.2   Gene-Disease Associations

Formally, let D be a set of diseases and G be a set of genes, such that $D = [d_1, d_2, ..., d_n]$ and $G = [g_1, g_2, ..., g_m]$, with n and m respectively the size of D and G.

IDDA performs the cross-referencing as the Cartesian product to build a domain for GDA:

$$\forall g \in G \ \exists d \in D : f(d, g) \rightarrow GDA$$

2.3   Disease-Disease Associations

Let $D = [d_1, d_2, ..., d_n]$ be a set of unique diseases, and $G = [g_1, g_2, ..., g_m]$ be a set of unique genes, respectively with a size of n and m.

Assuming $GDAs$ as the complete set of GDAs extracted by *IDDA*, and each GDA as pair (d_x, g_y) with $d_x \in D$, $g_y \in G$, $1 <= x <= n$ and $1 <= y <= m$.

Let denote $Gd_1d_2 = [cg_1, cg_2, ..., cg_k]$ the subset of k common genes (cg) identified for a specific DDA.

A DDA (d_1, d_2) is formally identified in accordance with the following conditions:

$$\forall cg_i \in Gd_1d_2 \ \exists (d_1, cg_i) \in GDAs, (d_2, cg_i) \in GDAs : Gd_1d_2 \subseteq (GDAs \cap d1, d2)$$

with $1 \leq i \leq k$, $d_1 \in D$, and $d_2 \in D$.

2.4   Score Evaluation

IDDA calculates an own score useful to provide a weight for each association.

Similarly to *DOSE*, Jiang is used by *IDDA* as default method to calculate the semantic similarity based on MF between genes.

Formally, Jiang is an Information Content (IC)-based score that can be defined as follows:

$$sim(d_1, d_2) = 1 - \min(1, IC(d_1) + IC(d_2) - 2 \cdot IC(MICA))$$

IDDA calculates its own score for a DDA, to evaluate an associative rank. Let *DDA* be the association between two diseases $D1$ and $D2$. Let $G1 = g_{11}, \ldots, g_{1n}$ be a set of genes with $n = |G1|$ related to $D1$, and $G2 = g_{21}, \ldots, g_{2m}$ be a set of genes with $m = |S|$ for $D2$ networks with $m = |G2|$. A *DDA* is evaluated when there exist one or more common genes between $D1$ and $D2$, formally if $G1 \cap G2 \neq \{\}$.

IDDA measures the pairwise gene similarity by using Jaccard index [12]. The latter computes the proportion of shared genes between $G1$ and $G2$ relative to the total number of genes of $D1$ and $D2$, normalizing the number of common genes in each DDA ($|G1 \cap G2|$).

Formally, the Jaccard index (J) between $G1$ and $G2$ is defined as:

$$J(G1, G2) = \frac{|G1 \cap G2|}{|G1| + |G2| - (|G1 \cap G2|)}$$

with $0 \leq J(G1, G2) \leq 1$. More generally, $J(G1, G2) = 1$ when $G1 = G2$, otherwise, $J(G1, G2) = 0$ when $G1 \cap G2 = \{\}$.

IDDA evaluates two main concepts that we denoted as internal and external similarity: IS and ES, respectively.

The former concerns the average semantic similarity among the common genes between $D1$ and $D2$, assuming $DDA(D1, D2)$, related to $DDA(D1, D2)$.

The latter concerns the average semantic similarity between the other genes belonging to $D1$ and $D2$. Both of these are normalized applying the Jaccard index (reported as J).

Formally, IS and ES are defined below, as well as the semantic similarity cross-function (f). The latter is based on the Jiang method; it is denoted below with $jiang(A)$ or $jiang(A, B)$, with A and B two generic sets of genes without duplicates. Note that $jiang(A)$ performs a score for each of the combinations of A, while $jiang(A, B)$ performs a score between all pairs of genes (A_i, B_j) with $1 \leq i \leq |A|$ and $1 \leq j \leq |B|$.

Formally, the semantic similarity function implemented in *IDDA* is defined as follows:

$$f(G) \longleftarrow \frac{\sum_{i=1}^{|G|-1} \sum_{j=i+1}^{|G|} G_{ij}}{n}$$

with $G = jiang((G1 \cup G2) \otimes (G1 \cup G2))$ (duplicates are discarded).

Internal Similarity (IS):

$$IS = f(G1 \cap G2) \cdot J$$

External Similarity (ES):

$$ES = jiang(G1 - (G1 \cup G2), G2 - (G1 \cap G2)) \cdot (1 - J)$$

***IDDA*'s Score:**

$$IDDA\_score(D1, D2) = 1 - (IS + ES)$$

The value related to "IDDA_score" is expressed within the range $[0, 1]$, where 0 represents a condition of no similarity while 1 represents perfect similarity (the latter can be obtained by comparing a disease with itself, or by comparing two concepts related to the same disease).

3   Results and Discussion

This section reports the results performed to evaluate the efficiency and the validity of *IDDA*.

In preprocessing, a list of $318,001$ diseases was acquired using the ClinVar, that was integrated with DO for extracting $102,851$ disease's identifiers. Based on preprocessed data, a set of $461,633$ GDAs are extracted from DisGeNET.

IDDA identified a preliminary set of $19,957,259$ DDAs with a high redundancy, that was processed producing $10,283,680$ DDAs. The latter are reported as associations (d_1, d_2, cg), with cg the number of common genes between d_1 and d_2. Additionally, these are linked to DO terms to allow a comparison with other methods that supports only DO as source for the information. Therefore, the resulting dataset consists of $5,705$ DDA associated to a unique term in DO.

Homogeneous subgroup are isolated to identify similarity within the samples in *IDDA*'s dataset; this task was performed by using K-Means [18] as clustering algorithm. Furthermore, the elbow method [10] is applied, to determinate the optimal no. of clusters (k). Briefly, the elbow method selects the number of clusters to be such that adding a cluster does not significantly reduce the within-group sum of squares.

In our experimentation, *IDDA* was compared with *DOSE* (see Sect. 1), to evaluate its performance.

Note that *DOSE* applies the Jiang score on the DO's graph, thus the result is not related to the no. of genes (or other genomic information), contrarily to *IDDA*. Furthermore, *DOSE* was used to map the *IDDA* results on DO graph, by applying the Wang method on pathways related to each pair of diseases and the Jiang method for evaluating gene similarities.

We performed One-Way ANOVA tests as statistical analysis, to check the following hypothesis:

- differences among clusters in testing dataset are statistical significant both for *IDDA* and *DOSE* based on Jiang score.
- for each disease exists a correlation between its *IDDA* and the related pathway in DO.

Table 1 shows results for the first hypothesis. The One-Way ANOVA test between *IDDA* and *DOSE* is statistically significant. This confirms that the clustering (for $k = 3$) produced relevant groups that are effectively able to identify subgroups for the testing dataset.

Furthermore, this test verifies that both *IDDA*'s score and *DOSE* are able to identify the degree for a DDA. This hypothesis suggests that (i) there is also a correlation between the two dependent variables, and (ii) *IDDA*'s score is able to evaluate a DDA in according to *DOSE* method. Note that the two methods use different approaches, respectively the first evaluates the gene similarity, while the second evaluates the similarity by using DO information which do not contain genomic data.

Table 1. First hypothesis. The One-Way ANOVA test was performed between *IDDA* and *DOSE*. It confirms that the clustering (with $k = 3$) has produced relevant groups that are effectively able to identify subgroups for the testing dataset. The result is statistically significant. *Note: Df is the degrees of freedom.*

		Sum of Squares	Df	Mean Square	F	sig.
IDDA	**Between Groups**	0.269	2	0.135	41.680	< 0.01
	Within Groups	0.446	138	0.003		
	Total	0.715	140			
DOSE	**Between Groups**	1.223	2	0.612	41.736	< 0.01
	Within Groups	2.022	138	0.015		
	Total	3.245	140			

The second hypothesis is evaluated by performing a Bivariate (Pearson, two-tailed) Correlation between *IDDA* and *DOSE*, as shown in Table 2. The result is statistically significant.

Table 2. Second hypothesis. Bivariate correlation (Person, two-tailed) between *IDDA* and *DOSE* scores.

		IDDA	DOSE
IDDA	**Pearson Correlation**	1	0.483**
	Sig. (2-tailed)		< 0.01
	N	141	141
DOSE	**Pearson Correlation**	0.483**	1
	Sig. (2-tailed)	< 0.01	
	N	141	141

Statistical analysis confirms that *IDDA* is able to identify a set of DDA that can be checked by other relevant methods applied to DO.

4   Conclusion

In this paper, we proposed *IDDA*, a solution to infer DDAs by integrating ontologies, gene set enrichment analysis, and semantic similarity among GO terms.

IDDA extracts DDAs based on an in-house score, which uses GDAs for its evaluations. In a preliminary step, it performs data enrichment to retrieve concepts both for diseases and genes, by integrating several curated biomedical ontologies and medical dictionaries: ClinVar, MedGen, DO, DisGenet and GO.

Our experimentation has been conducted to evaluate *IDDA*'s score validity, by comparing results with other relevant methods, as well as by mapping each DDA to DO.

Acknowledgements. This work was funded by the Next Generation EU - Italian NRRP, Mission 4, Component 2, Investment 1.5, call for the creation and strengthening of 'Innovation Ecosystems', building 'Territorial R&D Leaders' (Directorial Decree n. 2021/3277) - project Tech4You - Technologies for climate change adaptation and quality of life improvement, n. ECS0000009. This work reflects only the authors' views and opinions, neither the Ministry for University and Research nor the European Commission can be considered responsible for them.

References

1. Ashburner, M., et al.: Gene ontology: tool for the unification of biology. The Gene Ontology Consortium. Nat. Genet. **25**(1), 25–29 (2000)
2. Asim, M.N., Wasim, M., Khan, M.U.G., Mahmood, W., Abbasi, H.M.: A survey of ontology learning techniques and applications. Database (Oxford) **2018**, January 2018
3. Cinaglia, P., Cannataro, M.: Network alignment and motif discovery in dynamic networks. Netw. Model. Anal. Health Inform. Bioinform. **11** (2022). https://doi.org/10.1007/s13721-022-00383-1
4. Cinaglia, P., Cannataro, M.: Identifying candidate gene-disease associations via graph neural networks. Entropy (Basel) **25**(6) (2023)
5. Cinaglia, P., Cannataro, M.: A method based on temporal embedding for the pairwise alignment of dynamic networks. Entropy **25**(4) (2023). https://doi.org/10.3390/e25040665
6. Cinaglia, P., Guzzi, P.H., Veltri, P.: INTEGRO: an algorithm for data-integration and disease-gene association. In: 2018 IEEE International Conference on Bioinformatics and Biomedicine (BIBM), pp. 2076–2081 (2018). https://doi.org/10.1109/BIBM.2018.8621193
7. Cinaglia, P., Milano, M., Cannataro, M.: Multilayer network alignment based on topological assessment via embeddings. BMC Bioinform. **24**(1) (2023). https://doi.org/10.1186/s12859-023-05508-5. http://dx.doi.org/10.1186/s12859-023-05508-5
8. Cinaglia, P., Tradigo, G., Cascini, G.L., Zumpano, E., Veltri, P.: A framework for the decomposition and features extraction from lung DICOM images. In: Proceedings of the 22nd International Database Engineering & Applications Symposium, IDEAS 2018, pp. 31–36. Association for Computing Machinery (2018)
9. Cinaglia, P., Vázquez-Poletti, J.L., Cannataro, M.: Massive parallel alignment of RNA-SEQ reads in serverless computing. Big Data Cognit. Comput. **7**(2) (2023). https://doi.org/10.3390/bdcc7020098. https://www.mdpi.com/2504-2289/7/2/98

10. Fukuoka, Y., Zhou, M., Vittinghoff, E., Haskell, W., Goldberg, K., Aswani, A.: Objectively measured baseline physical activity patterns in women in the mPED Trial: cluster analysis. JMIR Public Health Surveill. **4**(1), e10 (2018)

11. Fung, K.W., Bodenreider, O.: Utilizing the UMLS for semantic mapping between terminologies. In: AMIA Annual Symposium Proceedings, pp. 266–270 (2005)

12. Fuxman Bass, J.I., Diallo, A., Nelson, J., Soto, J.M., Myers, C.L., Walhout, A.J.: Using networks to measure similarity between genes: association index selection. Nat. Methods **10**(12), 1169–1176 (2013)

13. Goh, K.I., Cusick, M.E., Valle, D., Childs, B., Vidal, M., Barabasi, A.L.: The human disease network. Proc. Natl. Acad. Sci. U.S.A. **104**(21), 8685–8690 (2007)

14. Jiang, J.J., Conrath, D.W.: Semantic similarity based on corpus statistics and lexical taxonomy. CoRR **cmp-lg/9709008** (1997)

15. Kibbe, W.A., et al.: Disease Ontology 2015 update: an expanded and updated database of human diseases for linking biomedical knowledge through disease data. Nucleic Acids Res. **43**(Database issue), D1071–1078 (2015)

16. Landrum, M.J., et al.: ClinVar: improving access to variant interpretations and supporting evidence. Nucleic Acids Res. **46**(D1), D1062–D1067 (2018)

17. Pinero, J., et al.: DisGeNET: a comprehensive platform integrating information on human disease-associated genes and variants. Nucl. Acids Res. **45**(D1), D833–D839 (2017)

18. Steinley, D., Brusco, M.J.: Initializing k-means batch clustering: a critical evaluation of several techniques. J. Classification **24**(1), 99–121 (2007)

19. Wang, J.Z., Du, Z., Payattakool, R., Yu, P.S., Chen, C.F.: A new method to measure the semantic similarity of GO terms. Bioinformatics **23**(10), 1274–1281 (2007)

20. Yu, G., Wang, L.G., Yan, G.R., He, Q.Y.: DOSE: an R/Bioconductor package for disease ontology semantic and enrichment analysis. Bioinformatics **31**(4), 608–609 (2015)

Network Model with Application to Allergy Diseases

Konrad Furmańczyk[1,4]([⊠]) [iD], Wojciech Niemiro[2] [iD], Mariola Chrzanowska[3,4] [iD], and Marta Zalewska[4] [iD]

[1] Institute of Information Technology, Warsaw University of Life Sciences, Warsaw, Poland
konrad_furmanczyk@sggw.edu.pl
[2] Faculty of Mathematics, Informatics and Mechanics University of Warsaw, Warsaw, Poland
[3] Institute of Economics and Finance, Warsaw University of Life Sciences, Warsaw, Poland
mariola_chrzanowska@sggw.edu.pl
[4] Department of Prevention of Environmental Hazards, Allergology and Immunology, Medical University of Warsaw, Warsaw, Poland
marta.zalewska@wum.edu.pl

Abstract. We propose a new graphical model to describe the comorbidity of allergic diseases. We present our model in two versions. First, we introduce a generative model that reflects the variables' causal relationships. Then, we propose an approximation of the generative model by a misspecified model, which is computationally more efficient and easily interpretable. In both versions of our model, we consider typical allergic disease symptoms and covariates. We consider two directed acyclic graphs (DAGs). The first one describes information about the coexistence of certain allergic diseases (binary variables). The second graph describes the relationships between particular symptoms and the occurrence of these diseases. In the generative model, the edges lead from diseases to symptoms, corresponding to causal relations. In the misspecified model, we reverse the direction of edges: they lead from symptoms to diseases. The proposed model is evaluated on a cross-sectional multicentre study in Poland (www.ecap.pl). An assessment of the stability of the proposed model is obtained using the bootstrap and jackknife techniques. Our results show that the misspecified model is a good approximation of the generative model and helps predict the incidence of allergic diseases.

Keywords: Network Model · Bayesian Network · Logistic Regression · Allergy Diseases

1 Introduction

Modelling dependence between different binary variables is an essential statistical task with many applications in medicine, life sciences, economics, and sociology. The basic statistical tools used in such situations are the autologistic (AL) model [2] and graphical network modelling [1,4,15]. General information on

L. Franco et al. (Eds.): ICCS 2024, LNCS 14835, pp. 105–112, 2024.
https://doi.org/10.1007/978-3-031-63772-8_9

graphical models for discrete data can be found in [12,13]. The classical AL model [2] has been applied in epidemiology, marketing, agriculture, ecology, forestry, geography, and image analysis [5–7,11,17]. The most common approach to estimation of the model parameters is the pseudo-likelihood [3] method. Zalewska et al. [21] recommended a heuristic estimation method. Recently, Shin et al. [17] invented and applied an AL network model for a disease progression study using pseudo-likelihood to estimate the model parameters.

Our paper proposes a new graphical model that is related to but different from the AL model. We aim to describe the interdependence of allergic diseases in contrast to most studies that do not consider dependences between allergies [8,9,20].

We present our model in two versions. First, we introduce a generative model that reflects the variables' causal relationships. Then, we propose an approximation of the generative model by a misspecified model, which is computationally more efficient and easily interpretable. We focus on the misspecified version, which we consider more practical. In both versions of our model, we consider typical allergic disease symptoms, family history of allergic disease, and control variables as covariates. We consider two directed acyclic graphs (DAGs), both based on experts' knowledge. The first one describes information about the coexistence of certain allergic diseases (binary variables). The second graph describes the relationships between particular symptoms and the occurrence of these diseases. In the generative model, the edges lead from diseases to symptoms, corresponding to causal relations. In the misspecified model, we reverse the direction of edges: they lead from symptoms to diseases. This trick significantly reduces computational costs. Our model was naturally divided into separate logistic models for individual allergy diseases. Each logistic regression is estimated by the standard generalized linear model (GLM) procedure. Our general approach is very flexible and can be applied to any dependence model for binary variables. We compare predictions based on two versions of our model. We argue that the misspecified model is a good approximation for the more logically consistent but computationally expensive generative model. The paper is organized as follows. Section 2 introduces the proposed methodology. In Sect. 3, we apply this methodology to construct a new model of comorbidity of allergic diseases, based on a big epidemiological data set (ECAP) [16]. At the end of this section, we present an evaluation of the proposed model. Section 4 and 5 contain discussion and conclusions. All computations are carried out with the R package (www.r-project.org). Below we provide a graphical user guide illustrating our methodology.

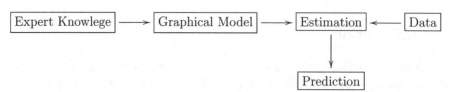

2   Hierarchical Logistic Network Models

2.1   Generative Model

Our proposed model contains four groups of variables. In the first group, we consider a random vector $\mathbf{Y} = (Y_1, \ldots, Y_p)^T$ with binary components. Each of these variables determines presence or absence of a given allergic disease for a patient. In our application we describe p allergic diseases. Taking into account the known co-occurrence of diseases, the relationships between them are described by a directed graph with the adjacency matrix $\mathbf{A} = (a_{ki})$ as follows: $a_{ki} = 1$ if Y_i is affected by Y_k and otherwise $a_{ki} = 0$.

In the second group, we have a random vector of symptoms of our diseases $\mathbf{S} = (S_1, \ldots, S_m)^T$. The remaining two groups consist of common factors $\mathbf{F} = (F_1, \ldots, F_l)^T$, which can affect all considered diseases (for example genetic features) and a vector of additional covariates $\mathbf{X} = (X_1, \ldots, X_r)^T$ such as gender, age, residence of a patient, etc. Symptoms S_i can be continuous or discrete random variables. It is usually known which symptoms are characteristic for each disease. This knowledge can be represented by a directed graph with adjacency matrix $\mathbf{B} = (b_{kj})$ such that: $b_{kj} = 1$ if Y_k causes S_j and otherwise $b_{kj} = 0$.

The full generative model includes diseases $\mathbf{Y}$, symptoms $\mathbf{S}$, common factors $\mathbf{F}$ and additional covariates $\mathbf{X}$. This graph has edges among $\mathbf{Y}, \mathbf{S}$ variables given by matrices $\mathbf{A}, \mathbf{B}$, and all edges leading from $\mathbf{F}, \mathbf{X}$ variables to all components of $\mathbf{Y}, \mathbf{S}$. We assume that the graph corresponding to the adjacency matrix $\mathbf{A}$ is acyclic. Consequently, the whole graph is a directed acyclic graph (DAG). The conditional probability distribution of $\mathbf{Y}, \mathbf{S}$ is given by

$$
\begin{aligned}
P(\mathbf{Y} = \mathbf{y}, \mathbf{S} = \mathbf{s} | \mathbf{F} = \mathbf{f}, \mathbf{X} = \mathbf{x}) &= \prod_{i=1}^{p} P(Y_i = y_i | \mathbf{Y}_{pa}(Y_i), \mathbf{F} = \mathbf{f}, \mathbf{X} = \mathbf{x}) \\
&\times \prod_{j=1}^{m} P(S_j = s_j | \mathbf{Y}_{pa}(S_j), \mathbf{F} = \mathbf{f}, \mathbf{X} = \mathbf{x}),
\end{aligned}
\tag{1}
$$

where $\mathbf{Y}_{pa}(Y_i) = \{Y_k : Y_k \rightarrow Y_i\}$ is a set of diseases which affect the occurrence of disease Y_i, $\mathbf{Y}_{pa}(S_j) = \{Y_k : Y_k \rightarrow S_j\}$ is a set of diseases which cause symptom S_j. We assume the following parametric form of conditional distributions:

$$
\log \frac{P(Y_i = 1 | \mathbf{Y}_{pa}(Y_i), \mathbf{F} = \mathbf{f}, \mathbf{X} = \mathbf{x})}{P(Y_i = 0 | \mathbf{Y}_{pa}(Y_i), \mathbf{F} = \mathbf{f}, \mathbf{X} = \mathbf{x})} = \omega_{0i} + \sum_{k=1}^{p} a_{ki} \omega_{ki} Y_k + \mathbf{x}^T \boldsymbol{\alpha}_i + \mathbf{f}^T \boldsymbol{\beta}_i, \tag{2}
$$

$$
\log \frac{P(S_j = 1 | \mathbf{Y}_{pa}(S_j), \mathbf{F} = \mathbf{f}, \mathbf{X} = \mathbf{x})}{P(S_j = 0 | \mathbf{Y}_{pa}(S_j), \mathbf{F} = \mathbf{f}, \mathbf{X} = \mathbf{x})} = \gamma_{0j} + \sum_{k=1}^{p} b_{kj} \gamma_{kj} Y_k + \mathbf{x}^T \boldsymbol{\delta}_j + \mathbf{f}^T \boldsymbol{\epsilon}_j. \tag{3}
$$

We thus have the following model parameters: $\omega_{0i} \in R, \omega_{ki} \in R, \boldsymbol{\alpha}_i \in R^r, \boldsymbol{\beta}_i \in R^l, \gamma_{0j} \in R, \gamma_{kj} \in R, \boldsymbol{\delta}_j \in R^r, \boldsymbol{\epsilon}_j \in R^l$. Since the conditional probability (1) consists of the product of $p + m$ probabilities, the parameters of each factor can be estimated separately by a standard logistic regression procedure. To improve prediction accuracy we also applied weighted logistic regression. However, the results obtained by both methods were almost identical (Supplement [19]: C3-C4).

2.2 Misspecified Model

Unfortunately, the model presented in the previous subsection is computationally demanding, and its parameters are difficult to interpret. We propose using another, misspecified model that does not reflect causal relations between variables but is computationally more accessible for a big network and has parameters with simple, intuitive meaning. We change the direction of edges joining symptoms and diseases. Entries of adjacency matrix $\mathbf{B}$ will now be interpreted as follows: $b_{ij} = 1$ indicates the presence of arrow $Y_i \leftarrow S_j$. We assume that the remaining edges of the graph are the same as in the generative model. In the misspecified model, Eq. (1) is replaced by Eq. (4), and Eqs. (2)–(3) are replaced by Eq. (5) as follows:

$$P(\mathbf{Y} = \mathbf{y}|\mathbf{S}, \mathbf{F}, \mathbf{X}) = \prod_{i=1}^{p} P(Y_i = y_i|\mathbf{Y}_{pa}(Y_i), \mathbf{S}_{pa}(Y_i), \mathbf{F}, \mathbf{X}), \qquad (4)$$

where $\mathbf{S}_{pa}(Y_i) = \{S_j : Y_i \leftarrow S_j\}$ is a set of symptoms related to occurrence of disease Y_i. Similarly as in generative model, we assume a log-linear form of conditional distributions. To simplify notation, we use the same symbols for the parameters for both models.

$$\log \frac{P(Y_i = 1|\mathbf{Y}_{pa}(Y_i), \mathbf{S}_{pa}(Y_i), \mathbf{F} = \mathbf{f}, \mathbf{X} = \mathbf{x})}{P(Y_i = 0|\mathbf{Y}_{pa}(Y_i), \mathbf{S}_{pa}(Y_i), \mathbf{F} = \mathbf{f}, \mathbf{X} = \mathbf{x})} = \omega_{0i} + \sum_{k=1}^{p} a_{ki}\omega_{ki}Y_k$$

$$+ \sum_{j=1}^{m} b_{ij}\gamma_{ij}S_j + \mathbf{x}^T\boldsymbol{\alpha}_i + \mathbf{f}^T\boldsymbol{\beta}_i. \qquad (5)$$

3   Application to Modelling Allergic Diseases

In this section we apply the proposed approach to investigate the prevalence of allergic diseases and their interdependences. Our model is based on a big epidemiological study in Poland (ECAP) [16]. More details can be found in the Supplement [18]-Section A.

3.1   The Structure of the Model

The first group of variables consists of 5 selected allergic diseases Y_1, Y_2, Y_3, Y_4, Y_5. The left panel of Fig. 1 illustrates the dependences between them, based on the literature and on discussions with medical doctors [10,14]. The second group of variables consists of typical symptoms of those diseases: S_1, S_2, S_3. Additionally we consider history of allergy diseases in the family:

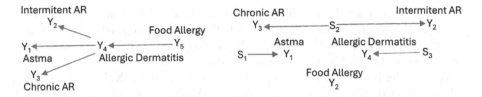

Fig. 1. The Graphs with adjacency matrices $\mathbf{A}, \mathbf{B}$

F_1, F_2, F_3, F_4, F_5. The right panel of Fig. 1 shows dependences between allergic diseases and their symptoms. The direction of arrows in Fig. 1 lead from symptoms to diseases which corresponds to the misspecified model. In the last group of variables, we consider control covariates: X_1, X_2, X_3, X_4 (they describe age, gender, residence of patients). More detailed description of all the variables can be found in the Supplement [18]-Section B).

3.2   Generative and Misspecified Models of Allergy Diseases

We recall the generative model in which diseases cause symptoms. Taking into account the structure of the graph with adjacency matrices $\mathbf{A}, \mathbf{B}$, we see that, conditionally on covariates $\mathbf{F}$ and $\mathbf{X}$, the conditional distribution of $\mathbf{Y}$ given symptoms $\mathbf{S}$ has the form

$$P(Y_1|Y_2,Y_3,Y_4)P(Y_2|Y_4)P(Y_3|Y_4)P(Y_4|Y_5)P(Y_5)P(S_1|Y_1)P(S_2|Y_2,Y_3)P(S_3|Y_4).$$

(We omitted $\mathbf{F}$ and $\mathbf{X}$ in this formula).

Now we turn to the misspecified model. Conditionally on covariates $\mathbf{F}$ and $\mathbf{X}$, the joint probability $P(\mathbf{Y}, \mathbf{S})$ is determined as:

$$P(\mathbf{Y}|\mathbf{S}) = P(Y_1|Y_2,Y_3,Y_4,S_1)P(Y_2|Y_4,S_2)P(Y_3|Y_4,S_2)P(Y_4|Y_5,S_3)P(Y_5).$$

We now formulate specific equations restricting attention to the misspecified model only. We assume the logistic form of the conditional probabilities (formulas (4)–(5)). We estimate each of them separately using standard R function 'glm'. The subsequent equations concern the logits for asthma Y_1, intermittent allergic rhinitis Y_2, chronic allergic rhinitis Y_3, allergic dermatitis Y_4. The equations are:

$$logit_1 = \omega_{01} + \sum_{j=1}^{4} \alpha_{j1}X_j + \sum_{j=1}^{5} \beta_{j1}F_j + \gamma_{11}S_1 + \sum_{j=2}^{4} \omega_{j1}Y_j.$$
$$logit_2 = \omega_{02} + \sum_{j=1}^{4} \alpha_{j2}X_j + \sum_{j=1}^{5} \beta_{j2}F_j + \gamma_{22}S_2 + \omega_{42}Y_4.$$
$$logit_3 = \omega_{03} + \sum_{j=1}^{4} \alpha_{j3}X_j + \sum_{j=1}^{5} \beta_{j3}F_j + \gamma_{32}S_2 + \omega_{43}Y_4.$$
$$logit_4 = \omega_{04} + \sum_{j=1}^{4} \alpha_{j4}X_j + \sum_{j=1}^{5} \beta_{j4}F_j + \gamma_{43}S_3 + \omega_{54}Y_5.$$

3.3   Comparison of Two Versions of Our Model

We compute the 'diagnostic' probabilities of diseases given symptoms for the generative and the misspecified model. It is worth noting that in the case of a large network, it would not be possible to calculate $P(\mathbf{Y}|\mathbf{S}, \mathbf{F}, \mathbf{X})$ or $P(Y_i|\mathbf{S}, \mathbf{F}, \mathbf{X})$ exactly in the generative model. In this situation, the misspecified model has an advantage over the generative model. The two models can be compared in the case of a small network as that considered here.

We consider five scenarios (different values of of covariates $\mathbf{X}, \mathbf{F}$, symptoms $\mathbf{S}$ and coexistent diseases Y_i). Let $p_1 = P(Y_1 = 1|Y_2 = 0, Y_3 = 0, Y_4 = 0, S_1), q_1 = P(Y_1 = 1|Y_2 = 1, Y_3 = 1, Y_4 = 1, S_1), p_2 = P(Y_2 = 1|Y_4 = 0, S_2), q_2 = P(Y_2 = 1|Y_4 = 1, S_2), p_3 = P(Y_3 = 1|Y_4 = 0, S_2), q_3 = P(Y_3 = 1|Y_4 = 1, S_2), p_4 = P(Y_4 = 1|Y_5 = 0, S_3), q_4 = P(Y_4 = 1|Y_5 = 1, S_3)$. The results for the first two scenarios are presented in Tables 1 (all 5 scenarios are given in the Supplement [18]-Section C). The difference between the two models obtained is negligible.

Table 1. Comparison between the generative model and misspecified model

Scenario	Model	p_1	p_2	p_3	p_4	q_1	q_2	q_3	q_4
1	generative	0.021	0.081	0.077	0.024	0.597	0.104	0.134	0.024
	misspecified	0.023	0.085	0.080	0.015	0.566	0.097	0.125	0.044
2	generative	0.103	0.282	0.322	0.208	0.886	0.326	0.461	0.524
	misspecified	0.088	0.270	0.307	0.081	0.842	0.299	0.421	0.216

3.4   Estimation of Parameters and Evaluation of the Model

We report the estimated coefficients of logistic regression, their standard errors, the odds ratios with confidence intervals (CI) in the Supplement [18]-Section C. The accuracy of estimators and robustness of our model is evaluated using the bootstrap and jackknife techniques. The dataset is divided into a learning and testing sample to assess if the proposed model is adequate. The ROC curve and average AUC on the testing sample are determined from 20 repetitions. Table 2 shows the AUC values for the averaged AUC values for bootstrap and jackknife. The ROC curves (Fig. 1–Fig. 16) are collected in the Supplement [18]-Section D as well as interpretation of the results from the medical point of view. Our results show good stability of the model.

Table 2. AUC for each logit

$logit_i$	$i = 1$	$i = 2$	$i = 3$	$i = 4$
bootstrap	0.8470	0.6986	0.7201	0.7931
jackknife	0.8165	0.6857	0.7215	0.7921

4   Discussion

Previous studies of multimorbidity in allergy [8–10,14,20] were based on fitting of single logistic models that did not take into account the correlations between the studied diseases. Our graphical model uses two DAGs to describe such dependences. The proposed model can be used in studies of associations of other diseases and, in general, in the study of correlations in complex systems.

5   Conclusions

Both versions of our model (generative and misspecified) produced similar results. The latter is computationally more efficient and easily interpretable. Evaluation of the model using bootstrap and jackknife techniques yielded average AUCs ranging from 0.67 to 0.84 (Table 2), indicating relatively high stability of the results. Both bootstrap and jackknife methods could be used to construct confidence intervals for the model parameters and classification metrics. Our model can help predict the incidence of allergic diseases and will allow for a better understanding of the complex co-occurrence of these diseases. It also sheds light on the impact of such covariates as gender, age, family history, etc. on allergic diseases. The proposed model can be easily extended by adding other potential factors influencing the occurrence of the diseases. Due to the nature of our task, we considered the low-dimensional case where the number of observations n is greater than the number of features p. Naturally, the proposed approach can be generalized to the high-dimensional case $p > n$ by adding the Lasso [19] or Ridge penalty for log-likelihood for each logit model separately. This will be the topic of further research.

Disclosure of Interests. The authors have no competing interests to declare that are relevant to the content of this article.

References

1. Abeyasinghe, P.M., et al.: Consciousness and the dimensionality of DOC patients via the generalized Ising model. J. Clin. Med. **9**(5), 1332 (2020)
2. Besag, J.E.: Nearest-neighbour systems and the auto-logistic model for binary data. J. R. Stat. B Stat. Methodol. **34**(1), 75–83 (1972)
3. Besag J.E.: Statistical analysis of non-lattice data. Statistician **24**(3), 179–195 (1975)
4. Briganti, G., Linkowski, P.: Exploring network structure and central items of the narcissistic personality inventory. Int. J. Methods Psychiatr. Res. **29**(1), e1810 (2000)
5. Caragea, P.C., Kaiser, M.S.: Autologistic models with interpretable parameters. JABES **14**, 281–300 (2009)
6. Gégout-Petit, A., Guérin-Dubrana, L., Li, S.: A new centered spatio-temporal autologistic regression model with an application to local spread of plant diseases. Spat. Stat. **31**, 100361 (2019)

7. He, F., Zhou, J., Zhu, H.: Autologistic regression model for the distribution of vegetation. JABES **8**(2), 205–222 (2003)
8. Jung, S., et al.: Risk factors and comorbidities associated with the allergic rhinitis phenotype in children according to the ARIA classification. Allergy, Asthma Immunol. Res. **12**(1), 72–85 (2020)
9. Kim, H.Y., et al.: Prevalence and comorbidity of allergic diseases in preschool children. Korean J. Pediatr. **56**(8), 338–342 (2013)
10. Krzych-Fałta, E., Furmańczyk, K., Piekarska, B., Tomaszewska, A., Sybilski, A., Samoliński, B.K.: Allergies in urban versus countryside settings in Poland as part of the epidemiology of the allergic diseases in Poland (ECAP) study-challenge the early differential diagnosis. Adv. Dermatol. Allergol. **33**(5), 359–368 (2016)
11. Koutsias, N.: An autologistic regression model for increasing the accuracy of burned surface mapping using Landsat Thematic Mapper data. Int. J. Remote Sens. **24**(10), 2199–2204 (2003)
12. Maathuis, M., Drton, M., Lauritzen, S., Wainwright, M. (eds.): Handbook of Graphical Models. Chapman & Hall/CRC Press (2019)
13. Madigan, D., York, J., Allard, D.: Bayesian graphical models for discrete data. Int. Stat. Rev. **63**(2), 215–232 (1995)
14. Raciborski, F., et al.: Dissociating polysensitization and multimorbidity in children and adults from a Polish general population cohort. Clin. Transl. Allergy **9**, 4 (2019)
15. Ravikumar, P., Wainwright, M.J., Lafferty, J.: High-dimensional Ising model selection using l1-regularized logistic regression. Ann. Statist. **38**, 1287–1319 (2010)
16. Samoliński, B., Raciborski, F., Lipiec, A., et al.: Epidemiologia Chorób Alergicznych w Polsce (ECAP). Alergol Pol. **1**(1), 10–18 (2014)
17. Shin, Y.E., et al.: Autologistic network model on binary data for disease progression study. Biometrics **75**(4), 1310–1320 (2019)
18. Furmańczyk, K., Niemiro, W., Chrzanowska, M., Zalewska, M.: Supplementary material to the paper 'network model with application to allergy diseases' (2024). https://github.com/kfurmanczyk/Network-_Allergy/blob/main/Supplement1.pdf
19. Tibshirani, R.: Regression shrinkage and selection via the lasso. J. R. Stat. **58**(1), 267–288 (1996)
20. Westman, M., et al.: Natural course and comorbidities of allergic and nonallergic rhinitis in children. J. Allergy Clin. Immunol. **129**(2), 403–408 (2012)
21. Zalewska, M., Niemiro, W., Samoliński, B.: MCMC imputation in autologistic model. Monte Carlo Methods Appl. De Gruyter **16**(3–4), 421–438 (2010)

TM-MSAligner: A Tool for Multiple Sequence Alignment of Transmembrane Proteins

Joel Cedeño-Muñoz[1], Cristian Zambrano-Vega[2], and Antonio J. Nebro[3,4](✉)

[1] Facultad de Ciencias Pecuarias y Biológicas, State Technical University of Quevedo, Quevedo, Los Ríos, Ecuador
jacedeno@uteq.edu.ec

[2] Facultad de Ciencias la Ingeniería, State Technical University of Quevedo, Quevedo, Los Ríos, Ecuador
czambrano@uteq.edu.ec

[3] ITIS Software, Edificio de Investigación Ada Byron, University of Málaga, Málaga 29071, Spain
ajnebro@uma.es

[4] Dept. de Lenguajes y Ciencias de la Computación, University of Málaga, Málaga 29071, Spain

Abstract. Transmembrane proteins (TMPs) are crucial to cell biology, making up about 30% of all proteins based on genomic data. Despite their importance, most of the available software for aligning protein sequences focuses on soluble proteins, leaving a gap in tools specifically designed for TMPs. Only a few methods target TMP alignment, with just a couple of the available to researchers. Considering that there are a few particular differences that ought to be taken into consideration aligning TMPs sequences, standard MSA methods are ineffective to align TMPs. In this paper, we present TM-MSAligner, a software tool designed to deal with the multiple sequence alignment of TMPs by using a multi-objective evolutionary algorithm. Our software include features such as transmembrane substitution matrix dynamically used according to the topology region, a high penalty to gap opening and extending, and two MSA quality scores, Sum-Of-Pairs with Topology Prediction and Aligned Segments, that can be optimized at the same time. This approach reduce the number of Transmembrane (TM) and non-Transmembrane (non-TM) broken regions and improve the TMP quality score. TM-MSAligner outputs the results in an HTML format, providing an interactive way for users to visualize and analyze the alignment. This feature allows for the easy identification of each topological region within the alignment, facilitating a quicker and more effective analysis process for researchers.

Keywords: Multiple sequence alignment · transmembrane proteins · multi-objective optimization · evolutionary algorithms · software framework

L. Franco et al. (Eds.): ICCS 2024, LNCS 14835, pp. 113–121, 2024.
https://doi.org/10.1007/978-3-031-63772-8_10

1   Introduction

The study of Transmembrane Proteins (TMPs) sequences has taken increasing attention in recent years due to their fundamental roles in various biological processes and their significance as potential drug targets and life science research [9,14,15]. Transmembrane proteins are involved in vital cellular functions, such as signal transduction, ion transport, and cell adhesion, making them key players in maintaining cellular homeostasis and energy production [11]. Sequence analysis methods for TMPs are of great interest in the biomedical and bioinformatics domains and understanding the structural and functional aspects of these proteins is crucial for unraveling the complexities of cellular mechanisms.

Multiple Sequence Alignment (MSA) remains one of the most powerful tools for assessing evolutionary sequence relationships and for identifying structurally and functionally important protein regions [11]. It serves as a foundational step for a range of further analyses in protein family studies, including homology modeling, predicting secondary structures, and understanding phylogenetic relationships. Transmembrane regions exhibit unique amino acid compositions and conservation patterns, differing significantly from soluble proteins. Traditional MSA methods fail to consider these distinctions when aligning TMPs, resulting in reduced accuracy of the alignments. Furthermore, there are few techniques available that can align TMPs while also optimizing for more than one MSA quality score, highlighting a gap that we address in this paper.

We introduce TM-MSAligner, a novel software tool aimed at finding multiple sequence alignments of TMPs using a multi-objective evolutionary algorithm [2]. These are stochastic nature-inspired search algorithms belonging to the family of metaheuristics [1] that do not guarantee to find optimal solutions but they usually provide accurate solutions in a reasonable amount of time. The alignment of TMPs is formulated as a bi-objective optimization where the Sum-Of-Pairs with Topology Prediction and Segments Aligned are defined as scores to be maximized, so its optimum is a set of trade-off solutions between the two objectives known. In the field of multi-objective optimization, this set is known as the Pareto set and their correspondence in the objective space is referred as to Pareto front. Due to the stochastic feature of multi-objective evolutionary algorithms, they provide as a result an approximation to the Pareto front.

The results obtained when TM-MSAligner is executed are generated in both CSV and HTML format, allowing the latter to plot the found alignments, so that the user can choose the solution that best meet defined criteria and the HTML page will show the alignment using the MSABrowser[1] viewer. In this way, the topology of the amino acids is shown in different colours, which facilitates the process of identifying the topological regions included in the alignments.

The rest of the paper is structured as follows. The package is described in Sect. 2 and an usage example is detailed in Sect. 3. The next section includes a discussion about our tool and, finally, Sect. 5 provides the conclusions and future works.

---

[1] MSABrowser: https://thekaplanlab.github.io/.

Table 1. Parameter space of TM-MSAligner. Types: (c)ategorical, (i)nteger, (r)eal. (CDA: crowdingDistanceArchive, IRG: insertRandomGap, MAGG: mergeAdjunt-edGapsGroups, SCG: shiftClosedGaps, SANGG: splitANonGapsGroup, ExtArch: external archive)

Parameter/Component	Type	Domain	Dependency
algorithmResult	c	{externalArchive, population}	
populationSizeWithArchive	i	[10, 200]	algorithmResult == ExtArch
externalArchive	c	{CDA, unbounded, hypevolume}	algorithmResult == ExtArch
offspringPopulationSize	i	[1, 400]	
selection	c	{tournament, random}	
selectionTournamentSize	i	[2, 10]	selection == tournament
replacement	c	{rankingAndDensityEstimator}	
ranking	c	{dominance, strength}	
densityEstimator	c	{crowdingDistance, knn}	
kValueForKNN	i	[1, 3]	densityEstimator == knn
variation	c	{crossoverAndMutation}	
crossover	c	{SPX}	
crossoverProbability	r	[0.0, 1.0]	
mutation	c	{IRG, MAGG, SCG, SANGG}	
mutationProbabilityFactor	r	[0.0, 2.0]	

2 Software Description

The core of TM-MSAligner is a multi-objective evolutionary algorithm that combines features of algorithms such as NSGA-II [3] and SPEA2 [16]. The package is implemented in Java as a Maven project and uses the jMetal framework for multi-objective optimization using metaheuristics [4,8] as a dependence. TM-MSAligner is an open source project under MIT license[2].

2.1 Software Architecture

The evolutionary algorithm of TM-MSAligner architecture is based on a workflow of components where each component is a step of the algorithm. Some of these components have more than one implementation and they can also have control parameters. The parameter space of TM-MSAligner is shown in Table 1.

The approach adopted to create the initial population is to pre-compute alignments by using existing tools (e.g., ClustalW2, T-Coffee, Muscle, Kalign, Mafft, Probcons, etc.) and recombining them to get the desired number of individuals for the population. This way the algorithm is able of providing accurate

---

[2] TM-MSAligner: https://github.com/jMetal/TM-MSAligner.

solutions faster than initializing the population with alignments obtained by filling the gaps randomly.

The evaluation of the population can be carried out sequentially or in parallel. The parallel model can be synchronous, where the behavior of the evolutionary algorithm does not change, or asynchronous [5], which can be more efficient than the synchronous one when using a large number of cores.

2.2   Transmembrane Proteins Features

TM-MSAligner includes some features to lead with TMPs sequences. The most relevant are:

- **Topology Prediction:** Topology prediction is used to identify transmembrane regions within the protein sequences. We have used DeepTMHMM [6], a deep learning protein language model-based algorithm to predict the topology of both alpha-helical and beta-barrels proteins. This software exports the results in a *.3line* format file, where each line represents the name of the sequence, the sequence of amino-acids and the TM topology, respectively. This information must be included in the input data file.
- **Transmembrane substitution matrix:** The substitution matrix is determined dynamically in our algorithm. It depends on consistent TM predictions over a column. The Sum-Of-Pairs MSA quality score applies the PHAT [10] (Predicted Hydrophobic and Transmembrane) substitution matrix on consistently predicted TM regions and BLOSUM [7] substitution matrix on non-TM regions.
- **Regions based Gap Penalty:** A TM-regions based Gap Penalty is incorporated in our proposal. The TM regions and non-TM regions are respectively denoted with different *Open Gap Penalties* and *Extended Gap Penalties*, so, with the aim of make TM regions harder to be broken during the aligning, the gap penalties in TM regions is higher than gap penalties in non-TM regions.
- **Aligned Segments:** The second fitness score to optimize is to generate alignments with the highest number of aligned regions with the same topology inside the MSA.

2.3   Execution Modes

TM-MSAligner allows to use a number of choices to configure and run the evolutionary algorithm:

- `TM_MSAligner`: All parameters can be set manually, allowing advanced users to fine-tune the settings. This execution mode is provided through a class that is typically executed in an integrated development environment (IDE), such as Eclipse or IntelliJ Idea.
- `ConfigurableTM_MSAligner`: The algorithm accepts as input a string with a particular combination of parameter values, what it is very convenient if we intend to run TM-MSAligner from the command line.

- **BAliBASETest:** This mode provides an adapted version of TM-MSAligner to solve the instances of the BAliBASE-ref7 benchmarking [12]. The unaligned sequences, the topology information of the proteins and the pre-computed alignments are saved in the **resources** folder of the project.

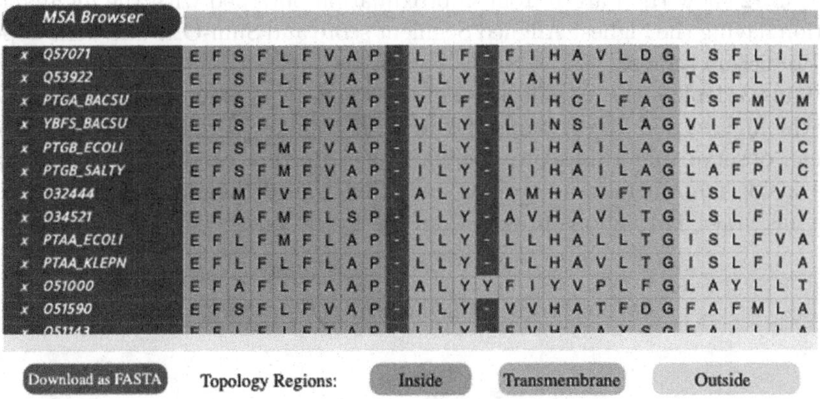

Fig. 1. Example of how MSAViewer displays the alignments, giving a different color to each topology for an easy identification of the regions.

2.4 Output Results

TM-MSAligner generates the following results:

- A Web page with the plot of the Pareto Front approximation and the visualization of the alignments selected by user, who can click the preferred point over the plot figure.
- List of alignments which represents the Pareto Front approximation. Each MSA solution is illustrated in HTML format using the viewer MSABrowser [13]. With the aim of identify the regions into the MSA, the topology of the aminoacids is colored with different values, as can be seen in the example of a MSA visualization shown in Fig. 1.
- The output of the *Observer* selected. It can be a plot figure or the best scores reached by the algorithm during its execution.

3   Illustrative Example

To illustrate the working of TM-MSAligner, we use it to solve the TMPs dataset of BAliBASE called reference 7 [12]. The original sequences, the transmembrane topology information, and the pre-computed alignments are saved in the *ref7* sub-folder of the *resources* directory.

Figure 2 show the Pareto front approximation obtained with the highlighted solution having the highest Aligned Segment score and Sum-Of-Pairs with Topology Prediction scores.

Figure 3 depicts the visualization of the alignments assigned to the solution with the highest and lowest Sum-Of-Pair with Topology Prediction score values. We can observe that this score penalizes the insertion of Opening Gaps, even more inside transmembrane regions, and that the alignment with the highest score has fewer TM regions broken than the lowest one.

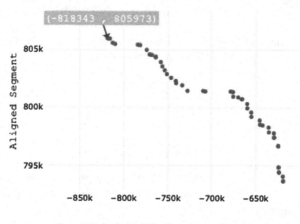

Fig. 2. Solution selected with the highest Segment Aligned score.

TM and non-TM regions in one block

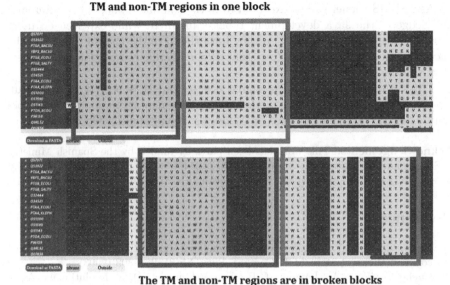

The TM and non-TM regions are in broken blocks

Fig. 3. Visualization of the Alignments with the highest (image above) and lowest (image below) Sum-Of-Pair with Topology Prediction scores.

4   Discussion

Nowadays, there are many MSA procedures that have been built and tested to align homologous soluble proteins, but only few have been adapted to lead with TMPs and only a pair are currently available. Given the biomedical importance of TMps and the large and growing gap between the number of solved TMP structures and the number of TMP sequences, sequence analysis techniques are crucial.

To address this challenge, TM-MSAligner is a software tool that allows users with a bioinformatics background to find sets of accurate alignments for TMPs representing trade-offs solutions according to the Sum-Of-Pairs with Topology Prediction and Aligned Segment objectives. The alignments performed by our software will have more conserved TM and non-TM regions. The parallelism features of TM-MSAligner can help to accelerate the optimization process by making use of the available cores of modern CPUs.

5   Conclusions

We have presented TM-MSAligner, a software tool developed to find multiple sequence alignment of transmembrane proteins by using a multi-objective evolutionary algorithm. For the purpose to improve the alignment accuracy for TMPs, some specific features have be taken, adapting our MSA method to lead with TMPs.

As TM-MSAligner can be executed from the command line, only a minimum knowledge of the Java development tools is required, while experienced users have access to the source code, so they have the have the chance to extend the package with new components (e.g., mutation and crossover operators) and new algorithms.

TM-MSAligner has an open source license and it is hosted in a GitHub repository containing the source code and the documentation. The software tool can be downloaded and executed on Windows, Linux and macOS computers.

Acknowledgements. This work has been partially funded by the Spanish Ministry of Science and Innovation via Grant PID2020-112540RB-C41 (AEI/FEDER, UE) and by the Junta de Andalucía, Spain, under contract QUAL21 010UMA.

References

1. Blum, C., Roli, A.: Metaheuristics in combinatorial optimization: overview and conceptual comparison. ACM Comput. Surv. **35**(3), 268–308 (2003)
2. Coello Coello, C.A., Lamont, G.B., Van Veldhuizen, D.A.: Evolutionary Algorithms for Solving Multi-Objective Problems, 2nd edn. Springer, New York (2007). https://doi.org/10.1007/978-0-387-36797-2. iSBN 978-0-387-33254-3
3. Deb, K., Pratap, A., Agarwal, S., Meyarivan, T.: A fast and elitist multiobjective genetic algorithm: NSGA-II. IEEE Trans. Evol. Comput. **6**(2), 182–197 (2002)
4. Durillo, J.J., Nebro, A.J.: jMetal: a Java framework for multi-objective optimization. Adv. Eng. Softw. **42**(10), 760–771 (2011)
5. Durillo, J.J., Nebro, A.J., Luna, F., Alba, E.: A study of master-slave approaches to parallelize NSGA-II. In: 2008 IEEE International Symposium on Parallel and Distributed Processing, pp. 1–8 (2008)
6. Hallgren, J., et al.: DeepTMHMM predicts alpha and beta transmembrane proteins using deep neural networks. bioRxiv (2022)
7. Henikoff, S., Henikoff, J.: Amino acid substitution matrices from protein blocks. Proc. Nat. Acad. Sci. **89**(22), 10915–10919 (1992)
8. Nebro, A.J., Durillo, J.J., Vergne, M.: Redesigning the jMetal multi-objective optimization framework. In: Genetic and Evolutionary Computation Conference, pp. 1093–1100 (7 2015)
9. Ng, D.P., Poulsen, B.E., Deber, C.M.: Membrane protein misassembly in disease. Biochimica et Biophysica Acta (BBA) - Biomembranes **1818**(4), 1115–1122 (2012). Protein Folding in Membranes
10. Ng, P.C., Henikoff, J.G., Henikoff, S.: PHAT: a transmembrane-specific substitution matrix. Bioinformatics **16**, 760–766 (2000)
11. Pirovano, W., Abeln, S., Feenstra, K.A., Heringa, J.: Multiple alignment of transmembrane protein sequences. In: Structural Bioinformatics of Membrane Proteins, pp. 103–122. Springer, Vienna (2010). https://doi.org/10.1007/978-3-7091-0045-5_6
12. Thompson, J.D., Koehl, P., Ripp, R., Poch, O.: BAliBASE 3.0: latest developments of the multiple sequence alignment benchmark. Proteins Struct. Funct. Bioinf. **61**(1), 127–136 (2005)
13. Torun, F.M., Bilgin, H.I., Kaplan, O.I.: MSABrowser: dynamic and fast visualization of sequence alignments, variations and annotations. Bioinf. Adv. **1**(1), vbab009 (2021)

14. Wallin, E., Heijne, G.V.: Genome-wide analysis of integral membrane proteins from eubacterial, archaean, and eukaryotic organisms. Protein Sci. **7**(4), 1029–1038 (1998)

15. Yin, H., Flynn, A.D.: Drugging membrane protein interactions. Ann. Rev. Biomed. Eng. **18**(1), 51–76 (2016). pMID: 26863923

16. Zitzler, E., Laumanns, M., Thiele, L.: SPEA2: improving the strength pareto evolutionary algorithm. Technical report, 103, Swiss Federal Institute of Technology (ETH), Zurich, Switzerland (2001)

Determining Mouse Behavior Based on Brain Neuron Activity Data

Anastasia Vodeneeva[1]([⊠]), Iosif Meyerov[1], Yury Rodimkov[1], Mikhail Ivanchenko[1], Vladimir Sotskov[2,3], Mikhail Krivonosov[1], and Konstantin Anokhin[4]

[1] Lobachevsky State University of Nizhny Novgorod, Nizhny Novgorod, Russia
alxndrvna@icloud.com
[2] Center of Interdisciplinary Research in Biology, Collège de France, Paris, France
[3] Institute of Biology, École Normale Supérieure, Paris, France
[4] Institute for Advanced Brain Studies, Lomonosov Moscow State University, Moscow, Russia

Abstract. The study of the relationship between brain neuron activity and behavioral responses of humans and other animals is an area of interest, although it has received relatively little attention from scientific biology and medical research centers. In this paper, we consider the problem of determining a mouse position in a circular track based on its neural activity data, and investigate the use of machine learning for solving this problem. The study is conducted in two parts: a classification task, where the model predicts which sector of the track the mouse is in at a particular time, and a regression task, where it predicts exact coordinates for each time step. We propose a neural network-based solution for both tasks, based on a graph of brain neuron activity. Accuracy results were obtained: 89% for classification and 93% for regression.

Keywords: Brain · Neural activity · Artificial intelligence

1 Introduction

The mechanisms underlying brain function and human and animal behavior comprise one of the most significant areas of research within modern science. Complexity, variability, and motivation are the most vital characteristics of the behavioral patterns of living organisms [1]. In this study, we explore the possibility of predicting an organism behavior based on neural impulses using machine learning (ML) tools. The experiment involves a mouse placed on a circular track and freely moving within it. Brain neuron impulses were recorded using a head-mounted NVista HD miniscope [2], which could detect calcium signals from neurons. Cell images were captured using a set of genetically engineered calcium indicators [3, 4]. The mouse with the miniscope was placed on a track that had been previously cleaned of foreign odors. There are four marks along the mutually perpendicular diameters of the track, which allow the animal to draw any conclusions about its current position. At each point in time during the experiment, the coordinates of the mouse position are recorded. The video recording frequency is 20 frames per second, and its total duration is 15 min and 39 s. Data for the experiment was obtained in article [5].

© The Author(s), under exclusive license to Springer Nature Switzerland AG 2024
L. Franco et al. (Eds.): ICCS 2024, LNCS 14835, pp. 122–129, 2024.
https://doi.org/10.1007/978-3-031-63772-8_11

The main scientific interest in this problem is the ability to determine the coordinates of a mouse location based on impulses from brain neurons using a graph of neural connections and ML methods. The potential of artificial intelligence to analyze and replicate the intelligence of living biological beings offers many opportunities for biological and medical research. This ability forms the basis for our work. In the course of our research, we answer the following questions:

1. Is it possible to construct such an artificial neural network architecture that allows tracking the mouse coordinates with acceptable accuracy based on calcium activity in mouse hippocampal neurons?
2. How well can models be trained using existing data of movement trajectories and neural activity during movements?
3. Which of the two mathematical formulations of the ML task is more suitable for solving this problem?

The article is organized as follows: Sect. 2 reviews background and related work; Sects. 3 and 4 describes the process of solving this problem through classification and regression, respectively. Section 5 concludes the article.

2   Background and Related Work

Despite widespread interest in this problem in the fields of biology and medicine, very little research has been published on this subject.

In [6], authors described an ML method for analyzing the behavior of mice kept in groups up to four individuals for several days in a controlled environment in real time. It was described how this method can be used to study the effects of mutations in genes linked to autism on mouse behavior. In [7], ML techniques were used to distinguish between different mouse conditions based on brain activity and camera data. The aim of the study is to develop a learning approach that could accurately reflect classification results and transfer those results to other mouse conditions.

The work [8] demonstrated that continuous behavioral data can be analyzed using approaches similar to natural language processing. This data supports further research into detecting complex pathophysiological alterations accompanied by changes in the behavioral profile.

The work [9] explored the solution to determining the movement of a mouse based on data from brain neurons activity using a statistical approach without prior knowledge. The authors hypothesized that, when combined with innovative techniques for estimating coordinates, a created Bayesian model could extract data about complex behavior [10, 11]. In [12], the authors solved this problem by reconstructing time series of brain cell activity and identifying fields that constitute cognitive maps. The data was used in the form of a three-dimensional graph of cellular connections, based on an algorithm for reconstructing the dynamic graph of calcium event distribution, with two dimensions being the number of cells in the studied part of the brain and the third being the number of studied time points [13]. The reconstruction of these graphs was done using calcium events from neurons detected using the algorithm described in [5], which was also used in our work to obtain the data.

3 Classification of Mouse Position on a Circular Track

In our study, we classify the mouse position on a circular track by dividing it into sectors. We solve this problem by determining whether a given mouse position belongs to a particular class (sector) of the track. The object of this task is the coordinates of the mouse position angle at different points in time, and the class is a specific sector of the track that the mouse travels along. The set of vertices in the graph is the number of brain cells, which is 562, and the set of edges represents the connections between these cells. The total number of graphs in our dataset is 18 775, corresponding to the number of measurements taken at different times. The response is the angle α of the mouse at each time point (see Fig. 3).

We use a convolutional neural network (CNN) to solve the classification problem of determining the sector of the mouse position on a circular track. The sum of the squared differences between the output signals from the network and their required values is used as a measure of how well the network performs (MSE, mean squared error):

$$R_{MSE} = \frac{1}{n} \sum_{i=1}^{n} \left(\alpha_i - \alpha_i^{predict} \right)^2, \tag{1}$$

where n is a number of classes, α_i is a real angle of the mouse position, and $\alpha_i^{predict}$ is a predicted angle. CNN is used to solve classification problem, based on an example from article [14]. CNN has a structure shown in Fig. 1a. The first 75% of data in time is taken for training, and the remaining 25% is taken for testing. For some points in time, a visualization of the neuron connections is created to obtain a clearer picture of what is happening (see Fig. 2). To generalize results, we use a function that calculates the error as the ratio of difference in real and predicted values to circumference (RGE, resulting generalized error):

$$R_{RGE} = \frac{|\alpha - \alpha^{predict}|}{2\pi} \times 100\%, \tag{2}$$

where r is a radius of the track, α is a real angle of the mouse position, $\alpha^{predict}$ is a predicted angle (all angles are taken in radians).

To compare results obtained in two different cases, classification results are converted to regression results by finding median value for each class:

$$R_i = \frac{\min\left\{ \left| Y_i - Y_i^{predict} \right|, n - \left| Y_i - Y_i^{predict} \right| \right\}}{n}, \tag{3}$$

where n is a number of classes, Y_i is a real number of the sector, $Y_i^{predict}$ is a predicted number. Here and below, all errors are given for the test set.

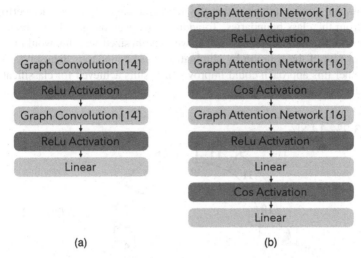

(a)   (b)

Fig. 1. Layer-by-layer structure of CNN (*a*) and GNN (*b*).

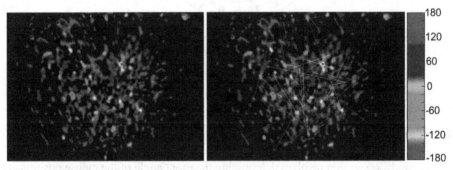

Fig. 2. Images of brain graphs at two moments of time (5 and 25 s from the start). Cells with the same color are activated at the position with that color on the scale (in degrees). (Color figure online)

We started by trying to locate the mouse by dividing the circle into two parts and identifying the halves. Here, we could achieve an RGE of 18%. Next, we attempted to predict the quarter in which the mouse is located, with an RGE of about 19%. For dividing into eight parts, the RGE was 22%; for dividing into twelve parts, it was 25%. Since the mouse size is approximately 8.3% of the circumference, solving the classification problem makes sense if the number of classes does not exceed 12 (see Fig. 3). Additionally, removing intervals where the mouse moves less than 8.3% of the way around the circle reduced the RGE to 14%. It was suggested that if all intervals with constant positions are removed from the dataset and CNN is trained on this new dataset, a smaller error could be achieved. We also considered that the network architecture was too simplistic for training on such a complex task. Therefore, we decided to increase the number of hidden layers and modify the activation functions accordingly. Additionally, the experimental results indicated that the main challenges in the model performance stemmed from the class boundaries. To address this, we changed the problem formulation from classification to regression, as we hypothesized that this would reduce the overall error rate of the solution by eliminating class boundaries themselves. In order to test whether this approach could improve the results achieved for classification, we formulated a regression task.

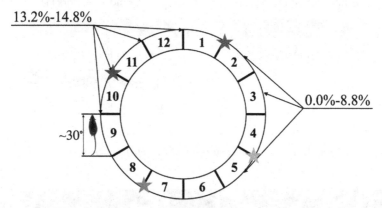

Fig. 3. The case of dividing a track into 12 sectors. The sector measure is 30°, and the length of a mouse is 30°. Colored dots indicate the physical markers for self-identification of mouse. (Color figure online)

4   Regression of Mouse Position on a Circular Track

For our study, regression involves determining the exact angle of the mouse position at each point in time. The set of brain neuron impulses at each time point is considered to be a set of features. The change in the mouse location on the track is the dependent variable. The angle of the mouse coordinate at each moment in time is the output. We chose the mean absolute error (MAE) as an error metric:

$$R_{MAE} = \frac{1}{n} \sum_{i=1}^{n} \left| \alpha_i - \alpha_i^{predict} \right|, \tag{4}$$

where n is a number of predicted values, α_i is a real mouse position angle, $\alpha_i^{predict}$ is a predicted angle [14]. The structure of a graph neural network (GNN) used to solve a regression problem is shown in Fig. 1b [15, 16]. MSE (1) and MAE (4) are used as loss functions. The error obtained using MAE is smaller than the error using MSE. After that, GNN is used for classification and produces better results than CNN.

The plot of error changes on test data is shown in Fig. 4a. As can be seen from this plot, the maximum regression RGE does not exceed 12%, which is a better result than for classification. In addition, by looking at areas of constant mouse position, we concluded that error peaks occur exactly at moments when the mouse stops or starts moving. During all other time intervals, RGE does not exceed 7%. Based on results from solving regression task, plot of predicted mouse coordinate was drawn in Fig. 5b. As expected, because of absence of dividing the track into sectors errors at boundaries disappeared, but because of unpredictable behavior during constant coordinates intervals, maximum RGE equals to 7%. For comparative analysis, dynamic of RGE changes for classification task was also plotted. Plot of classification error changes is shown in Fig. 4b. Here we show that the maximum RGE for classification is greater than for regression, being approximately 15% (which is better than previously obtained). Additionally, it is clear that error reaches this maximum not only at constant positions, but also between them. This occurs at the boundaries between classes when mouse moves from one class to another. Therefore, for classification, the maximum RGE is 11% and for regression 7%. By changing the problem formulation from classification to regression, accuracy increases by 4%. A plot shows the mouse movement curve based on original data and predicted data in regression (see Fig. 5b). The predicted coordinate generally follows the real coordinate dynamics, but has larger fluctuations at intervals with small changes in real coordinates. For more detailed conclusions, a trajectory of the average coordinate was plotted in a sliding window of 100 frames (5 s). This plot is presented in Fig. 5a. The moving average is calculated using an interval $[t - w, t]$, where w is the window size, and t is the averaged data argument. The maximum RGE here is 8%. We conclude that in a window of this width, the predicted trajectory closely follows the actual one.

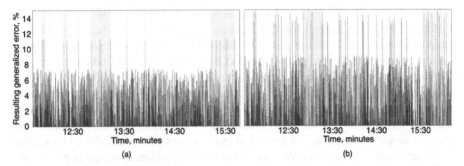

Fig. 4. Error plot for regression (a) and classification (b). The vertical axis represents RGE in percentage terms, the horizontal axis represents time from the start in minutes. The largest errors are shown in red; the intervals where the mouse moves less than 8.3% of the circumference are shown in yellow. (Color figure online)

Based on all the results obtained, Table 1 was compiled showing the values of maximum RGE for two ML problem formulations on two network types (see Fig. 1) using the train and test datasets.

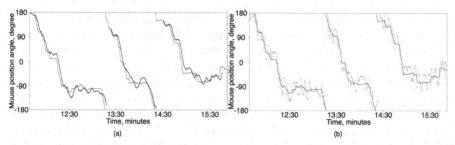

Fig. 5. Average coordinate in a sliding window of 5 s (100 frames wide) (*a*) and the coordinates at each time point (*b*). The vertical axis represents mouse position angle, horizontal axis represents time from the beginning in minutes. Actual coordinates are green; predicted coordinates are blue on *a*, are red on *b*. (Color figure online)

Table 1. Generalizing table of the RGE errors for two formulations of problems when solving them using two methods for constructing a neural network with train and test data.

	Classification		Regression	
	CNN (Fig. 1a) (%)	GNN (Fig. 1b) (%)	CNN (Fig. 1a) (%)	GNN (Fig. 1b) (%)
Train	13	8	9	4
Test	22	15	13	7

5   Conclusions

As a result of the experiments, GNN was found to solve a classification task with an RGE (2) of 11% and a regression task with an RGE (2) of 7%, respectively. Thus, it would be advisable to formulate and solve this problem in terms of regression analysis. Given that the problem addressed in this work has not been widely studied, it is impossible to determine with certainty the minimum error that could be achieved with this data. Additionally, it remains uncertain whether we have all the necessary information available to construct a high-quality neural network. It is now essential to interpret these findings from a neurobiological standpoint and develop a strategy for improvement. This approach should be based on real biological processes, and its results should be applied to a larger number of mice in order to test whether the patterns observed in one mouse apply to others. Does the network trained on the first day of the experiment produce the desired level of accuracy when re-run on the second and third days? What patterns exist between the graphs of neuron activity on different days? We plan to investigate these questions as part of our future research.

Acknowledgments. This research was funded by the "Center of Photonics" funded by the Ministry of Science and Higher Education of the Russian Federation (contract no. 075-15-2022-293).

Data and Code Availability. All data, code and launch scripts used for the article is provided as part of the replication package. It is available at https://github.com/nastyalabs/mouseBrain.

Disclosure of Interests. The authors have no competing interests to declare that are relevant to the content of this article.

References

1. Andrianov, V.: Neurons, brain and behavior. Bull. Int. Acad. Sci. Russ. Sect. **2**, 25–29 (2007)
2. Ghosh, K., et al.: Miniaturized integration of a fluorescence microscope. Nat. Methods **8**, 871–878 (2011)
3. Barykina, N., et al.: FGCaMP7, an improved version of fungi-based ratiometric calcium indicator for in vivo visualization of neuronal activity. Int. J. Mol. Sci. **21**, 3012 (2020)
4. Subach, O., et al.: Novel genetically encoded bright positive calcium indicator NCaMP7 based on the mNeonGreen fluorescent protein. Int. J. Mol. Sci. **21**, 1644 (2020)
5. Sotskov, V., et al.: The rapid formation of CA1 hippocampal cognitive map in mice exploring a novel environment. In: Advances in Cognitive Research, Artificial Intelligence and Neuroinformatics: Proceedings of the 9th International Conference on Cognitive Sciences, Intercognsci-2020, 10–16 Oct 2020, Moscow, Russia 9. Springer International Publishing (2021)
6. De Chaumont, F., et al.: Real-time analysis of the behaviour of groups of mice via a depth-sensing camera and machine learning. Nat. Biomed. Eng. **3**(11), 930–942 (2019)
7. Mantripragada, S., Dionne, E., Chen, J.: Identifying behavioral movements of a mouse using machine learning. Columbia Junior Sci. J. Spring **2020**, 3 (2020)
8. Gharagozloo, M., et al.: Machine learning in modeling of mouse behavior. Front. Neurosci. **15**, 700253 (2021)
9. Weissenberger, Y., King, A.J., Dahmen, J.C.: Decoding mouse behavior to explain single-trial decisions and their relationship with neural activity. bioRxiv, 567479 (2019)
10. Kingma, D.P., Welling, M.: Auto-Encoding Variational Bayes. Iclr 1–14 (2014)
11. Rezende, D.J., Mohamed, S., Wierstra, D.: Stochastic Backpropagation and 316 Approximate Inference in Deep Generative Models (2014)
12. Sotskov, V., et al.: Calcium imaging reveals fast tuning dynamics of hippocampal place cells and CA1 population activity during free exploration task in mice. Int. J. Mol. Sci. **23**(2), 638 (2022)
13. Mitroshina, E., et al.: Novel algorithm of network calcium dynamics analysis for studying the role of astrocytes in neuronal activity in Alzheimer's disease models. Int. J. Mol. Sci. **23**(24), 15928 (2022)
14. Defferrard, M., Bresson, X., Vandergheynst, P.: Convolutional neural networks on graphs with fast localized spectral filtering. Adv. Neural Inf. Process. Syst. **29** (2016)
15. Nunez, E., Steyerberg, E.W., Nunez, J.: Regression modeling strategies. Revista Española de Cardiología (English Edition) **64**(6), 501–507 (2011)
16. PaddlePaddle. https://github.com/PaddlePaddle/PGL. Last accessed 17 Feb 2024

Fact-Checking Generative AI: Ontology-Driven Biological Graphs for Disease-Gene Link Verification

Ahmed Abdeen Hamed[1,2(✉)], Alessandro Crimi[1], Byung Suk Lee[3], and Magdalena M. Misiak[4]

[1] Sano Centre for Computational Medicine, Cracow, Poland
a.hamed@sanoscience.org
[2] Complex Adaptive Systems and Computational Intelligence Laboratory, Binghamton University, Binghamton, NY, USA
[3] Department of Computer Science, University of Vermont, Burlington, VT, USA
[4] Department of Physiology and Biophysics, Washington DC, USA

Abstract. Since the launch of various generative AI tools, scientists have been striving to evaluate their capabilities and contents, in the hope of establishing trust in their generative abilities. Regulations and guidelines are emerging to verify generated contents and identify novel uses. we aspire to demonstrate how ChatGPT claims are checked computationally using the rigor of network models. We aim to achieve fact-checking of the knowledge embedded in biological graphs that were contrived from ChatGPT contents at the aggregate level. We adopted a biological networks approach that enables the systematic interrogation of ChatGPT's linked entities. We designed an ontology-driven fact-checking algorithm that compares biological graphs constructed from approximately 200,000 PubMed abstracts with counterparts constructed from a dataset generated using the ChatGPT-3.5 Turbo model. In 10-samples of 250 randomly selected records a ChatGPT dataset of 1000 "simulated" articles , the fact-checking link accuracy ranged from 70% to 86%. This study demonstrated high accuracy of aggregate disease-gene links relationships found in ChatGPT-generated texts.

Keywords: ChatGPT · fact-checking · generative AI · biological graphs · biological ontology · network medicine

1 Introduction

The rise of new generative AI technologies holds both potential and concerns. Particularly, the emergence of ChatGPT [1] caused scientists to raise various concerns related to the capabilities and the inauthentic contents of such tools. Van Dis et al. identified five key priorities aimed at educating the general public about the potential of ChatGPT and formulating an effective response to this transformative AI tool. Among the five guidelines, fact-checking and human verification of ChatGPT contents were highlighted [2]. Inspired by such guidelines,

L. Franco et al. (Eds.): ICCS 2024, LNCS 14835, pp. 130–137, 2024.
https://doi.org/10.1007/978-3-031-63772-8_12

here we present our work on computational fact-checking of biological networks we constructed from ChatGPT-generated content. The utilization of biological ontology (i.e., Disease Ontology, Gene Ontology, Gene Ontology Annotations) give credibility to the biological terms that make up the nodes of the graphs. Using biological entities from ontology to extract and construct biological graphs from the biomedical literature offers trustworthy ground truth. Using network models and algorithms offer the rigor needed to perform fact-checking at the aggregate level. This study assumes a closed-world assumption [3–6], which sets the fact-checking scope within the knowledge embedded in the literature dataset and not beyond.

Knowledge graphs have been instrumental in advancing fact-checking methodologies, enabling structured and nuanced analyses of claims and assertions. For example, Tchechmedjiev et al. introduced ClaimsKG, a comprehensive knowledge graph that houses fact-checked claims, allowing informed queries on truth values and related aspects [7]. Vedula and Parthasarathy's work stood out by introducing FACE-KEG, a knowledge graph tailored to expound whether a statement is true or false, addressing the transparency gap in fact-checking [8]. Lin et al. made strides with ontology-based subgraph patterns, constructing graph fact-checking rules that integrate intricate patterns, capturing both topological and ontological constraints [9–11]. Notably, Ciampaglia et al. laid a foundation for fact-checking by leveraging knowledge graphs to scrutinize claims, drawing from reliable sources like Wikipedia [12].

Wang et al. harnessed entity category information, using prototype-based learning to enhance verification accuracy and reasoning capabilities in knowledge graph-based fact-checking, marking a significant advancement in this domain [13]. Khandelwal et al.'s approach encompassed structured and unstructured data from knowledge graphs to address the challenge of evaluating facts amidst growing data and misinformation [14]. Orthlieb et al.'s attention-based path ranking model exhibited promise in automating fact-checking through knowledge graphs, emphasizing interpretability and competitive results [15]. Another notable contribution came from Shi, who introduced ProjE, a neural network model that improved the completion of knowledge graphs and the accuracy of fact-checking [16].

Recent advancements further underpin the significance of knowledge graphs in fact-checking. The approach of Wang et al. leveraged category hierarchy and attribute relationships, showcasing the potential of knowledge structure in fact verification [17]. Amidst the COVID-19 outbreak, Mengoni's extended knowledge graph enabled enhanced claim validation through leveraging existing fact-checking reports [18]. Kim introduced weighted logical rules mining and evidential path identification in knowledge graphs, enhancing computational fact-checking [19,20]. Zhu et al. designed a knowledge-enhanced fact-checking system, tapping into both unstructured document knowledge bases and structured graphs to robustly identify misinformation [21].

2   Methods

In this section, we present a comprehensive methodology for constructing a reliable knowledge framework to assess the quality of content generated using Chat-GPT. Our approach is centered around the utilization of biological graphs as rigorous models that offer quantitative analysis of objective outcomes. Graphs as a tool is also being investigated for the advancement of Large Language Models (LLMs) [22] and ChatGPT technologies [23].

The proposed approach consists of six key steps which as a whole contribute to verification of the authenticity and accuracy of AI-generated biomedical text: (1) ChatGPT prompt-engineering and simulated-articles generation, (2) partial-match ontology term chunking to increase the recall of term matching, (3) ontology feature extraction, where partial terms are used as the means to feature identification in the literature and ChatGPT text, (4) proximity-based biological graphs construction for capturing the strongest links among the biological terms, (5) biological graph topological analysis, by analyzing the structural properties of each type and comparing them accordingly, and (6) algorithmic fact-checking to assert the facts.

2.1   Prompt-Engineering ChatGPT for Simulated-Articles Generation

Using the ChatGPT APIs, we engineered a prompt that has two roles: (1) the system role which is to command the ChatGPT engine to generate biomedical abstracts and (2) a user role which is to command ChatGPT explicitly to perform the task shown in Algorithm 1, repeating it as needed until a dataset of the desired size is produced.

Algorithm 1. ChatGPT Prompt Engineering for Article Generation

---

Require: The number n of simulated articles.
Require: The number w of words in each article.
1: Generate a list of n simulated PubMed-style abstracts.
2: For each abstract containing three fields: GPT-ID, Title, and Abstract, make it w words.
3: Make the GPT-ID random, containing at most five letters and numbers.
4: Return the abstracts in a valid JSON format as an array of JSON records.
5: Investigate the biology of human disease-gene associations.
6: Provide details related to diseases, genes, cells, organisms, and any FDA-approved drugs, and state any relationships.

2.2   Feature Extraction and Biological Graph Construction

Ontology terms are inherently detailed and lengthy. In biomedical literature, these long names are frequently abbreviated for convenience. For instance, the term "female breast cancer" is often referred to as "breast cancer" in the text. Importantly, we maintained a connection between these bigrams and their corresponding original term IDs in the ontology, while also tracking their positions. Constructing the knowledge graphs required the following steps: (1) feature extraction using the diseases and gene ontology and (2) establishing the links among the terms extracted. The process of ontology feature extraction from text records is as follow: (1) it takes as in put a collection of abstract texts and an ontology containing terms, (2) reads each textual record in the collection to identifies mentions of ontology terms (and related bigrams) within the text, (3) checks if the term appears in the text. If the term is a single word, the algorithm records the position of the match, the term itself, and other relevant information. For terms with more than one word, the algorithm generates bigrams (pairs of adjacent words) and checks for their presence in the text. If found, it records the position, term, bigram, and additional information. The process terminates by producing a set of matches for each record, indicating where ontology terms and bigrams were found within the text. Concretely, we constructed two different undirected but weighted graphs of disease and genes nodes. The first type one was constructed publication-driven from the mentions of disease and genes occurring in a dataset of biomedical abstracts extracted from PubMed Central [24]. A disease and a gene are connected if they occur in the same abstract. Then the link is weighted with the distance among the terms. Both gene and diseases names are ontology terms from the Human Disease Ontology (DOID) [25,26], and the Gene Ontology and Annotation (GOA) [27–29].

2.3 Fact-Checking ChatGPT Biological Graphs

The purpose of this step is to investigate the authenticity of contents gathered from ChatGPT and other generative AI models, and to test whether such contents may bridge the disease–genes gap in our understanding. In this regard, we propose a computational approach that captures how much true knowledge is stated in ChatGPT graphs and also identifies what may be considered noise or novelties. The idea is to compare the various link types (disease-gene, gene-gene, disease-disease) and determine how much they overlap with those in the ground truth literature graph. This offers fact checking at an aggregate level without having to verify the link semantics. Specifically, from 10 graphs constructed earlier, we implemented a process that systematically computes the number of edges in a ChatGPT-generated graph that coincide with edges in the corresponding graph derived from literature abstracts. While being in the search space, the algorithm also tracks the link to discern each type and evaluates the balance in the facts founds. It extracts all links before it also processes one link at a time, and checks it against the ground-truth graph constructed from the literature.

3   Results

We used various network metrics that compare the ChatGPT graphs with literature Graphs objectively. Table 1 encapsulates the essential metrics pertaining to each type of the knowledge graphs (i.e., literature and ChatGPT counterparts).

Table 1. The statistical result of comparing 10 GPT graphs with 10 literature graphs generated from the same number of records.

Source	Metric	G1	G2	G3	G4	G5	G6	G7	G8	G9	G10
GPT	No of Nodes	70	80	80	86	80	74	66	75	75	79
PubMed		137	63	100	104	116	118	101	113	95	154
GPT	No. of Edges	110	138	113	141	120	116	108	116	124	139
PubMed		297	124	165	214	251	240	207	366	147	393
GPT	N/E Ratio	0.64	0.58	0.71	0.61	0.67	0.64	0.61	0.65	0.60	0.57
PubMed		0.46	0.51	0.61	0.49	0.46	0.49	0.49	0.31	0.65	0.39
GPT	No. of Diseases	54	64	65	67	60	57	54	60	59	61
PubMed		117	54	87	88	97	101	79	97	75	131
GPT	No. of Genes	16	16	15	19	20	17	12	15	16	18
PubMed		20	9	13	16	19	17	22	16	20	23
GPT	Gene-Gene Link No	46	67	59	64	51	48	54	56	56	62
PubMed		229	86	120	151	196	192	148	311	94	316
GPT	Disease-Gene Link No	54	50	45	58	50	47	47	48	49	58
PubMed		57	34	40	55	45	36	44	43	42	61
GPT	Disease-Disease Link No	10	19	9	19	19	20	6	11	19	19
PubMed		11	4	5	5	10	10	14	10	11	16

Each consecutive two rows embody a distinct scenario for a given statistic, while the columns reference the dataset selected randomly by a given seed. The "No. of Nodes" 2-row denotes the count of all nodes, which symbolize diseases, genes. The "No. of Edges" 2-row unit quantifies the interconnections between nodes, reflecting relationships (e.g., disease - gene) or interactions (protein - protein). The "N/E Ratio" 2-row unit computes the balance between nodes and edges, potentially demonstrating the network complexity of each graph. The "No. of Diseases" 2-row enumerates disease-related nodes, while "No. of Genes" 2-row unit does the same for genes. The "No. of Disease-Gene Links" 2-row unit indicates associations between diseases and genes. "No. of Disease-Disease Links" underscores connections between different diseases. Lastly, "Number of Gene-Gene Links" 2-row unit unveils interactions among gene nodes. Collectively, this table provides an intricate glimpse into the network's composition, connectivity, and relationships within the biological and medical framework, fostering a deeper understanding of its underlying dynamics of each type. Figures 1a and 1a demonstrate the comparisons of nodes and edges between ChatGPT and literature, respectively.

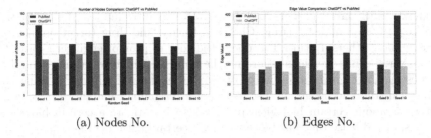

(a) Nodes No.   (b) Edges No.

Fig. 1. shows two subfigures: (a) on the left, the Number of Nodes, and (b) on the right the Number of Edges comparisons of 10 chatGPT graphs against literature, respectively.

4 Discussion

The discussion of our study results revolves around several key observations and findings that shed light on the comparison and potential utility of the constructed knowledge graphs. Our approach involved the comparison of two distinct types of graphs, both constructed from randomly selected datasets. This sampling strategy ensured an unbiased and fair basis for comparison between the two sources. In terms of topological analysis, it was our expectation to observe a less number of nodes and edges exhibited in ChatGPT. It was also our expectations to observe that the network generated from literature to be rich and complex, which was demonstrated by lower ratios of nodes to edges. However, we also observed an anomalous behavior among the 10 graph. Particularly, G9 has surpassed its literature counterpart in the ratio of number of nodes to edges. Such an observation indicate complexity of certain ChatGPT graphs which warrant further pursuing.

One of the main pursuits of this work was to perform an unbiased fact-checking and verification of a truth graph constructed using ontologies for their credible terminology, and biomedical literature of publications that are funded by the National Institute of Health to ensure high quality and credibility of work. We ensure that the fact-checking process is bounded by a closed-world assumption to make our work possible. The outcome of the process yielded promising results: the precision of link overlaps ranged from [70% to 86%] which is significantly high given the close-world assumption. This finding gives a certain measure of confidence to cautiously consider investigating data generated by ChatGPT using careful prompt-engineering.

5 Conclusion and Future Direction

As we continue to refine our work, the next steps involve further investigation of the proximity distance among biomedical terms and test if they hold in other domain and research areas. The study of disease-gene can be further instantiated in precise complex disease such as Alzheimer's and comorbidities where little is known. Such investigations may necessitate the introduction of new ontologies (e.g., Gene ontology, Drug, Chemical Entity, and drug target ontologies) among many others. In turn, this opens the door to prompt-engineer ChatGPT to answer specific questions regarding the repurposeability of a drug. Another interesting direction is to entirely *retrain* the engines of ChatGPT using the confirmed-true knowledge and use its massive reasoning capabilities to answers questions about certain biological pathways to investigate a certain biological targets, or a disease that maybe caused by a certain clusters of genes.

Acknowledgments. This research is supported by the European Union's Horizon 2020 research and innovation programme under grant agreement Sano No 857533 and carried out within the International Research Agendas programme of the Foundation for Polish Science, co-financed by the European Union under the European Regional Development Fund, Additionally, is created as part of the Ministry of Science and Higher Education's initiative to support the activities of Excellence Centers established in Poland under the Horizon 2020 program based on the agreement No MEiN/2023/DIR/3796'

References

1. OpenAI. ChatGPT: Conversational ai assistant. OpenAI Platform (2023). Accessed 14 Aug 2023
2. Van Dis, E.A., Bollen, J., Zuidema, W., Van Rooij, R., Bockting, C.L.: ChatGPT: five priorities for research. Nature **614**(7947), 224–226 (2023)
3. Przymusinski, T.C.: An algorithm to compute circumscription. Artif. Intell. **38**(1), 49–73 (1989)
4. Query rewriting for ontology-mediated conditional answers (2020)
5. Torralba, A., Efros, A.A.: Unbiased look at dataset bias (2011)
6. Minker, J.: On indefinite databases and the closed world assumption. In: Loveland, D.W. (ed.) CADE 1982. LNCS, vol. 138, pp. 292–308. Springer, Heidelberg (1982). https://doi.org/10.1007/BFb0000066
7. Tchechmedjiev, A., et al.: CLAIMSKG: a knowledge graph of fact-checked claims. Semant. Web - ISWC **11779**, 2019 (2019)
8. Vedula, N., Parthasarathy, S.: Face-keg: fact checking explained using knowledge graphs. In: Proceedings of the 14th ACM International Conference on Web Search and Data Mining, pp. 526–534 (2021)
9. Lin, P., Song, Q., Shen, J., Wu, Y.: Discovering graph patterns for fact checking in knowledge graphs. In: Pei, J., Manolopoulos, Y., Sadiq, S., Li, J. (eds.) DASFAA 2018. LNCS, vol. 10827, pp. 783–801. Springer, Cham (2018). https://doi.org/10.1007/978-3-319-91452-7_50
10. Lin, P., Song, Q., Yanhong, W.: Fact checking in knowledge graphs with ontological subgraph patterns. Data Sci. Eng. **3**, 341–358 (2018)
11. Lin, P., Song, Q., Yinghui, W., Pi, J.: Discovering patterns for fact checking in knowledge graphs. J. Data Inf. Qual. (JDIQ) **11**(3), 1–27 (2019)
12. Ciampaglia, G.L., Shiralkar, P., Rocha, L.M., Bollen, J., Menczer, F., Flammini, A.: Computational fact checking from knowledge networks. PloS One **10**(6), e0128193 (2015)

13. Wang, S., Wang, L., Mao, W.: A kg-based enhancement framework for fact checking using category information. In: 2020 IEEE International Conference on Intelligence and Security Informatics (ISI), pp. 1–6. IEEE (2020)
14. Khandelwal, S., Kumar, D.: Computational fact validation from knowledge graph using structured and unstructured information. In: Proceedings of the 7th ACM IKDD CoDS and 25th COMAD, pp. 204–208 (2020)
15. Orthlieb, T., Abdessalem, H.B., Frasson, C.: Checking method for fake news to avoid the twitter effect. In: Cristea, A.I., Troussas, C. (eds.) ITS 2021. LNCS, vol. 12677, pp. 68–72. Springer, Cham (2021). https://doi.org/10.1007/978-3-030-80421-3_8
16. Shi, B., Weninger, T.: Proje: embedding projection for knowledge graph completion. In: Proceedings of the AAAI Conference on Artificial Intelligence, AAAI 2017, pp. 1236–1242. AAAI Press (2017)
17. Wang, S., Mao, W., Wei, P., Zeng, D.D.: Knowledge structure driven prototype learning and verification for fact checking. Knowl.-Based Syst. **238**, 107910 (2022)
18. Mengoni, P., Yang, J.: Empowering covid-19 fact-checking with extended knowledge graphs. In: International Conference on Computational Science and its Applications, pp. 138–150. Springer, Heidelberg (2022). https://doi.org/10.1007/978-3-031-10536-4_10
19. Kim, J., Choi, K.S.: Unsupervised fact checking by counter-weighted positive and negative evidential paths in a knowledge graph. In: Proceedings of the 28th International Conference on Computational Linguistics (2020)
20. Kim, J.S., Choi, K.S.: Fact checking in knowledge graphs by logical consistency. Semantic Web J. **swj2721** (2021)
21. Zhu, B., Zhang, X., Gu, M., Deng, Y.: Knowledge enhanced fact checking and verification. IEEE/ACM Trans. Audio Speech Lang. Process. **29**, 3132–3143 (2021)
22. Pan, S., Luo, L., Wang, Y., Chen, C., Wang, J., Wu, X.: A roadmap, Unifying large language models and knowledge graphs (2023)
23. Yang, L., Chen, H., Li, Z., Ding, X., Wu, X.: Chatgpt is not enough: enhancing large language models with knowledge graphs for fact-aware language modeling (2023)
24. Pubmed central (pmc). Accessed 2 Sept 2023
25. Hofer, P., Neururer, S., Goebel, G.: Semi-automated annotation of biobank data using standard medical terminologies in a graph database, vol. 228 (2017)
26. Sow, A., Guissé, A., Niang, O.: Enrichment of medical ontologies from textual clinical reports: towards improving linking human diseases and signs. In: Bassioni, G., Kebe, C.M.F., Gueye, A., Ndiaye, A. (eds.) InterSol 2019. LNICST, vol. 296, pp. 104–115. Springer, Cham (2019). https://doi.org/10.1007/978-3-030-34863-2_10
27. Huntley, R.P., et al.: The GOA database: gene ontology annotation updates for 2015. Nucleic Acids Res. **43**(D1), D1057–D1063 (2015)
28. Gene ontology annotations and resources. Nucleic Acids Res. **41** (2013)
29. Camon, E., et al.: The gene ontology annotation (goa) project: implementation of go in swiss-prot, trembl, and interpro (2003)

Identification of Domain Phases in Selected Lipid Membrane Compositions

Mateusz Rzycki[✉] [iD], Karolina Wasyluk, and Dominik Drabik [iD]

Department of Biomedical Engineering, Wroclaw University of Science and Technology, 50-370 Wroclaw, Poland
mateusz.rzycki@pwr.edu.pl

Abstract. Lipid microdomains are specialized structures that play crucial roles in various physiological and pathological processes, such as modulating immune responses, facilitating pathogen entry, and forming signaling platforms. In this study, we explored the dynamics and organization of lipid membranes using a combination of molecular dynamics simulations and a suite of machine learning (ML) techniques. Using ML algorithms, we accurately classified membrane regions into liquid order, liquid-disordered, or interfacial states, demonstrating the potential of computational methods to predict complex biological organizations. Our investigation was mainly focused on two lipid systems: POPC/PSM/CHOL, and DPPC/DLIPC/CHOL. The study underscores the dynamic interaction between ordered and disordered phases within cellular membranes, with a pivotal role of cholesterol in inducing domain formation.

Keywords: microdomains · lipid membranes · machine learning · molecular dynamics · gel domains

1 Introduction

The cell membrane is a complex and dynamic structure that plays a pivotal role in maintaining integrity and regulating various cellular processes [3]. The distribution of lipids within these membranes is heterogeneous and certain lipids aggregate to form distinct domains [12]. These include structured gel phases (S_o) and less structured liquid disordered phases (L_d), with a special case of structured liquid ordered phases (L_o) induced by sterols like cholesterol [15]. Lipid rafts, a notable L_o domain, contain sphingomyelin and cholesterol, affecting the membrane's structure and function due to its composition and interactions. Lipid microdomains are highly ordered regions within membranes, crucial for signal transduction, endocytosis, and membrane trafficking, due to their ability to stabilize and organize membrane proteins [5,16]. In contrast, the S_o domains exhibit a different narrative and are rare in living organisms. The L_d regions,

L. Franco et al. (Eds.): ICCS 2024, LNCS 14835, pp. 138–146, 2024.
https://doi.org/10.1007/978-3-031-63772-8_13

composed of disordered lipids, provide a dynamic matrix. This matrix surrounds the ordered domains and is essential for protein diffusion and cellular responsiveness [6]. Lipid rafts, specialized microdomains, play crucial roles in various physiological and pathological processes such as modulating immune responses, facilitating pathogen entry, and forming signaling platforms [7]. Therefore, the role of lipid domains in cell signaling and organization emphasizes the importance of understanding their impact on cell homeostasis.

Accurate identification of microdomains is particularly crucial, given their unique mechanical properties, which may render them potential targets for specific antimicrobial agents [14]. A promising method for identifying microdomains involves the use of machine learning (ML) techniques on molecular dynamics (MD) membrane trajectories, although research in this area remains limited. This study synthesizes methodologies from previous research [4,11], employing supervised learning in modeled and spontaneously formed microdomains in ternary lipid mixtures in MD, thus improving the current knowledge base. The primary advantage of this approach is its precision in predicting molecular organization, domain, nondomain, and interface, accurately reflecting experimental observations.

2   Methods

2.1   System Preparation and Simulation

Using the CHARMM-GUI membrane builder [18], we constructed several membrane systems with various lipid compositions, each designed to represent different molecular organizations. Six training systems were created, each containing 150 lipid molecules, representing different domain stages. Lipid compositions included phosphatidylcholine (POPC), sphingomyelin (PSM), cholesterol (CHOL), as well as other phosphatidylcholines (DPPC and DLIPC), mixed in specific ratios to mimic nondomain, interface, and domain-like phases. Lipid ratios were set as follows: 8:1:1 for nonraft (POPC/PSM/CHOL), 2:1:1 and 4:3:3 for interface stages (POPC/CHOL/PSM), and varying ratios such as 1:2:2, 2:1, and 1:1 for raft-like stages (CHOL/PSM). Analogously, for the second approach (spontaneous domain formation), we replaced PSM with DPPC and POPC with DLIPC in training systems, following established literature protocols [4,12,19].

Additionally, two specialized testing systems were devised, containing 1140 and 900 lipid molecules, respectively, with equal proportions of POPC/PSM/CHOL and DPPC/DLIPC/CHOL. The first system was structured with a central circular configuration of PSM and CHOL surrounded by POPC, designed to replicate an idealized raft-like domain [4,12,15]. The second system featured a stochastic distribution of DPPC, DLIPC, and CHOL, aimed at exploring lipid behavior in a nonordered environment. These models facilitate the study of lipids dynamics and the structural properties of different membrane configurations.

Molecular dynamics simulations were performed with Gromacs software (v. 2022) [1] and the CHARMM36 force field [18]. The simulation protocol involved energy minimization, NVT and NPT equilibration (constant number of particles, volume/pressure, and temperature), ending with a production run. All simulations were performed at 295.15 K, using standard CHARMM protocol. A detailed description of the simulation procedure is described in our previous work [14]. The production run involved at least 500ns.

2.2 Lipid Features and Machine Learning Techniques

In our molecular dynamics simulations, we analyzed the last 50 ns of each trajectory to extract ten different lipid features that characterize the local membrane environment. These features included lipid type (1), area per lipid (APL, 1), total lipid length (2), the number of surrounding lipids by each type, first shell (3), and composition of the second shell (3).

APL was determined using a custom MATLAB script. For each lipid, a position point was identified at the midpoint between the phosphorus atom (P) and the second carbon atom (C2) in the phospholipids, and between the C3 and O3 atoms for CHOL. The local environment of each lipid was further quantified by Voronoi tessellation to determine the composition of the first and second lipid shells. Additionally, the order parameter (S_{cd}) for acyl chains was calculated using the MEMBPLUGIN [8] S_{cd} addon in Visual Molecular Dynamics (VMD) software, averaging the values for both acyl chains of each lipid. The lengths of the lipids were measured as the geometric distance between the phosphorus and last carbon atoms in the acyl chain.

Table 1. The architectures and accuracy of tested ML models

ML model	Architecture	Best accuracy
KNN	5 closest neighbors	91% ± 3%
DNN	$10 \times 128 \times 128 \times 32 \times 8 \times 4 \times 3$, ReLu activation, 20% dropout, softmax without dropout (last layer)	87% ± 5%
RF	entropy criterion, 1,000 estimators, max. depth 15	88% ± 2%
SVC	polynomial kernel	85% ± 3%

Statistical analysis was performed using the Python Scipy library, employing a one-way ANOVA test with a significance threshold of 0.05 and the Tukey test for post hoc analysis to ensure the reliability of our findings. We employed a suite of ML techniques to unravel the complex organization of microdomains and lipid membranes, using different algorithms to categorize membrane compositions as raft/domain, nondomain, or interface. Initially, we normalized all input data and applied one-hot encoding to categorical variables, ensuring that our dataset was optimized for ML processing. Subsequently, several selected ML algorithms from the scikit-learn package were adopted: Support Vector Classifier (SVC), Random Forest (RF), K-Nearest Neighbors (KNN), and Deep Neural Network

(DNN). The models were trained on a random selection of 80% of data, and the remaining 20% of data was used to determine an accuracy metric for the method. The choice of an equal lipid composition (1:1:1) ensures a balanced distribution of features in the test dataset. In this article, we have selected and presented one ML technique for each approach with the highest accuracy of all the methods listed. These were KNN and RF for the first and second approaches, respectively. All architectures are presented in Table 1.

3   Results and Discussion

In this study, we explored two scenarios of lipid microdomain formation using molecular dynamics simulations. The first scenario involved an idealized domain with a core of CHOL and PSM, surrounded by unsaturated POPC lipids, showing slow lipid migration and interfacial mixing [4,12]. The second scenario explores the spontaneous segregation of saturated DPPC and unsaturated DLIPC lipids facilitated by the addition of CHOL. We used the L_o and L_d nomenclature for these domains, observing how DPPC's high phase-transition temperature influenced behavior differently from the POPC/PSM/CHOL mixture. Our simulations, which lasted approximately 2 μs, highlighted the complex and nuanced domain formation process, involving a more probabilistic gel domain (S_o) formation compared to the previous model. Using ML techniques, the selected parameters of all lipids were used to classify them into three distinct clusters representing raft-like/domain, nonraft/nondomain, and interface regions. The results of selected ML approaches are shown in Fig. 1.

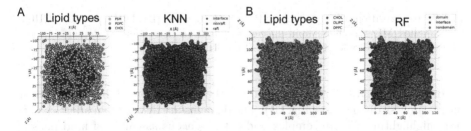

Fig. 1. Classification of (A) POPC/PSM/CHOL and (B) DPPC/DLIPC/CHOL membrane regions using KNN and RF methods, respectively based on selected lipid parameters (top view). Each left panel illustrates the structural composition of the lipid membrane and the right panel depicts the ML lipid classification.

The organization of the idealized raft system is presented in Fig. 2A. The raft, nonraft, and interfacial areas constitute 55%, 31%, and 13% of the system, respectively. This distribution is consistent with experimental studies on POPC/PSM/CHOL vesicles [13]. The formation of these domains is influenced by temperature variations, which often leads to a predominance of L_o regions due to the affinity of CHOL for PSM [20]. The interface mainly comprises PSM, CHOL (30%), and POPC (22%), while nonraft areas are dominated by CHOL and PSM with minimal POPC presence (see Fig. 2C). The nonraft region is predominantly composed of POPC (92%), with minor contributions from PSM (5%) and CHOL (3%). These distributions deviate slightly from the literature, where typically the L_d phase contains about 71% DOPC, 24% PSM, and 5% CHOL [3]. Our findings for the L_o phase align with the trends for CHOL and PSM but with a slightly lower PC representation. The discrepancies may be attributed to variations in lipid ratios (43:32:25) in both our study and the presence of more unsaturated DOPC, which supports phase separation [3]. It is worth noting that the experimental models hardly quantify the transition phase, thus a binary classification in ML might mirror the experimental outcomes closely.

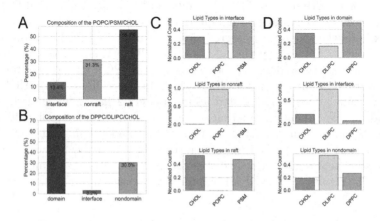

Fig. 2. Composition of (A) POPC/PSM/CHOL and (B) DPPC/DLIPC/CHOL with applied ML classifiers, lipid species distribution in recognized domains in (C) POPC/PSM/CHOL and (D) DPPC/DLIPC/CHOL systems.

In a DPPC/DLIPC/CHOL mixture, the distribution of lipids and their phases are less distinct (see Fig. 1B). The domains are dispersed in a lattice pattern, highlighting a more complex and heterogeneous assembly of lipid phases. Interestingly, the location of CHOL molecules can be seen mainly at the interface between DPPC and DLIPC (see Fig. 2B). Here, about 67% of the system is L_o, 30% is L_d, and the interfacial area comprises 3.2%, indicating the challenges in identifying transition zones. Domain compositions show a balanced distribution, with the nondomain phase mainly consisting of DLIPC, supplemented by about 30% DPPC and 20% CHOL. In contrast, the L_o phase predominantly includes DPPC and CHOL, with 15% of DLIPC (see Fig. 2D). This distribution and composition of these domains align with other experimental studies [2,15].

Additionally, we used ML predictions to analyze specific lipids based on their molecular organization, focusing on the area per lipid (APL) and order parameters. In the rafts, typically enriched with PSM and CHOL, we found a decrease in APL. This suggests a tighter lipid packing and a consequent increase in the order parameter, reflecting the dense lipid packing. The L_d phases, which contain more unsaturated lipids such as POPC, tend to indicate higher APL values and lower order parameters due to looser acyl chain packing and increased disorder [17]. Our findings, illustrated in Fig. 3A, confirm these patterns. Lipids in the raft phase consistently exhibited the lowest APL values, whereas those in the nondomain regions showed the highest. Interestingly, CHOL in the nonraft phase displayed unusually low APL values, probably distorted due to its limited occurrence and not entirely indicative of the overall behavior. Intermediate APL values in the interface region support its role as a transitional area between the L_o and L_d phases, reflecting the characteristics of both phases.

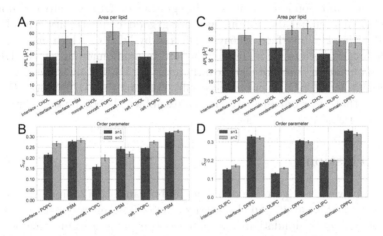

Fig. 3. Quantitative analysis of lipid organization in (A, B) POPC/PSM/CHOL and (C, D) DPPC/DLIPC/CHOL. Changes in APL (A, C) and order parameters (B, D) are displayed for all identified regions.

DPPC/DLIPC/CHOL observations are consistent with previous studies, indicating the highest APL values in nondomain areas and the lowest in domain areas [15]. The slight effect of the APL values of the interface region on the ML predictions suggests a diminished role for this parameter in the phase identification process.

The order parameter reflects the orientation of the lipid acyl chains in the membrane, with higher values indicating rigid, straight chains typical of the L_o phases. Our results show the highest order parameters for PSM and POPC in the raft domains and the lowest in the nondomain areas, with values in the interfacial zones falling between, showing a smooth transition from ordered to disordered regions (see Fig. 3B). Within the raft phase, there is a notable reduction in APL for POPC. This pattern is mirrored by the PSM, where the decrease in APL corresponds to an increase in the order of the lipid tails, highlighting the effect of CHOL on the behavior of associated lipids. This supports previous findings of the limited effect of PC on disorder in the L_o phase [4]. The interface serves as a transition zone with reduced POPC and PSM order parameters. It should be noted that the $sn1$ tail of POPC shows a significant deviation in order from its unsaturated $sn2$ tail. The APL values for POPC in nondomain areas align with those of pure PC systems, whereas an increase is observed in PSM [10]. These variations imply a significant influence of PC on neighboring lipids, accompanied by a decrease in PSM order and a shift between the $sn1$ and $sn2$ tails. The role of CHOL in these organizational changes appears to be minimal. In the second system, we observe a consistent trend in lipid tail ordering and APL with the highest order of DPPC in the domain phase and the lowest in the L_d phase (see Fig. 3 C, D). This trend is also apparent with DLIPC. These differences are more nuanced than previously noted, particularly in distinguishing between the interface and nondomain phases. Integration of DLIPC into the domain phase showed minimal disruption, indicating good compatibility with DPPC and CHOL [9]. The interface is heavily affected by DLIPC, which makes it akin to a nondomain because of its liquid fraction. In nondomain areas, the DPPC-CHOL combination does not significantly alter the DLIPC tail ordering. These findings highlight CHOL's pronounced effect on ordering DPPC over DLIPC, suggesting that the domain more closely resembles the L_o phase, contrary to initial expectations of a S_o phase.

4 Conclustions

In this study, we investigate the dynamics and organization of lipid membranes, focusing on the formation and characterization of lipid microdomains. We used molecular dynamics simulations and a suite of machine learning techniques to analyze and classify membrane regions into ordered, disordered, or interfacial states. Our findings highlight the dynamic interaction between the L_o and L_d phases, with cholesterol playing a crucial role in the formation of the L_o domains. We examine the organization of the membrane in different mixtures: POPC/PSM/CHOL and DPPC/DLIPC/CHOL to understand the behavior in varying compositional contexts. Idealized circular rafts were identified better than spontaneously induced ones. However, it is worth noting that in both systems the domain compositions were consistent with literature reports. The dependence on the initial training system's composition is a main limitation that could be reduced through experimentally supported training preparations. Further development of more sophisticated machine learning models capable of integrating multiscale data will enhance the ability to predict membrane organization and dynamics under various physiological conditions.

Acknowledgement. M.R. thanks the National Science Centre, Poland for the financial support (grant number 2022/45/N/NZ9/02130).

References

1. Abraham, M.J., Murtola, T., Schulz, R.: Gromacs: high performance molecular simulations through multi-level parallelism from laptops to supercomputers. SoftwareX **1–2**, 19–25 (2015)
2. Barnoud, J., Rossi, G., Marrink, S.J., Monticelli, L.: Hydrophobic compounds reshape membrane domains. PLoS Comput. Biol. **10**(10), e1003873 (2014)
3. Bezlyepkina, N., Gracià, R.S., Shchelokovskyy, P., Lipowsky, R., Dimova, R.: Phase diagram and tie-line determination for the ternary mixture DOPC/eSM/Cholesterol. Biophys. J . **104**(7), 1456–1464 (2013)
4. Canner, S.W., Feller, S.E., Wassall, S.R.: Molecular organization of a raft-like domain in a polyunsaturated phospholipid bilayer: a supervised machine learning analysis of molecular dynamics simulations. J. Phys. Chem. B **125**(48), 13158–13167 (2021)
5. Chichili, G.R., Rodgers, W.: Cytoskeleton-membrane interactions in membrane raft structure. Cell. Mol. Life Sci. **66**(14), 2319–2328 (2009)
6. Cournia, Z., Allen, T.W., Andricioaei, I., Antonny, B., Bondar, A.N.: Membrane protein structure, function, and dynamics: a perspective from experiments and theory. The J. Membr. Biol. **248**(4), 611–640 (2015)
7. Drabik, D., Drab, M., Penič, S., Iglič, A., Czogalla, A.: Investigation of nano- and microdomains formed by ceramide 1 phosphate in lipid bilayers. Sci. Rep. **13**(1), 1–14 (2023)
8. Guixà-González, R., et al.: MEMBPLUGIN: studying membrane complexity in VMD. Bioinformatics **30**(10), 1478–1480 (2014)
9. Keller, F., Heuer, A.: Chain ordering of phospholipids in membranes containing cholesterol: what matters? Soft Matter **17**(25), 6098–6108 (2021)
10. Leftin, A., Molugu, T.R., Job, C., Beyer, K., Brown, M.F.: Area per lipid and cholesterol interactions in membranes from separated local-field 13C NMR spectroscopy. Biophys. J . **107**(10), 2274 (2014)
11. López, C.A., Vesselinov, V.V., Gnanakaran, S., Alexandrov, B.S.: Unsupervised machine learning for analysis of phase separation in ternary lipid mixture. J. Chem. Theory Comput. **15**(11), 6343–6357 (2019)
12. Peter, C., Kremer, K., Carbone, P., Niemel, P.: Concerted diffusion of lipids in raft-like membranes. Faraday Discuss. **144**, 411–430 (2009)
13. Pokorny, A., Yandek, L.E., Elegbede, A.I., Hinderliter, A., Almeida, P.F.F.: Temperature and composition dependence of the interaction of d-lysin with ternary mixtures of sphingomyelin/cholesterol/POPC. Biophys. J . **91**, 2184–2197 (2006)
14. Rzycki, M., Drabik, D., Szostak-Paluch, K., Hanus-Lorenz, B., Kraszewski, S.: Unraveling the mechanism of octenidine and chlorhexidine on membranes: does electrostatics matter? Biophys. J . **120**(16), 3392–3408 (2021)
15. Sodt, A.J., Sandar, M.L., Gawrisch, K., Pastor, R.W., Lyman, E.: The molecular structure of the liquid-ordered phase of lipid bilayers. J. Am. Chem. Soc. **136**(2), 725–732 (2014)
16. Staubach, S., Hanisch, F.G.: Lipid rafts: signaling and sorting platforms of cells and their roles in cancer. Expert Rev. Proteomics **8**(2), 263–277 (2011)
17. Veatch, S.L., Keller, S.L.: Miscibility phase diagrams of giant vesicles containing sphingomyelin. Phys. Rev. Lett. **94**(14), 148101 (2005)
18. Wu, E.L., et al.: CHARMM-GUI membrane builder toward realistic biological membrane simulations. J. Comput. Chem. **35**(27), 1997–2004 (2014)

19. Yang, J., Martí, J., Calero, C.: Pair interactions among ternary DPPC/POPC/ cholesterol mixtures in liquid-ordered and liquid-disordered phases. Soft Matter **12**(20), 4557–4561 (2016)
20. Yasuda, T., Tsuchikawa, H., Murata, M., Matsumori, N.: Deuterium NMR of raft model membranes reveals domain-specific order profiles and compositional distribution. Biophys. J . **108**(10), 2502–2506 (2015)

MonoWeb: Cardiac Electrophysiology Web Simulator

Lucas Marins Ramalho de Lima[1] , Rafael Rocha Ribeiro[1] ,
Lucas Arantes Berg[2] , Bernardo Martins Rocha[1] ,
Rafael Sachetto Oliveira[3] , Rodrigo Weber dos Santos[1] ,
and Joventino de Oliveira Campos[1( )]

[1] Federal University of Juiz de Fora, Juiz de Fora, MG, Brazil
joventino.campos@ufjf.br
[2] University of Oxford, Oxford, UK
[3] Federal University of São João del-Rei, São João del-Rei, MG, Brazil

Abstract. Computational modeling emerged to address scientific problems by developing mathematical models for their description and creating computational codes to obtain solutions. Employing this technique in studying cardiac electrophysiology enables a better understanding of heart function, which requires considerable time and technological expertise. *MonoWeb* is a structured platform for simulating electrophysiological activity in cardiac tissues, using the monodomain model in a browser-based manner. This tool provides not only an accessible platform to simulate cardiac electrical activity but also integrates visualization and flexible configuration with an intuitive interface. Through communication with the *MonoAlg3D* simulator, it allows the input of advanced parameters, and different cellular models, including selecting arrhythmia examples, and stimuli, with the goal of making this experience easier and practical for electrophysiology professionals.

Keywords: Computational Electrophysiology · Web Simulator · Educational simulator for cardiologists

1 Introduction

Computational modeling of cardiac electrophysiology has advanced significantly in recent decades by studying the electrical activity of the heart under healthy and pathological conditions [2,10]. Several studies were conducted to investigate the generation of arrhythmias, providing an in-depth view of the influence of various physiological conditions, such as fibrosis, ischemia, hypoxia, and myocarditis [10]. This approach allows for a more comprehensive understanding of the underlying mechanisms of these conditions, providing valuable information for the diagnosis and treatment of patients with heart disease.

The usage of high-performance computing and efficient numerical methods allow simulations to become suitable for clinical applications and several works

L. Franco et al. (Eds.): ICCS 2024, LNCS 14835, pp. 147–154, 2024.
https://doi.org/10.1007/978-3-031-63772-8_14

have presented results in the context of patient-specific simulations [2,8,10]. The process involves the acquisition of anatomical data, image segmentation, geometric model reconstruction, discretization and simulations to study cardiac behavior under different conditions.

However, performing these simulations is computationally expensive and requires powerful machines to run the analysis. Furthermore, the simulator usually is developed for researcher use, and it does not present an easy interface for physicians' understanding, which is a barrier to clinical translation.

In this context, the work presented in [3] proposed a simulator based on the *WebGL* library that runs in the web browser, which solves the mathematical model using the graphical process unit (GPU). The *OpenCARP* software [6] has also a tool named *carputilsGUI*, which is a web-based interface to run and visualize simulations, such as single cell action potential and tissue reentry induction. By developing such an interface, the aim is not only to facilitate access to advances in computational modeling of the cardiac system but also to empower healthcare professionals to apply these tools in their daily practice.

Therefore, this work presents *MonoWeb*, a user interface designed to simplify the execution and visualization of electrophysiology simulations, eliminating the need for advanced programming knowledge on the part of users. This tool aims to bridge the gap between the complexity of computational models and the accessibility for physicians and educators, allowing a broader and more effective application of these technologies.

2   Methods

2.1   Monodomain Model

To model the electrophysiological activity of the cardiac tissue, the monodomain model was used, which represents the propagation of the transmembrane potential through the tissue using a partial differential equation. It is described by a reaction-diffusion equation coupled with a system of ordinary differential equations (ODEs) that describes the cellular-level excitation [7].

The solution of the monodomain equation in a wide range of cellular models was addressed using *MonoAlg3D* [7]. This is an adaptive simulator that employs the finite volume method for solving the monodomain model. Its implementation is accelerated through parallel programming techniques on GPUs.

The *MonoAlg3D* configuration is done through files in the INI format, in which the user specifies the details of the desired simulation. These files require comprehensive information, including aspects such as domain, discretizations, linear systems, GPU optimization, stimuli, and any other pertinent data to the simulation process. This level of detail can pose challenges for individuals lacking expertise in scientific computing.

2.2   Trame

Trame is a visual analytics application developed by *Kitware* [4]. The framework has integration with various data analysis tools, including *VTK/ParaView*,

PyVista, and *Plotly*. The API gets the data from *Paraview* and allows the user to interact with the loaded model without the need to install anything. The processing functions and the state that must be shared with the user are built in plain Python using Model-View-ViewModel (MVVM). This software architecture separates the user interface from the business logic, facilitating maintenance, testing, and application development, promoting an organized and reusable structure.

2.3 Paraview

ParaView is an open-source data visualization and analysis application designed to handle extremely large datasets [1]. Users import data in several formats and then apply filters and visualization techniques to explore and analyze the data interactively. Paraview also provides *pvpython* (Paraview Python), a way for users to access *ParaView's* extensive library of filters, algorithms, and visualization techniques directly from their Python scripts. This tool was used to interpret the VTU files generated by *MonoAlg3D* and displayed in the *MonoWeb* interface.

3 MonoWeb

MonoWeb is structured into three key areas: a toolbar located at the top of the interface, a parameter input section to the left of the program, and a visualization area, as shown in Fig. 1.

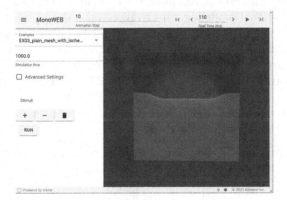

Fig. 1. *MonoWeb* interface with an example experiment being visualized, with the advanced options disabled. There is a toolbar to control the simulation time step, a menu with examples to run, and a section of stimuli configurations.

The visualization area leverages *ParaView* libraries to read *MonoAlg3D's* output files in PVD and VTU formats. This allows the user to interact with the simulation, for example, by adjusting the camera position using the mouse, which facilitates a more dynamic experience during result analysis.

The toolbar in the visualization area enables control over the displayed simulation frame. Additionally, users can select a specific moment in the simulation and adjust the time step size for the animation function.

The simulation parameters can be adjusted on the left side of the window, as depicted in Fig. 1. In the first dropdown menu, it is possible to select from four simulation examples with different arrhythmia cases. The following input field represents the total simulation time in milliseconds.

Below, there is a checkbox to display advanced settings, where there are three fields indicating the electrical conductivity of the tissue (σ) in the three Cartesian directions x, y, and z, respectively. There is also a dropdown menu to create the simulation domain. Domain functions define the shape of the tissue for the simulation based on certain parameters. All functions require spatial discretization of the domain, therefore, there are three fields for the values of dx, dy, and dz, respectively.

There are ten domain definition functions available, allowing users to build meshes in spherical, cubic, prismatic, as well as square and rectangular planar shapes. Moreover, within the domain functions, it is possible to configure fibrosis zones of various shapes.

3.1   Configuring the Cellular Model

Understanding the case studies requires some concepts of action potential generation in cardiac cells, as described by cellular models. Action potentials are initiated by the opening and closing of ion channels within the cell, typically sensitive to the transmembrane potential or the concentration of specific ions inside the cell. Computational modeling aims to represent the transport of these ionic currents through channels, capturing the general behavior of an excitable cell and its action potential. A key characteristic of action potentials is the refractory period, occurring during repolarization and extending slightly after the cell returns to its resting state, preventing the initiation of a new action potential in that cell (Fig. 2).

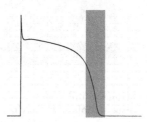

Fig. 2. Example of action potential of the Ten-Tusscher 3 cellular model for endocardial cells [9] showing the refractory period highlighted in red. (Color figure online)

The behavior of an action potential is described according to the cellular model. Some models focus only on a general phenomenological description with few ODEs, while others provide detailed descriptions of ion transport across the membrane and specialized organelles. The first three examples available in *MonoWeb* use the Ten-Tusscher 3 cellular model [9] for endocardial cells while the last example uses Mitchell-Shaeffer [5], but it can be changed to other available cellular models.

3.2 Configuring the Stimuli

There is a section for applying stimuli in the computational cardiac tissue, where there are three buttons: one to add a new stimulus, another to remove the last added stimulus, and the third to remove all the included stimuli. *MonoWeb* allows the user to add up to 10 stimuli in a customizable way.

For each stimulus, it is possible to configure the start time, duration and applied current. Additionally, for each stimulus, the user can choose a function to facilitate its placement. Several functions are available to stimulate specific regions of the cardiac tissue, such as those enabling the delineation of boundaries along cartesian directions, facilitating the stimulation of areas within these bounds. Additionally, there are functions that enable the stimulation of spherical regions around a point, with the option to adjust the radius.

4   Case Studies

MonoWeb presents four different examples with distinct scenarios, in order to provide preliminary simulations that may be customized and to serve as educational material on computational electrophysiology.

The initial example, named *EX01 3d wedge healthy*, consists of a simulation of a prism-shaped healthy tissue with a rectangular base. This simulation aims to illustrate the propagation of a single stimulus through healthy tissue, highlighting the generation of a planar wave, as presented in Fig. 3. For this purpose, a stimulus is applied at time $t = 0.0$ ms with a duration of $t_{stim} = 2.0$ ms at one edge of the domain, as presented in Fig. 3(a). As time progresses in the simulation, it is observed that the shape of the wavefront is preserved due to the uniformity of the tissue in the domain, as shown in Fig. 3(b).

The second example, *EX02 plain mesh S1S2 protocol*, depicts the application of the S1–S2 protocol of electrical stimulation in a healthy square tissue. The S1–S2 protocol is a commonly performed study to induce arrhythmias in virtual environments and medical examinations.

In this scenario, an initial stimulus is applied at the domain's edge to trigger a planar wave as observed in Fig. 4(a). After 360 ms from the application of this first stimulus, a second stimulus is applied in one-quarter of the domain, as shown in Fig. 4(b). This second intervention results in unidirectional blockage in the horizontal direction due to the refractory period of cells in that region, which were stimulated more recently than the cells in the vertical direction where propagation occurs.

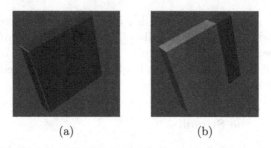

(a)   (b)

Fig. 3. Example 1 visualized in *MonoWeb*, demonstrating the application of a single stimulus in a three-dimensional parallelepiped-shaped tissue.

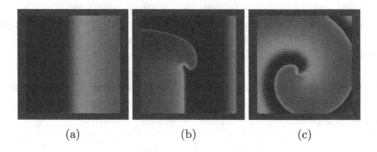

(a)   (b)   (c)

Fig. 4. Example 2 visualized in *MonoWeb*, demonstrating the induction of an arrhythmia in a plain healthy tissue by applying the S1–S2 protocol.

As the stimulus propagates vertically, the adjacent tissue segment exits the refractory period, becoming susceptible to stimulation once more. Consequently, the stimulus spreads horizontally and, subsequently, into the region previously subjected to unidirectional blockage. This sequence gives rise to a self-sustaining spiral wave, representing a case of arrhythmia where tissue deviates from its expected behavior, assuming a disordered pattern, as presented in Fig. 4(c).

The third example, named *EX03 plain mesh with ischemia*, demonstrates the application of the S1–S2 protocol in ischemic tissue. The domain of this example is a flat mesh with a circular region of ischemia in the center. To model ischemia, certain conditions are applied to the ionic channels in the cellular model, within the circular region, causing a delay in stimulus propagation in that region. The first stimulus propagates through the tissue, stimulating the ischemic region slightly slower than normal, as illustrated in Fig. 5(a).

When applying a new stimulus identical to the first one, initially, it propagates around the ischemic region without depolarizing it, indicating unidirectional blockage (Fig. 5(b)–(c)). When the stimulus reaches the upper part of the tissue, as shown in Fig. 5(d), it stimulates the upper region of the central circle of ischemia, inducing reentry.

The last example, *EX04 plain mesh spiral breakup*, aims to illustrate, using the Mitchell-Shaeffer cellular model [5], the transition from arrhythmia to a fibrillation case. To achieve this, stimuli from the S1-S2 protocol are applied to

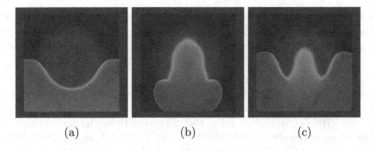

(a)   (b)   (c)

Fig. 5. Simulation result of example 3 visualized in *MonoWeb*, demonstrating the induction of an arrhythmia caused by the application of the S1-S2 protocol from the electrophysiological study in flat tissue with circular ischemia.

a square tissue, following precisely what was demonstrated in example 2, with the purpose of inducing the formation of a spiral wave, as shown in Fig. 6(a). After a period of time, the manifestation of the phenomenon known as Alternans becomes evident. This phenomenon is characterized by variations in the action potential duration (APD), oscillating between long and short periods.

As time progresses, Alternans results in the breakup of the spiral into multiple others, as shown in Fig. 6(b)–(c), outlining a fibrillation scenario, which is a more severe form of arrhythmia.

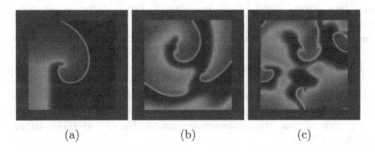

(a)   (b)   (c)

Fig. 6. Simulation result of example 4 visualized in *MonoWeb*, demonstrating the induction of fibrillation caused by the application of the S1–S2 protocol from the electrophysiological study in flat tissue.

5   Conclusion

In summary, this work highlights the value of computational simulations of heart electrical function for educational and clinical purposes. With the development of *MonoWeb*, we aim to overcome the challenges of complexity and resource demands associated with computational electrophysiology simulations, making its execution and interpretation more accessible.

We intend to improve *MonoWeb* in future work, overcoming limitations related to predefined computational domains and web access, besides providing new case studies in the software in order to present a more robust web-based simulator.

Acknowledgements. This work was supported by the Federal University of Juiz de Fora (UFJF), the "Coordenação de Aperfeiçoamento de Pessoal de Nível Superior" (CAPES) 88881.708850/2022-01, CNPq, the "Empresa Brasileira de Serviços Hospitalares" (Ebserh) grant numbers 423278/2021-5, 310722/2021-7, and 315267/2020-8, and by the Minas Gerais State Research Support Foundation (FAPEMIG) - Brazil TEC APQ 01340-18 and APQ 00748-18.

References

1. Ahrens, J., Geveci, B., Law, C.: Paraview: an end-user tool for large data visualisation. In: Visualization Handbook (2005)
2. Camps, J., Berg, L.A., Wang, Z.J., Sebastian, R., et al.: Digital twinning of the human ventricular activation sequence to clinical 12-lead ecgs and magnetic resonance imaging using realistic purkinje networks for in silico clinical trials. Med. Image Anal. 103108 (2024)
3. Kaboudian, A., Cherry, E.M., Fenton, F.H.: GPU load balancing using sparse cartesian grids: making interactive webgl simulations of complex ionic models even faster on 3d heart structures. In: 2023 Computing in Cardiology (CinC), vol. 50, pp. 1–4 (2023). https://doi.org/10.22489/CinC.2023.136
4. Kitware Inc. Trame Guide Overview (2021). https://kitware.github.io/trame/. Accessed 19 Feb 2024
5. Mitchell, C.C., Schaeffer, D.G.: A two-current model for the dynamics of cardiac membrane. Bull. Math. Biol. **65**(5), 767–793 (2003)
6. Plank, G., et al.: The opencarp simulation environment for cardiac electrophysiology. Comput. Methods Prog. Biomed. **208**, 106223 (2021)
7. Sachetto Oliveira, R., Martins Rocha, B., Burgarelli, D., et al.: Performance evaluation of gpu arallelization, space-time adaptive algorithms, and their combination for simulating cardiac electrophysiology. Int. J. Numer. Methods Biomed. Eng. **34**(2), e2913 (2018)
8. Soares, T.D.J., et al.: Studying arrhythmic risk with in-silico programmed ventricular stimulation and patient-specific computational models. In: Gervasi, O., et al. (eds.) ICCSA 202. LNCS, vol. 14112, pp. 41–51. Springer, Heidelberg (2023). https://doi.org/10.1007/978-3-031-37129-5_4
9. Ten Tusscher, K.H., Panfilov, A.V.: Cell model for efficient simulation of wave propagation in human ventricular tissue under normal and pathological conditions. Phys. Med. Biol. **51**(23), 6141 (2006)
10. Trayanova, N.A., Lyon, A., Shade, J., Heijman, J.: Computational modeling of cardiac electrophysiology and arrhythmogenesis. Physiol. Rev. **104**, 1265–1333 (2023)

Enhancing Breast Cancer Diagnosis: A CNN-Based Approach for Medical Image Segmentation and Classification

Shoffan Saifullah[1,2]([✉]) [iD] and Rafał Dreżewski[1] [iD]

[1] Faculty of Computer Science, AGH University of Krakow, Krakow, Poland
{saifulla,drezew}@agh.edu.pl
[2] Department of Informatics, Universitas Pembangunan Nasional Veteran
Yogyakarta, Yogyakarta, Indonesia
shoffans@upnyk.ac.id

Abstract. This study introduces a novel Convolutional Neural Network (CNN) approach for breast cancer diagnosis, which seamlessly integrates segmentation and classification. The segmentation process achieves high precision, with Jaccard Index (JI) values of 0.89, 0.92, and 0.87 for Normal, Benign, and Malignant regions, respectively, resulting in an overall JI of 0.896. Similarly, the Dice Similarity Coefficient (DSC) values are notably high, with 0.94, 0.96, and 0.92 for the corresponding regions, yielding an overall DSC of 0.943. The CNN model exhibits high accuracy, specificity, precision, recall, and F1 score across all classes, establishing its reliability for clinical applications. This research comprehensively evaluates the model's performance metrics, addressing challenges in breast cancer diagnostics and proposing an innovative CNN-based solution. Beyond immediate applications, it lays a robust foundation for future medical imaging advancements, enhancing diagnostic accuracy and patient outcomes.

Keywords: Breast Cancer Detection · Medical Image Analysis · CNN · Image Segmentation · Image Classification

1 Introduction

Breast cancer poses a health challenge globally [5], particularly impacting women, with advancements in its understanding and treatment. However, precise and timely diagnosis remains a formidable hurdle. These challenges underscore the urgent need for advancements in breast cancer diagnosis to navigate its intricate landscape with heightened accuracy and efficiency [14].

The complexities of breast cancer diagnosis arise from its diverse manifestations and progression patterns, influenced by genetic, molecular, and environmental factors [20]. Early symptoms often present subtly, complicating timely detection. Clinical examination and medical imaging interpretation further complicate matters, leading to delayed diagnoses and missed opportunities for early

© The Author(s), under exclusive license to Springer Nature Switzerland AG 2024
L. Franco et al. (Eds.): ICCS 2024, LNCS 14835, pp. 155–162, 2024.
https://doi.org/10.1007/978-3-031-63772-8_15

intervention [7,12,22]. Beyond clinical complexities, breast cancer diagnosis presents societal and healthcare challenges [25], including emotional distress from false positives and negatives [4,24]. Healthcare systems struggle to meet diagnostic demands [6], exacerbating disparities in access to advanced resources and early detection.

Researchers worldwide explore innovative solutions, such as advanced imaging, molecular diagnostics, and AI applications like Convolutional Neural Networks (CNNs) [23]. This study introduces a CNN-based approach to breast cancer diagnosis, aiming to improve outcomes by addressing existing challenges. The article's structure includes a review of related works (Sect. 2), detailed methodology (Sect. 3), comprehensive results analysis (Sect. 4), and conclusions with future research directions (Sect. 5).

2 Related Works

Breast cancer diagnosis has witnessed transformative advancements [3,15] with integration of Convolutional Neural Networks (CNNs) into imaging technologies [16]. CNNs have revolutionized the field by significantly enhancing accuracy and efficiency across various diagnostic processes [10]. Traditionally, interpretations of medical images [11], such as mammograms and ultrasounds, were susceptible to human subjectivity, leading to inconsistent results [13]. However, recent studies have demonstrated the power of CNNs augmenting classification accuracy by leveraging extensive datasets containing diverse breast tissue images [8]. For instance, Li et al. [9] achieved a remarkable 94.55% accuracy in distinguishing malignant from benign lesions using mammography images, significantly reducing false positives and false negatives.

Furthermore, CNNs have played a crucial role in significantly improving medical image segmentation [16], a critical aspect of breast cancer diagnosis [2,14]. Automated segmentation ensures consistency and expedites treatment accessibility by accurately delineating tumor boundaries within medical images. Despite these advancements, challenges persist, including the subtle nature of early-stage symptoms and the complexities involved in image interpretation [16]. Innovative concepts like artificial intelligence (AI) and deep learning techniques have been pivotal in addressing these challenges [12,20], promising more accurate and efficient breast cancer diagnosis.

In summary, CNN integration in breast cancer diagnosis marks a paradigm shift, reshaping diagnostics and improving outcomes [22]. These advances highlight CNN's transformative potential in breast cancer management.

3 Methods

This section details our methodology, using CNNs for breast cancer diagnosis. Our approach improves accuracy and efficiency by combining image classification and segmentation. We cover data collection, preprocessing, CNN architecture, and workflow.

3.1 Datasets of Breast Ultrasound Images and Preprocessing

Our dataset comprises 1,578 breast ultrasound images from Kaggle.com [21], categorized into Normal (266 images), Benign (891 images), and Malignant (421 images), each with ground truth labels (Table 1). Preprocessing steps ensured dataset consistency and readiness for CNN model training and evaluation. Standard normalization techniques and resizing were applied for uniformity, while data augmentation techniques, including rotation and flipping, enhanced model generalization [17].

Table 1. Samples of the dataset.

Category	Normal	Benign	Malignant
Breast Ultrasound Image			
Ground Truth (Mask)			

3.2 Convolutional Neural Networks (CNNs) Architecture

Our approach integrates two key components: U-Net for segmentation and CNN for classification, both pivotal for accurate breast cancer diagnosis. The U-Net architecture (Fig. 1), renowned for its effectiveness in segmentation tasks, operates with an $256 \times 256 \times 3$ input layer. It consists of a 5×5 Convolutional Block with ReLU activation, followed by Maxpooling and dropout to prevent overfitting. The bottleneck section employs a $16 \times 16 \times 1024$ feature map, which is crucial for capturing intricate details. The expansive path restores the feature map's original size, maintaining spatial details with upsampling layers and incorporating skip connections to preserve fine-grained details [1].

In parallel, the CNN classification network sequentially processes the segmented regions. Following the U-Net segmentation, the CNN operates on features extracted from the segmented images. Utilizing 5×5 convolutional blocks, the CNN discerns intricate patterns indicative of different breast cancer classes. Subsequent layers flatten the feature representation, followed by dense layers with dropout to learn complex relationships within the data. A softmax activation assigns probability scores to each class (Normal, Benign, Malignant), guided by categorical cross-entropy loss [2].

This integrated architecture optimizes segmentation and classification by leveraging the strengths of U-Net [18,19] for precise delineation and CNN for accurate classification. Our commitment to reliable breast cancer diagnosis is evident in this comprehensive approach, promising better patient outcomes.

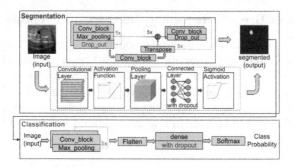

Fig. 1. CNN architecture used in the proposed approach.

3.3   Proposed Method

Our breast cancer detection method integrates segmentation and classification, inspired by [17,18]. We preprocess breast cancer images to ensure dataset uniformity and quality, laying the groundwork for model training. Our approach utilizes a modified U-Net for lesion segmentation, coupled with a CNN classifier trained to identify malignancy patterns within segmented regions. Quantitative evaluation measures segmentation accuracy using Dice Similarity Coefficient (DSC) and Jaccard Index (JI). Comprehensive validation assesses accuracy, precision, recall, and F1-score metrics, distinguishing between benign and malignant cases. Model outputs, including segmentation and classification results, are visually interpreted to understand discriminative features aiding accurate diagnosis. The proposed method undergoes validation with real-world clinical data to ensure clinical relevance, aligning predictions with actual outcomes.

4   Results and Discussion

This section presents the performance of our CNN model in segmenting and classifying breast cancer images. We analyze the results, discuss the implications, and compare them with existing approaches in the field.

4.1   Breast Cancer Segmentation Results

Our proposed CNN-based model was meticulously evaluated to ensure its efficacy in segmenting breast cancer images. The training process, depicted in Fig. 2(a), illustrates convergence at epoch 1000, with high accuracy (0.9875) and minimal loss (0.0281) indicating optimal performance. Successful assessment using DSC (0.9063) and JI (0.8307) metrics confirm the model's segmentation efficacy, as illustrated in Fig. 2(b).

Performance metrics, detailed in Table 2, underscored the precision of our model in segmenting breast cancer regions. The DSC and JI scores, including 0.94 (DSC) and 0.89 (JI) for Normal, 0.96 (DSC) and 0.92 (JI) for Benign,

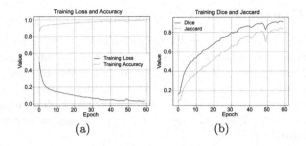

(a)   (b)

Fig. 2. Training performance of CNN: (a) model approach evaluated on accuracy and loss function, and (b) segmentation results assessed by DSC and JI.

Table 2. Evaluation metrics for segmented classes and overall performance.

No.	Segmentation Metrics	Predicted			Overall
		Normal	Benign	Malignant	
1	Jaccard Index (JI)	0.89	0.92	0.87	0.896
2	Dice Similarity Coefficients (DSC)	0.94	0.96	0.92	0.943

and 0.92 (DSC) and 0.87 (JI) for Malignant regions, indicated a high degree of accuracy in delineating tissue classes. Furthermore, the consolidated DSC of 0.943 and JI of 0.896 affirmed the model's balanced accurate segmentation.

Figure 3 shows the segmentation outcomes, demonstrating the model's accuracy in isolating distinct regions within breast images. This visualization aids in assessing segmentation accuracy and identifying areas for further analysis. The model's ability to delineate cancerous and non-cancerous regions enhances diagnostic reliability and treatment planning.

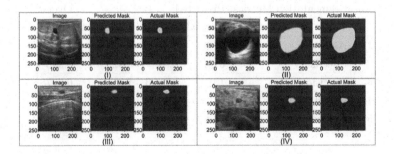

Fig. 3. Segmentation testing results in breast images, comparing model delineations with Ground Truth Masks for accuracy assessment.

4.2   Breast Cancer Classification Results

The breast cancer classification results demonstrate the exceptional performance of our CNN model in accurately categorizing breast tissue into Normal, Benign, and Malignant classes. Figure 4 depicts the model's training progress, showcasing the convergence of accuracy to near perfect levels and minimal loss, indicating successful adaptation to the dataset's complexities. The high accuracy of 0.9972 and low loss of 0.0022 highlight the model's proficiency in learning intricate tissue features during training.

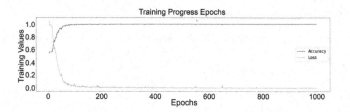

Fig. 4. Training progress depicting breast cancer classification performance, accuracy and loss metrics, showcasing model learning dynamics and convergence.

The model underwent testing with a separate dataset to evaluate generalization capabilities, following an 80%:20% split. Table 3 presents the testing phase's summary. With an accuracy of 0.896 and specificity of 0.943, the model effectively classifies breast cancer cases while minimizing false positives, essential for clinical reliability.

Table 3. Breast cancer classification efficacy metrics.

No.	Segmentation Metrics	Predicted			Overall
		Normal	Benign	Malignant	
1	Accuracy	0.89	0.92	0.87	0.896
2	Specificity	0.94	0.96	0.92	0.943
3	Precision	1.0	0.9333	1.0	0.9778
4	Recall	0.9286	0.9333	0.85	0.904
5	F1-Score	0.963	0.9333	0.9189	0.9387

Precision values indicate the reliability of the model's predictions, with a near-perfect precision of 1.0 for the Benign class, ensuring confidence in classifying benign cases. However, there is a minor scope for improvement in the recall for the Malignant class, suggesting opportunities for enhancing the model's ability to capture all true positive instances of malignant cases. The F1-score of 0.9387 signifies a balanced assessment, reflecting the model's effectiveness in making accurate predictions across all classes and ensuring robust performance in breast cancer classification.

5 Conclusions

This study demonstrates the efficacy of a CNN for breast cancer diagnosis, integrating medical image segmentation and classification. Precise segmentation identifies Normal, Benign, and Malignant tissues, while classification validates robustness across all classes. It contributes to medical image analysis by offering a reliable framework for accurate diagnosis, enhancing clinical decision-making. Future research may explore techniques (dataset split included) to improve generalization capabilities, advancing breast cancer diagnosis using deep learning.

Acknowledgement. This research was supported by the Polish Ministry of Science and Higher Education funds assigned to AGH University of Krakow and by PLGrid under grant no. PLG/2023/016757.

References

1. Alshehri, M.: Breast cancer detection and classification using hybrid feature selection and DenseXtNet Approach. Mathematics **11**(23), 4725 (2023). https://doi.org/10.3390/math11234725
2. Balasubramaniam, S., Velmurugan, Y., Jaganathan, D., Dhanasekaran, S.: A modified LeNet CNN for breast cancer diagnosis in ultrasound images. Diagnostics **13**(17), 2746 (2023). https://doi.org/10.3390/diagnostics13172746
3. Chopra, S., Khosla, M., Vidya, R.: Innovations and challenges in breast cancer care: a review. Medicina **59**(5), 957 (2023). https://doi.org/10.3390/medicina59050957
4. Gadaleta, E., Thorn, G.J., Ross-Adams, H., Jones, L.J., Chelala, C.: Field cancerization in breast cancer. J. Pathol. **257**(4), 561–574 (2022). https://doi.org/10.1002/path.5902
5. Hamdy, S., Nye, C.: Comics and revolution as global public health intervention: the case of Lissa. Glob. Public Health **17**(12), 4056–4076 (2022). https://doi.org/10.1080/17441692.2019.1682632
6. Hunleth, J., Steinmetz, E.: Navigating breast cancer screening in rural Missouri: from patient navigation to social navigation. Med. Anthropol. **41**(2), 228–242 (2022). https://doi.org/10.1080/01459740.2021.2015347
7. Jaiswal, V., Suman, P., Bisen, D.: An improved ensembling techniques for prediction of breast cancer tissues. Multimedia Tools Appl. (2023). https://doi.org/10.1007/s11042-023-16949-8
8. Labrada, A., Barkana, B.D.: A comprehensive review of computer-aided models for breast cancer diagnosis using histopathology images. Bioengineering **10**(11), 1289 (2023). https://doi.org/10.3390/bioengineering10111289
9. Li, H., Zhuang, S., Li, D.A., Zhao, J., Ma, Y.: Benign and malignant classification of mammogram images based on deep learning. Biomed. Signal Process. Control **51**, 347–354 (2019). https://doi.org/10.1016/j.bspc.2019.02.017
10. Liew, X.Y., Hameed, N., Clos, J.: A review of computer-aided expert systems for breast cancer diagnosis. Cancers **13**(11), 2764 (2021). https://doi.org/10.3390/cancers13112764

11. Lozano, A., Hassanipour, F.: Infrared imaging for breast cancer detection: an objective review of foundational studies and its proper role in breast cancer screening. Infrared Phys. Technol. **97**, 244–257 (2019). https://doi.org/10.1016/j.infrared.2018.12.017

12. Najjar, R.: Redefining radiology: a review of artificial intelligence integration in medical imaging. Diagnostics **13**(17), 2760 (2023). https://doi.org/10.3390/diagnostics13172760

13. Ortiz, M.M., Andrechek, E.R.: Molecular characterization and landscape of breast cancer models from a multi-omics perspective. J. Mammary Gland Biol. Neoplasia **28**(1), 12 (2023). https://doi.org/10.1007/s10911-023-09540-2

14. Panico, A., Gatta, G., Salvia, A., Grezia, G.D., Fico, N., Cuccurullo, V.: Radiomics in breast imaging: future development. J. Pers. Med. **13**(5), 862 (2023). https://doi.org/10.3390/jpm13050862

15. Pulumati, A., Pulumati, A., Dwarakanath, B.S., Verma, A., Papineni, R.V.L.: Technological advancements in cancer diagnostics: improvements and limitations. Cancer Rep. **6**(2) (2023). https://doi.org/10.1002/cnr2.1764

16. Rahman, H., Naik Bukht, T.F., Ahmad, R., Almadhor, A., Javed, A.R.: Efficient breast cancer diagnosis from complex mammographic images using deep convolutional neural network. Comput. Intell. Neurosci. **2023**, 1–11 (2023). https://doi.org/10.1155/2023/7717712

17. Saifullah, S., Drezewski, R.: Modified histogram equalization for improved CNN medical image segmentation. Procedia Comput. Sci. **225**(C), 3020–3029 (2023). https://doi.org/10.1016/j.procs.2023.10.295

18. Saifullah, S., Suryotomo, A.P., Dreżewski, R., Tanone, R., Tundo, T.: Optimizing brain tumor segmentation through CNN U-Net with CLAHE-HE image enhancement. In: Proceedings of the 2023 1st International Conference on Advanced Informatics and Intelligent Information Systems (ICAI3S 2023), pp. 90–101. Atlantis Press (2024). https://doi.org/10.2991/978-94-6463-366-5_9

19. Saifullah, S., Yuwono, B., Rustamaji, H.C., Saputra, B., Dwiyanto, F.A., Drezewski, R.: Detection of chest X-ray abnormalities using CNN based on hyperparameters optimization. Eng. Proc. **52**, 1–7 (2022). https://doi.org/10.3390/ASEC2023-16260

20. Sebastian, A.M., Peter, D.: Artificial intelligence in cancer research: trends, challenges and future directions. Life **12**(12), 1991 (2022). https://doi.org/10.3390/life12121991

21. Shah, A.: Breast Ultrasound Images Dataset (2020). https://www.kaggle.com/datasets/aryashah2k/breast-ultrasound-images-dataset

22. Singh, N., Srivastava, M., Srivastava, G.: Enhancing the deep learning-based breast tumor classification using multiple imaging modalities: a conceptual model. Commun. Comput. Inf. Sci. **1546**, 329–353 (2022). https://doi.org/10.1007/978-3-030-95711-7_29

23. Ting, F.F., Tan, Y.J., Sim, K.S.: Convolutional neural network improvement for breast cancer classification. Expert Syst. Appl. **120**, 103–115 (2018). https://doi.org/10.1016/j.eswa.2018.11.008

24. Topol, E.: Deep medicine: how artificial intelligence can make healthcare human again. Hachette UK (2019)

25. Travado, L., Rowland, J.H.: Supportive care and psycho-oncology issues during and beyond diagnosis and treatment. In: Gentilini, O., Partridge, A.H., Pagani, O. (eds.) Breast Cancer in Young Women, pp. 197–214. Springer, Cham (2020). https://doi.org/10.1007/978-3-030-24762-1_17

Integration of Self-supervised BYOL in Semi-supervised Medical Image Recognition

Hao Feng[2], Yuanzhe Jia[2], Ruijia Xu[3], Mukesh Prasad[1], Ali Anaissi[2], and Ali Braytee[1(✉)]

[1] School of Computer Science, University of Technology Sydney, Ultimo, Australia
`ali.braytee@uts.edu.au`
[2] School of Computer Science, The University of Sydney, Camperdown, Australia
[3] North University of China, Taiyuan, China

Abstract. Image recognition techniques heavily rely on abundant labeled data, particularly in medical contexts. Addressing the challenges associated with obtaining labeled data has led to the prominence of self-supervised learning and semi-supervised learning, especially in scenarios with limited annotated data. In this paper, we proposed an innovative approach by integrating self-supervised learning into semi-supervised models to enhance medical image recognition. Our methodology commences with pre-training on unlabeled data utilizing the BYOL method. Subsequently, we merge pseudo-labeled and labeled datasets to construct a neural network classifier, refining it through iterative fine-tuning. Experimental results on three different datasets demonstrate that our approach optimally leverages unlabeled data, outperforming existing methods in terms of accuracy for medical image recognition.

Keywords: self-supervised learning · semi-supervised learning · medical image recognition · limited labels

1 Introduction

Artificial Intelligence (AI) has significant potential in medical applications but still faces challenges in medical image recognition [1]. Medical image datasets are rich in spatial resolution, color channels, and diverse representations of organs, diseases, and anatomical structures. Obtaining relevant labels for medical classification is a primary hurdle, often requiring the expertise of medical professionals, leading to a time-consuming and logistically complex annotation process [2]. Semi-supervised learning offers a potential solution for limited labeled data. This approach can improve model performance by incorporating additional unlabeled data. Similarly, self-supervised learning has emerged as a robust strategy for exploiting the inherent patterns in unlabeled data and acquiring representations that capture the underlying semantics [3].

L. Franco et al. (Eds.): ICCS 2024, LNCS 14835, pp. 163–170, 2024.
https://doi.org/10.1007/978-3-031-63772-8_16

We propose a method to address the challenge of limited labeled data in medical image classification by combining the advantages of self-supervised and semi-supervised learning. In our method, Bootstrap Your Own Latent (BYOL) [4], a pre-training model for semi-supervised learning is employed to acquire useful representations from large amounts of unlabeled data. These representations are then fine-tuned using smaller labeled datasets to achieve high performance. Our method offers two key advantages: firstly, BYOL enhances the model generalization by capturing structural and semantic information from unlabeled medical data. Secondly, semi-supervised learning optimizes the model performance by leveraging labeled and unlabeled data.

2   Related Work

Medical image recognition heavily relies on labeled datasets. Many AI techniques, such as deep convolutional neural networks, perform optimally when abundant data and high-quality annotations are available [5]. Challenges persist when faced with limited labeled data [6]. To tackle this issue, unsupervised learning techniques like Sparse Coding (SC), Auto-encoders (AE), and Restricted Boltzmann Machines (RBMs) have been proposed. However, they are often limited to learning basic image features and struggle with the intricate characteristics present in medical images. Fortunately, leveraging unlabeled data through semi-supervised and self-supervised learning provides a promising avenue to reduce dependence on labeled data in medical image recognition.

Semi-supervised learning combines supervised and unsupervised methods, utilizing labeled and unlabeled data for enhanced performance. When dealing with brain images that lack precise diagnostic information, semi-supervised support vector machines provide feature labels to overcome classification challenges [7]. This approach is also effective on imbalanced data. Multi-Curriculum Pseudo-Labelling (MCPL) evaluates the learning progress and adjusts the threshold for each category. This adaptation improves the efficiency of semi-supervised learning, especially for imbalanced medical image classification [8]. Furthermore, semi-supervised learning can extract additional semantic information from unlabeled data [9]. Overall, by leveraging unlabeled data, semi-supervised learning yields valuable information and training examples, thereby reducing labeling costs, addressing data imbalances, and enhancing model performance.

Although semi-supervised learning has reduced the expenses associated with medical image labeling, it still necessitates labeled data, a formidable challenge in medical images. There is a pressing need to diminish the model's reliance on annotated data. Recent studies have shown that incorporating self-supervised learning into the anomaly detection framework can significantly improve the accuracy by exploiting valuable information from the original unlabeled data [10]. Moreover, a novel self-supervised learning algorithm, Bootstrap Your Own Latent (BYOL), was proposed and achieved a high classification accuracy on ImageNet, with the potential for further improvement using deeper architectures [4]. Unlike other self-supervised learning algorithms, such as Contrastive

Predictive Coding (CPC) and SimCLR, BYOL distinguishes itself by eliminating the need to generate negative samples for comparison. Therefore, the BYOL algorithm is a promising choice for augmenting semi-supervised models.

3   A Brief on BYOL

BYOL is a self-supervised learning method for feature extraction in semi-supervised learning tasks. One of its key merits lies in its independence from labeled data, as it exclusively leverages unlabeled data during the training process. This approach enables the model to derive meaningful representations from a large volume of unlabeled data, thereby enhancing its efficacy in semi-supervised learning applications.

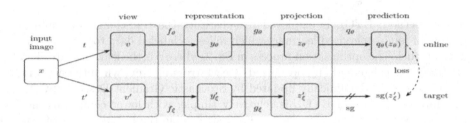

Fig. 1. BYOL architecture overview [4]

As shown in Fig. 1, the core idea of BYOL is to develop two neural networks that learn from each other, the online network θ and the target network ξ, where v is the input of θ and v' is the input of ξ. v and v' are obtained by applying stochastic image augmentation methods t and t' on the input image x. f_θ and f_ξ are encoders that share the same structure with different parameters so that v and v' get their respective corresponding representations y_θ and y'_ξ after passing through f_θ and f_ξ, where y_θ is considered as a feature of the image that will be used eventually. After this, the online network θ and the target network ξ map y_θ and y'_ξ to the latent space to get z_θ and z'_ξ using g_θ and g'_ξ, respectively. The prediction result $q_\theta (z_\theta)$ can be obtained from the online network θ by adding a layer of network structure q_θ. The target of BYOL is trying to make $q_\theta (z_\theta)$ as close to z'_ξ as possible using the loss function in Eq. 1

$$\mathcal{L}_{\theta,\xi} \triangleq \left\| \bar{q_\theta} \left(z_\theta \right) - \bar{z}'_\xi \right\|_2^2 = 2 - 2 \bullet \frac{\left\langle q_\theta \left(z_\theta \right), z'_\xi \right\rangle}{\left\| q_\theta \left(z_\theta \right) \right\|_2 \bullet \left\| z'_\xi \right\|_2} \tag{1}$$

4   Proposed Method

Our proposed method for medical image recognition is designed by utilising BYOL in a semi-supervised setting, which mainly consists of two steps: pre-training and fine-tuning (Fig. 2).

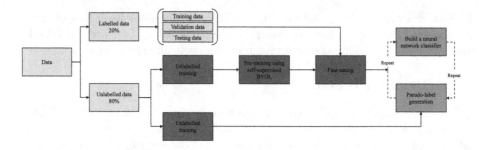

Fig. 2. The proposed semi-supervised learning with BYOL.

4.1   Pre-training

In our method, BYOL is employed to substitute the traditional pre-training process of semi-supervised learning, which relies on labeled data to generate the initial model. The optimisation process in Eq. 2, where η is the learning rate. The loss $L(\theta, \xi)$ is calculated and optimized at each training step, while the network parameters θ are dynamically updated by Stochastic Gradient Descent (SGD). The implementation of BYOL is relatively simple, with lower computing and memory requirements. As a result, we can streamline the initial phases of raw data preprocessing and feature extraction.

$$\theta \leftarrow optimizer\left(\theta, \nabla_\theta \mathcal{L}_{\theta,\xi}^{BYOL}, \eta\right), \ \xi \leftarrow \tau\xi + (1 - \tau)\,\theta, \tag{2}$$

4.2   Fine-Tuning

The feature extraction model generated by BYOL is fine-tuned with labeled data to construct a neural network classifier. This implementation is divided into three parts: constructing a neural network classifier, generating pseudo-labels, and model training.

4.2.1   Constructing a Neural Network Classifier.
The pre-trained BYOL model is fine-tuned using a smaller dataset chosen for image classification. This entails substituting the prediction layer with one customized for the target task and applying supervised learning on labeled data. The network parameters are initialized with the pre-trained weights from the prior task, capitalizing on weight inheritance to enhance the initial state, accelerate training convergence, and exploit features acquired by the pre-trained model. The prediction layer is crafted with input dimensions aligned with the output of the projection layer in the online network and is concurrently updated with the online network. In addition, the loss function of BYOL transitions from image similarity to cross-entropy to provide intuitive insights and explanations. The cross-entropy loss function is defined in Eq. 3, where y represents the one-hot encoded ground truth labels for N samples, and p denotes the predicted class probabilities.

$$H(y,p) = -\sum_{i=1}^{N} y_i \log \left(\frac{e^{p_i}}{\sum_{j=1}^{N} e^{p_j}} \right) \tag{3}$$

4.2.2   Generating Pseudo-labels. Pseudo-labels are formulated by leveraging unlabeled data and generating predictions using the pre-trained model. These synthetic labels are then integrated into the original dataset, effectively updating the sample labels within the unlabeled data. Subsequently, the predicted labels are appended to the pseudo-label repository of the neural network classifier. This combination of pseudo-labels with a limited set of labeled data enlarges the dataset, encompassing both authentic and synthetic labels, albeit with potential noise. This expanded training dataset introduces a broader spectrum of data distributions, increased sample diversity, and enhanced informational content. Consequently, it empowers the model to acquire a more extensive understanding, elevating its capacity for generalization.

4.2.3   Model Training. The modified BYOL model undergoes retraining, leveraging both labeled and pseudo-labeled data. This fusion facilitates improved generalization and heightened accuracy in addressing the target task. Each iteration involves fine-tuning the model with labeled data, generating pseudo-labeled data based on the current model, and retraining the model with a combined dataset comprising both labeled and pseudo-labeled instances. This iterative optimization consistently elevates the model performance, enhancing its resilience and generalization capabilities. Striking a balance between the objectives of model improvement is also crucial throughout this process. Determining the optimal point for iterations, guided by specific requirements and constraints, is essential in selecting the most effective model.

5   Experiments and Results

5.1   Datasets

Three datasets are used in our experiments: OCT2017 [11]; COVID-19 X-ray [12]; and Kvasir [13]. OCT2017[1] consists of 109,312 images, of which 108,312 are used for training, 32 for validation and 968 for testing. COVID-19 X-ray[2] contains 2,905 posterior-to-anterior chest X-ray images, including positive cases, viral pneumonia cases and normal cases. Kvasir[3] collects manually annotated images covering several classes, including anatomical landmarks, pathological findings and endoscopic procedures, each class consisting of hundreds of images.

---

[1] https://tinyurl.com/OCT2017dataset.
[2] https://www.kaggle.com/datasets/tawsifurrahman/covid19-radiography-database.
[3] https://datasets.simula.no/kvasir/.

5.2   Hyperparameter Tuning

Due to space constraints, we only demonstrate the tuning process on the OCT2017 dataset. Three hyperparameters are tested: the number of epochs, the learning rate, and the number of pseudo-labels. We first searched for the best combination of epochs and learning rates, exploring a range of epochs from 50 to 250 and a range of learning rates from 0.001 to 0.03. We employ accuracy as an evaluation metric. The model achieved its highest accuracy of 0.95 with 250 epochs and a learning rate of 0.001, although the computation time increased with a larger epoch count (see Table 1). We then searched for the optimal number of pseudo-labels using grid search with values of 500, 1000, and 2,000, which attained accuracy scores of 0.961, 0.962, and 0.966, respectively. This allowed us to establish the optimal hyperparameters for our model: 250 epochs, a learning rate of 0.001, and 2,000 pseudo-labels.

Table 1. Selecting the optimal epoch and learning rate on OCT2017.

Learning Rate/Epoch	50	150	250
0.03	0.51	0.86	0.83
0.01	0.62	0.81	0.90
0.001	0.83	0.93	**0.95**
Computation Time	15 min	42 min	94 min

5.3   Results

Table 2. Classification results using accuracy on three datasets.

Method/Dataset	OCT2017	COVID-19	Kvasir
MeanTeacher [14]	0.92	0.93	0.92
MixMatch [15]	0.93	0.94	0.91
FixMatch [16]	0.92	0.93	0.92
VAT [17]	0.92	0.91	0.91
VATNM [18]	0.93	0.92	0.92
GLM [19]	0.94	0.93	0.93
VTS [20]	0.95	0.95	0.93
NNM [21]	0.95	0.96	0.93
Ours	**0.966**	**0.987**	**0.976**

The experiments aim to evaluate the model performance across various datasets, including OCT2017, COVID-19 X-ray and Kvarsir. Table 2 displays a comparative analysis of results across three distinct datasets, assessing the proposed model against various semi-supervised counterparts such as MeanTeacher, Mix-Match, FixMatch, VAT, VATNM, GLM, VTS and NNM. Evaluation metrics focus on test set accuracy. For OCT2017, benchmark models achieved scores between 0.92 and 0.95. Similarly, accuracy scores fell within the 0.91 to 0.96 range on COVID-19 X-rays, and performance was maintained within the 0.91 to 0.93 range on Kvasir. Noteworthy is the exceptional performance of our proposed method across all three datasets, achieving scores of 0.966 for OCT2017, 0.987 for COVID-19, and 0.976 for Kvasir. These results underscore the effectiveness and competitiveness of our approach when compared to existing methods.

6 Conclusion

This paper introduces a method to improve the performance of semi-supervised learning on medical image data by integrating self-supervised BYOL. Experimental findings demonstrate that BYOL markedly boosts semi-supervised learning, diminishing the dependence on labeled data and setting it apart from existing semi-supervised approaches.

Acknowledgement. We acknowledge Kexin Liu, Qian Deng, Ruyi Jin, Shuting Li, and Yongxin Dai for their invaluable contributions to the project.

References

1. Rajkomar, A., Dean, J., Kohane, I.: Machine learning in medicine. New Engl. J. Med. **380**, 1347–1358 (2019). https://doi.org/10.1056/NEJMra1814259
2. Litjens, G., et al.: A survey on deep learning in medical image analysis. Med. Image Anal. **42**, 60–88 (2017). https://doi.org/10.1016/j.media.2017.07.005
3. Chaturvedi, K., Braytee, A., Li, J., Prasad, M.: SS-CPGAN: self-supervised cut-and-pasting generative adversarial network for object segmentation. Sensors **23**(7), 3649 (2023)
4. Grill, J.B., et al.: Bootstrap your own latent: a new approach to self-supervised learning. arXiv.org (2020)
5. Zhou, Y., et al.: Vgg-fusionnet: a feature fusion framework from ct scan and chest x-ray images based deep learning for covid-19 detection. In: 2022 IEEE International Conference on Data Mining Workshops (ICDMW), pp. 1–9. IEEE (2022)
6. Esteva, A.: Dermatologist-level classification of skin cancer with deep neural networks. Nature (London) **542**(7639), 115–118 (2017). https://doi.org/10.1038/nature21056
7. Filipovych, R., Davatzikos, C.: Semi-supervised pattern classification of medical images: application to mild cognitive impairment (mci). NeuroImage (Orlando, Fla.) **55**(3), 1109–1119 (2011). https://doi.org/10.1016/j.neuroimage.2010.12.066
8. Peng, Z., et al.: Faxmatch: multi-curriculum pseudo-labeling for semi-supervised medical image classification. Med. Phys. (Lancaster) **50**(5), 3210–3222 (2023). https://doi.org/10.1002/mp.16312
9. Liu, Q., Yu, L., Luo, L., Dou, Q., Heng, P.A.: Semi-supervised medical image classification with relation-driven self-ensembling model. IEEE Trans. Med. Imaging **39**(11), 3429–3440 (2020). https://doi.org/10.1109/TMI.2020.2995518

10. Zhao, H., et al.: Anomaly detection for medical images using self-supervised and translation-consistent features. IEEE Trans. Med. Imaging **40**(4), 1404–1415 (2021). https://doi.org/10.1109/TMI.2020.3000458

11. Kermany, D., Zhang, K., Goldbaum, M.: Labeled optical coherence tomography (oct) and chest x-ray images for classification. Mendeley Data (2018). https://doi.org/10.17632/rscbjbr9sj.2

12. Sahoo, P., Roy, I., Ahlawat, R., et al.: Potential diagnosis of covid-19 from chest x-ray and ct findings using semi-supervised learning. Phys. Eng. Sci. Med. **45**(1), 31–42 (2022). https://doi.org/10.1007/s13246-021-01075-2

13. Liu, P., Qian, W., Cao, J., Xu, D.: Semi-supervised medical image classification via increasing prediction diversity. Appl. Intell. (Dordrecht, Netherlands) **53**(9), 10162–10175 (2023). https://doi.org/10.1007/s10489-022-04012-2

14. Tarvainen, A., Valpola, H.: Mean teachers are better role models: weight-averaged consistency targets improve semi-supervised deep learning results. Adv. Neural Inf. Process. Syst. **30** (2017)

15. Berthelot, D., Carlini, N., Goodfellow, I., Papernot, N., Oliver, A., Raffel, C.A.: Mixmatch: a holistic approach to semi-supervised learning. Adv. Neural Inf. Process. Syst. **32** (2019)

16. Sohn, K., et al.: Fixmatch: simplifying semi-supervised learning with consistency and confidence. Adv. Neural. Inf. Process. Syst. **33**, 596–608 (2020)

17. Miyato, T., Maeda, S.I., Koyama, M., Ishii, S.: Virtual adversarial training: a regularization method for supervised and semi-supervised learning. IEEE Trans. Pattern Anal. Mach. Intell. **41**(8), 1979–1993 (2018)

18. Cui, S., Wang, S., Zhuo, J., Li, L., Huang, Q., Tian, Q.: Towards discriminability and diversity: Batch nuclear-norm maximization under label insufficient situations. In: Proceedings of the IEEE/CVF Conference on Computer Vision and Pattern Recognition, pp. 3941–3950 (2020)

19. Gyawali, P.K., Ghimire, S., Bajracharya, P., Li, Z., Wang, L.: Semi-supervised medical image classification with global latent mixing. In: Martel, A.L., et al. (eds.) MICCAI 2020. LNCS, vol. 12261, pp. 604–613. Springer, Cham (2020). https://doi.org/10.1007/978-3-030-59710-8_59

20. Wang, X., Chen, H., Xiang, H., Lin, H., Lin, X., Heng, P.A.: Deep virtual adversarial self-training with consistency regularization for semi-supervised medical image classification. Med. Image Anal. **70**, 102010 (2021)

21. Liu, P., Qian, W., Cao, J., Xu, D.: Semi-supervised medical image classification via increasing prediction diversity. Appl. Intell. **53**(9), 10162–10175 (2023)

Computational Health

Local Sensitivity Analysis
of a Closed-Loop *in Silico* Model
of the Human Baroregulation

Karolina Tlałka[1,2]([✉])[iD], Harry Saxton[3], Ian Halliday[2,5], Xu Xu[4,5][iD],
Daniel Taylor[2,5], Andrew Narracott[2,5], and Maciej Malawski[1][iD]

[1] Sano Centre for Computational Medicine, Nawojki 11, 30-072 Kraków, Poland
k.tlalka@sanoscience.org
[2] Division of Clinical Medicine, School of Medicine and Population Health,
University of Sheffield, Sheffield, UK
[3] Materials and Engineering Research Institute, Sheffield Hallam University,
Howard Street, Sheffield S1 1WB, UK
[4] Department of Computer Science, University of Sheffield, Sheffield S1 4DP, UK
[5] Insigneo Institute for in silico Medicine, University of Sheffield, Sheffield, UK
https://sano.science/

Abstract. Using a minimal but sufficient closed-loop encapsulation and
the theoretical framework of classical control, we implement and test the
mathematical model of the baroregulation due to Mauro Ursino [24]. We
present and compare data from a local relative sensitivity analysis and
an input parameter orthogonality analysis from a regulated and then an
equivalent unregulated cardiovascular model with a single ventricle and
"CRC" Windkessel representation of the systemic circulation. We conclude: (i) a basic model of the closed-loop control is intrinsically stable;
(ii) regulation generally (but not completely) suppresses the sensitivity
of output responses on mechanical input parameters; (iii) with the sole
exception of the regulation set-point, the mechanical input parameters
are more influential on system outputs than the regulation input parameters. This work is the initial step for further analysis of more complex
and computationally expensive models of the cardiovascular system, with
baroreflex control, with possible applications in space-flight medicine or
research on exercise intolerance.

Keywords: Baroreflex · Sensitivity Analysis · Digital Twin

1 Introduction

The cardiovascular (CV) system is not an independent entity. Its function relies
on external stimuli like posture shifts, exercise state and oxygen levels [2]. Blood
pressure regulation is achieved by several long and short time-scale control mechanisms. The most important is the baroreflex [6]. A model incorporating a physiologically reasonable description within the framework of control theory, of the

L.-Franco et al. (Eds.): ICCS 2024, LNCS 14835, pp. 173–187, 2024.
https://doi.org/10.1007/978-3-031-63772-8_17

coupling with the CV system is critical to digital twin development. Such a model would support a range of emerging applications - CV system response to gravitational acceleration, haemorrhage and arrhythmia (where experimental observation is challenging), to name a few.

The baroreflex is a short-term (seconds to minutes response) neurological mechanism, regulating blood pressure by adapting the CV system response, most importantly the heart period, ventricular contractility, venous tone, and systemic resistance [6]. We describe baroregulation within the framework of control theory [9], as a negative feedback problem. The feedback and feedforward elements form a single-input, multiple-output sub–system (Fig. 1). Sensors (i.e., the baroreceptors) in the aortic arch and carotid sinus transduce mechanical strains to electrical signals which are transmitted via afferent nerves to the central nervous system. There, information is processed and a response signal is directed to local effectors, via sympathetic ("fight-or-flight") and parasympathetic ("rest-and-digest") nerves.

Several models of the baroreflex exist [7,8,17,24]. The Ursino model [24] represents the control system with succinct mathematical descriptions of particular neurological physiology. Ottesen et al. [17] present similar solutions - a simple, computationally inexpensive model with pressure changes as inputs for baroreflex function; also a more complex model with wider applications detailing down to nervous activity. Heldt et al. [7,8] developed a more empirical approach, based on DeBoer's earlier work [4] and making limited appeal to control theory concepts.

A model, a set of clinical hypotheses and an appropriate methodology should co-evolve. An initially parsimonious base model is advanced by increased complexity in some aspect of its function, motivated by a need to test particular hypothesis relating to, e.g. treatments, involving this function. A relevant example is the work of Gee et al. [5], where the authors evolve Ursino's model [24], with Park et al.'s modifications [18], extending the model to describe the intrinsic cardiac nervous system, aiming to study respiratory sinus arrhythmia.

To understand baroreflex operation, an appropriate model of flow and biomechanics (termed a mechanical model) must be coupled with the regulation model and suitable test scenarios devised. Long scenario timescales and high computational costs militate for a reduction in dimensionality. Mechanical model complexity can be systematically reduced to one-dimensional (1D) formulations [13], e.g. to describe pulse-wave propagation phenomena, or further, to zero-dimensional (0D) formulations (also called lumped-parameter models) [22] which was the approach used in this work. A 0D formulation (Fig. 2) was chosen because of the specificity and provenance of the baroreflex model available and used in this work [24], which provides lumped parameters values- no counterpart parameterisation is currently available (to our knowledge) in models with higher dimensionality. Because of this reduction, some output information was lost (e.g. arterial cross-sectional pressures and detailed flow patterns), but in this application such data are not essential. Advantages and weaknesses of the 0D method are described in [22].

Personalised medicine is focused on considered, substantive resource management [1,25]. Personalisation is perhaps the central problem in the field, in which a model is calibrated to provide a digital representation of a specific patient. This task increases in difficulty with an increase in the number of input parameters and so it is desirable to minimise this number, under the constraint that key behaviours must emerge from the model. The mathematical tool that facilitates the study of the impact of changes in model inputs on model outputs - and the interaction between them - is sensitivity analysis (SA), for which there is an extensive literature in the context of medical applications [15,16,21]. Information from SA may be supplemented with orthogonality analysis [15]. By analysis of the significance and orthogonality of inputs, one can systematically identify a minimal subset of parameters, to be used in the personalisation task [16]; the simplified model is then said to have been *reduced*. The result of a model reduction depends of course upon one's initial model.

We aim to assess the unmodified baroregulation model of Ursino [24] within: (i) a minimal but physiologically sufficient, closed-loop encapsulation and (ii) the framework of classical control theory. To achieve this, we perform local relative sensitivity analysis (LSA) using "one-at-the-time" (OAT) formulation and orthogonality analysis both of a regulated and an equivalent unregulated, model. The latter is defined as one with parameters R_{sys}, E_{LVmax} and τ_0 set to the values emerging from the periodic state of that regulated model. This comparison quantifies any shift in the relative influence of model input factors on a chosen subset of discrete, derived outputs. See Fig. 3. Interactions between parameters and higher-order effects are not considered in LSA- they are the subject of global sensitivity analysis (GSA) and future work. LSA was performed because of its simplicity, low computational cost and as a preliminary method in our investigations of regulated models, preceding a GSA. We return to this point in our Conclusions. While we are aware that LSA overlooks non-linear model properties, we assume, here, that in the periodic state (a physiological state of rest), the non-linear nature of the model will play only a negligible part, based upon findings with unregulated, purely mechanical models [19].

2 Methods

Here we describe our model formulation with emphasis on the baroreflex, we outline closed–loop simulations using the model and describe methods to perform a local relative sensitivity analysis and orthogonality analysis.

2.1 Baroreflex Model

Ursino's formulation [24] transparently represents key physiological functions. Formally, our baroreflex model is a set-point controlled closed-loop regulator. It is presented in the block diagram form in Fig. 1. The Laplace transfer functions shown re-cast Ursino's ordinary differential equation formulation. The associated mechanical model has a single ventricle representation (see Fig. 2) for which the formulation is reported elsewhere [3, 21]. These two sub-models were combined in ODE or state-space form [9] as follows:

$$\frac{d}{dt}\underline{P}(t; \underline{\theta}) = \underline{f}\left(\underline{P}; \underline{\theta}\right).$$

Above, $\underline{P}(t; \underline{\theta})$ is a vector of compartmental pressure time series (or pressure surrogates [24], in the case of the control sub-system), $\underline{\theta}$ is an input parameter vector and t is time. The ODE or state-space formulation of the cardiovascular model is mathematically described by Saxton et al., [21] and our baroreflex sub-model by Ursino [24].

A single mechanical model output of aortic pressure (a surrogate for Ursino's carotid sinus pressure), provides the input for the baroreceptor block and is compared to a pre-defined pressure set-point to evolve an error signal with units of pressure, a proxy for nervous electrical pulsation. The overall regulation algorithm drives central nervous system (CNS) and autonomic nervous system (ANS) regulatory responses, designed to minimise this error. This process has defined effector dynamics expressed here in the form of Laplace transfer functions. Note, signals from the effectors are superposed with a constant, base value of each system factors in the effectors' evolution. This imparts an effective integral action to the control, as recognised by Heldt [8]. In fact, it may be shown to conform with integral action control such as that posited by Heldt et al. [8], complicated somewhat because of a presence of time delays with electro-physiological and bio-chemical origins. Our regulated CV model, described in Sect. 2.2, was applied in a rest state (i.e. without any representation of a physiological perturbation). Regulation was applied from $t = 0$ and:

1. no cycle–averaging or smoothing was necessary for stable results,
2. beat-to-beat sampling of cardiac control parameters after Heldt et al. [7] is required for a physiologically feasible and mathematically stable solution. Specifically, stable results are impossible when evolving regulation HR effectors for heart period continuously, as it is easy to show that period within a beat can cause *reduction* of the value τ- which should increase monotonically. We update heart period and maximal ventricular elastance only at the beginning of a beat; systemic resistance was adjusted at every time step.

Input parameter values and associated sources (references) are summarised in Table 1.

2.2   Cardiovascular Circulation Model

To expose the operation of heart regulation, we choose to use the simplest feasible mechanical model, capable of exposing regulation phenomena. This simplicity allows one to uncover the structure and key details of the sensitivities' of the baroreflex model, which is our main interest. Specifically, we use a 0D single ventricle model, with a Shi double cosine elastance function [12], coupled to a passive "CRC" Windkessel model, representing the systemic circulation; the latter is shown in electrical analogue form in Fig. 2.

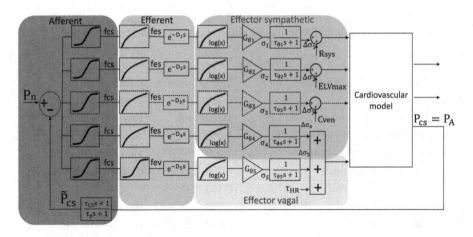

Fig. 1. Block diagram of our closed-loop baroreflex control mechanism [24]. The corresponding regulated mechanical model is represented in Fig. 2. For the baroreceptors, the input is the carotid sinus pressure, P_{CS}, and the output is a surrogate pressure $\tilde{P}_{CS}$ with units of spike-rate. The regulation set point, defined by Ursino as the pressure at the central point of the sigmoid describing carotid sinus pressure [24], is P_n (far left). Variables $f_{cs}(t)$ (carotid sinus frequency), $f_{es}(t)$ (efferent sympathetic frequency) and $f_{ev}(t)$ (efferent vagal frequency) have units of spikes per second. The delay blocks can be commuted with the signal compression blocks and represent the cumulative effect of ANS and CNS processing. Solid colour blocks locate the control functions to afferent, efferent or organ nervous activity. Afferent processing is described by a first-order ODE with 2 time constants -a first order Laplace transfer function (LTF)- followed by a sigmoidal functional compression to describe the conversion spike rate. In the efferent arc, depending on the unit of spiking activity, the efferent frequency is calculated for the sympathetic (vagal) arc using an exponential (sigmoid) functional block. In the effectors (blue and red regions of the diagram) processing involves logarithmisation and multiplication by a particular gain factor and first-order LTF. Sympathetic and parasympathetic effector processing modulates the base values of the mechanical system input parameters: systemic resistance, ventricular contractility, venous compliance, sympathetic heart period and the vagal heart period. The change of the heart period is the effect of summation of the sympathetic and parasympathetic influence. Of course, blocks in the effector part can also be commuted. (Color figure online)

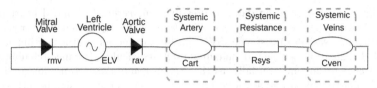

Fig. 2. Our single ventricle, mechanical model in electrical analogue form. The elastance of the left ventricle, E_{lv}, is a Shi-double cosine model [12], which relates chamber pressure and volume. The valves are assumed to have Ohmic behaviour, under both forward and reverse bias, with the regurgitating resistance set very large. Our notation for the resistances (haemodynamic dissipation) and capacitances (vessel compliance) etc. and their numerical values are declared in Table 1.

Table 1. Parameter values of the unregulated CV system. Adapted from [21]. For the corresponding regulation and Windkessel parameters, see Table 3 of reference [24].

Parameters	Symbols	Values	Units	Sources
Mean circulatory filling pressure	$mcfp$	8.000	$mmHg$	Zucker et al. [26]
Heart period	τ_0	0.580	s	Ursino [24]
Initial left ventricular volume	V_{LV}	160	ml	Kawel-Boehm et al. [11]
Minimal left-ventricular elastance	E_{LVmin}	0.060	$\frac{mmHg}{ml}$	Simaan et al. [23]
Maximal left-ventricular elastance	E_{LVmax}	2.000	$\frac{mmHg}{ml}$	Simaan et al. [23]
Time of systolic phase peak	τ_{S1LV}	0.300	s	Bjørdalsbakke et al. [3]
Time of systolic phase end	τ_{S2LV}	0.450	s	–
Aortic valve resistance	r_{av}	0.033	$\frac{mmHg \cdot s}{ml}$	Bjørdalsbakke et al. [3]
Mitral valve resistance	r_{mv}	0.060	$\frac{mmHg \cdot s}{ml}$	Bjørdalsbakke et al. [3]
Arterial compliance	C_{art}	1.130	$\frac{ml}{mmHg}$	Bjørdalsbakke et al. [3]
Systemic resistance	R_{sys}	1.663	$\frac{mmHg \cdot s}{ml}$	Kamoi et al. [10]
Venous compliance	C_{ven}	11.000	$\frac{ml}{mmHg}$	Bjørdalsbakke et al. [3]

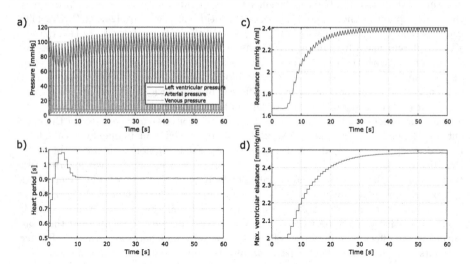

Fig. 3. Time series simulation data: a) pressure time series; b) applied heart period changes, shown as a time series (the *quantised* form of this regulation effector is apparent); c) systemic vascular resistance (SVR) evolution time series (the continuous form of the SVR regulation is clear); d) maximum left ventricular elastance (chamber contractility) changes in time. Initially, as the system equilibrates, some fluctuations of regulated values are apparent; as the system enters a periodic state, significantly smaller changes of regulation parameters occur.

2.3   Sensitivity and Orthogonality Analysis

We perform LSA of the unregulated and the equivalent closed-loop regulated CV models to determine the relative influence of input parameters on the chosen derived outputs and to investigate the impact of regulation on system sensitivities, and input parameter orthogonality to determine which inputs influence outputs in a similar way.

Local Relative Sensitivity Analysis. Although LSA represents a quasi–linearisation of input parameter effects about an operating point, a low computational cost means that LSA remains a canonical first step in understanding our model's input parameter effects to: (i) verify correct interaction between the CV and regulation models and (ii) identify the non-influential input parameters. Relative sensitivity matrices were calculated using a central difference method (Eq. 1), perturbing inputs one at a time, about a reference state $\underline{\theta}_0$:

$$s_{i,j} = 2 \left(\frac{X_j(t^*; \underline{\theta}^+) - X_j(t^*; \underline{\theta}^-)}{X_j(t^*; \underline{\theta}^+) + X_j(t^*; \underline{\theta}^-)} \right) \left(\frac{\theta_i}{\Delta\theta_i} \right). \tag{1}$$

Above, t^* represents a discrete sample time, $X_j(t^*; \underline{\theta}^+)$ the j-th output with $\theta_i \to (\theta_i + 0.5\Delta\theta_i)$ and $X_j(t^*; \underline{\theta}^-)$ the j-th output with $\theta_i \to (\theta_i - 0.5\Delta\theta_i)$. In the simulation, there were 36 inputs required to describe both the CV and baroreflex models. See Table 1 for elastance and valve factors and Table 3 of Ursino [24] for the following regulation and Windkessel parameters. Perturbed parameters were: R_{sys}, C_{art}, C_{ven}, τ_0, r_{av}, r_{mv}, τ_{S1LV}, τ_{S2LV}, $E_{LV_{max}}$, $E_{LV_{min}}$, P_n, k_a, f_{min}, f_{max}, τ_z, τ_p, $f_{es,\infty}$, $f_{es,0}$, $f_{es,min}$, k_{es}, $f_{ev,0}$, $f_{ev,\infty}$, $f_{cs,0}$, k_{ev}, $G_{T,v}$, $\tau_{T,v}$, $D_{T,v}$, $G_{T,s}$, $\tau_{T,s}$, $D_{T,s}$, $G_{E_{max,lv}}$, $\tau_{E_{max,lv}}$, $D_{E_{max,lv}}$, $G_{R,sp}$, $\tau_{R,sp}$, $D_{R,sp}$. The heart period, compliance and systemic resistance of the base mechanical model were chosen so that the emergent, regulated state was a plausible representation of a normal individual (heart period = 0.58). For the LSA presented below, the above factors were varied by ±5 % and ±10 % from the reference values. Two cases are considered: the periodic steady-state with and without regulation. For parity, the unregulated model's parameterisation was chosen so that, as far as possible, the regulated and unregulated periodic states are matched. In the regulated model, we analysed the influence of **36 inputs** on the following **10 outputs**: minimal and maximal values of: left ventricular pressure, arterial pressure, venous pressure and left ventricular volume, and heart period and cardiac output. In the unregulated model, heart period is excluded from the outputs and the following **10 inputs** were considered R_{sys}, C_{art}, C_{ven}, τ_0, r_{av}, r_{mv}, τ_{S1LV}, τ_{S2LV}, $E_{LV_{max}}$ and $E_{LV_{min}}$.

Orthogonality Analysis. A LSA helps to identify from the full input parameter list an optimal subset of inputs for use in model personalisation. Different optima exist. For example, one might select, using the criterion of influence, those model input parameters which, when changed, move the outputs the most.

However, two input parameters which, when changed, displace all output metrics in a parallel direction cannot act together to increase the dimensionality of the output space explored- the personalisation subspace. To identify such redundancy between the action of inputs we define a convenient metric of the orthogonality between two input parameters, which is based upon the sensitivity vectors

$$\hat{S}_i = (s_{i,1}, s_{i,2}, .., s_{i,N_0}), \quad i \in [1, 36],$$

defined for each input parameter. Above, $N_0 = 9$ (10) for the unregulated (regulated) system. This metric measures the displacement action about the base state, due to input i, across all outputs by comparing sensitivity vectors $\hat{S}_i$ and $\hat{S}_{i'}$. Measure $d_{ii'}$ is an inner product measure for the orthogonality between any two input parameters θ_i and $\theta_{i'}$

$$d_{i'i} = \sin\left[\cos^{-1}\left(\frac{\hat{S}_{i'}^T \cdot \hat{S}_i}{||\hat{S}_{i'}||\,||\hat{S}_i||}\right)\right], \quad i, i' = 1, .., n, \quad d_{i'i} \in [0, 1], \qquad (2)$$

Above, $||.||$ denotes the Euclidean norm and the sin function simply ensures that $d_{i'i} \in [0, 1]$. Following the work of Olsen et al. [14], the Fisher information matrix

$$\mathbf{F} = \mathbf{s} \cdot \mathbf{s}^T,$$

encapsulates the collective properties of influence and orthogonality of the sensitivity vectors, $\hat{S}$. Above, $\mathbf{s}$ is the 36×10 matrix with elements s_{ij}, see equation (1). By seeking both sensitive and orthogonal inputs, one can obtain an optimally effective set of inputs for model personalisation.

3   Results

The system was solved numerically using the MATLAB (The MathWorks, Inc., Natick, Massachusetts, USA) ode15s implicit Euler solver recommended for stiff differential equations (variable-order, variable-step method). The relative and absolute solver tolerance was $1e^{-7}$ and maximum step size was 0.01. An output function, called at each iteration, was implemented to accumulate and interpolate the emerging solution history inside the solver, to facilitate regulation delays. All regulation input parameter values were taken from Ursino [24]. The mechanical model parameters are declared in Table 1.

3.1   Local Sensitivity Analysis

The results of the LSA of the unregulated model are presented in Fig. 4. To facilitate comparison, the unregulated system heart period, contractility and SVR were chosen to produce outputs close to the regulated equivalent ($\tau_0 = 0.58$, $E_{LVmax} = 2.48$, $R_{sys} = 2.386$). Two heatmaps are presented for $\pm 5\%$ and $\pm 10\%$ perturbation of the model inputs. Despite the highly non-linear character of the model, these results show a similar pattern. In case of the cardiac output, the most influential parameter in both cases is the heart period, but it is not highly influential on the other outputs. The least influential parameters are venous compliance and aortic valve resistance; mitral valve resistance is more important (than aortic) which may point to the significance of diastolic filling. The arterial compliance is much more significant than venous compliance.

The LSA results for the CV system with regulation are presented in Fig. 5. As in the case of the un-regulated model, two heatmaps are presented, corresponding to a ±5% and ±10% perturbation of the model inputs. Again, despite the highly non-linear character of the model, these results show a similar pattern. In general, mechanical model inputs are more influential than regulation model inputs. Applying regulation does not significantly change the relative importance of the CV parameters. Unsurprisingly, the impact of the heart period on the cardiac output is more visible than in the unregulated system and the most influential control model factor is the set-point, (P_n). There is higher sensitivity on parameters bounding neural activity $(f_{min}, f_{max}, f_{es,0}, f_{es,min}, f_{ev,0}, f_{ev,\infty})$ compared with rate parameters (k_a, k_{es}, k_{ev}). Cardiac output is dominated by elastance parameters (E_{LVmin}, E_{LVmax}) and the heart period.

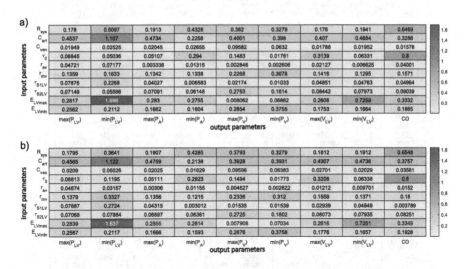

Fig. 4. Heatmaps of the local relative sensitivities of the unregulated base mechanical model, in Fig. 2, with parameters perturbed by a) ±5%; b) ±10%. The unregulated model was parameterised to generate outputs close to the regulated equivalent in Fig. 5. The chosen model outputs were for these tabulations, maximum and minimum pressures (left ventricular: P_{LV}, arterial: P_A and venous: P_V), maximum and minimum left ventricular volume (V_{LV}) and cardiac output (CO).

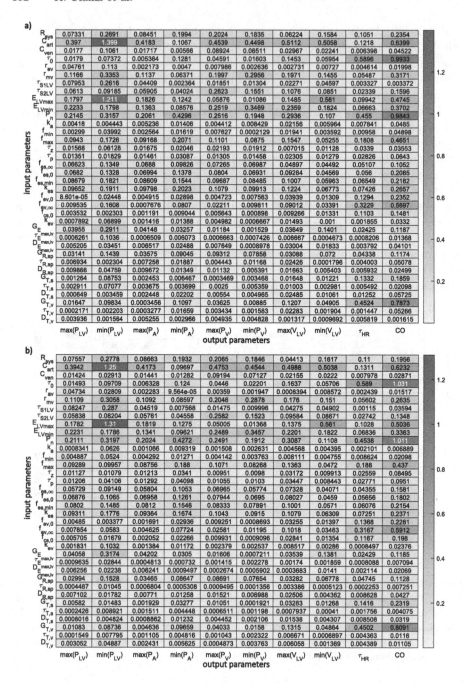

Fig. 5. Heatmap of the local relative sensitivities of the closed-loop, regulated model defined in Figs. 1 and 2, with parameters perturbed by ±5% and ±10%. The chosen model outputs were for these tabulations, maximum and minimum pressures (left ventricular: P_{LV}, arterial: P_A and venous: P_V), maximum and minimum left ventricular volume (V_{LV}), cardiac output (CO) and heart period τ_{HR}.

3.2   Orthogonality Analysis

The orthogonality analysis results presented here are for the regulated model, Saxton et al. have previously reported an orthogonality analysis of the unregulated model [21]. Sensitivity vector orthogonality is presented as a heatmap of the Fisher information matrix (Fig. 6). Interestingly, there appears to be more variation between the Fisher matrices in Fig. 6 than between the parent LSA plots, in Fig. 5. Some persistent structure is notable– e.g. the block diagonal, corresponding to the union of afferent and efferent parameters (confined between P_n and k_{ev}). In this region all factors have similar mutual orthogonality with some smaller *islands* being observed, corresponding to afferent activity between P_n and τ_p, sympathetic activity between $f_{es,\infty}$ and k_{es} and vagal activity between $f_{ev,0}$ and $f_{ev,0}$. Inside these islands, linear independence is very low. Between vagal and sympathetic parameters, the orthogonality is approximately 0.6.

The most interesting results are observed for the gains corresponding to regulated quantities– it can be seen that each regulated parameter has similar impact on the output as the corresponding gain. Again it can be seen that vagal activity dominates (with similar impact on outputs to afferent and vagal neural parameters). Heart parameters (between r_{av} and $E_{LV_{min}}$) do not have a common effect when compared with control parameters. The eigenspace of **F** contains valuable information but in the interest of a compact account, we chose to consider an intuitive re-interpretation of the orthogonality data, by examining the distribution of orthogonality, in the form of a histogram of the $d_{i,j}, i \leq j$ (upper triangular matrix elements of **F**) (Fig. 7). The interpretation of this figure is considered further in the Discussion section.

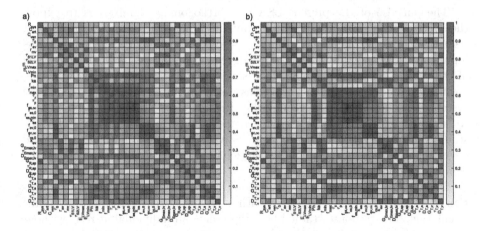

Fig. 6. Heatmaps of the orthogonality (Fisher information) in the closed-loop, regulated model defined in Figs. 1 and 2, with parameters perturbed by a) $\pm 5\%$; b) $\pm 10\%$.

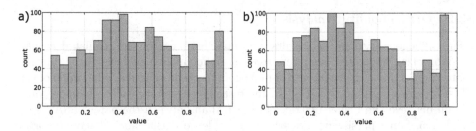

Fig. 7. The distribution of the orthogonality in the closed-loop regulated model defined in Figs. 1 and 2, with parameters perturbed by a) ±5%; b) ±10%. These are the distributions of the $d_{i,j}, i \leq j$, i.e., the elements of Fisher information, $\mathbf{F}$.

4   Discussion

The LSA of the baroreflex model of Ursino [24] performed here is, to our knowledge, the first on a closed-loop model. Our model is a faithful representation of the Ursino model and does not implement averaging of the regulatory signal over a cardiac cycle. We also present the corresponding LSA of a *shadow*, unregulated system, to provide context.

The overall distribution of the LSA heatmap is similar for both regulated and unregulated models and for results for input perturbations at ±5% and ±10%. This consistency supports the tentative conclusion that our results in Figs. 4 and 5 are truly characteristic of the system. However, the changes in the numerical values of the relative sensitivities between Figs. 4 and 5 suggest non-linear interactions and the need for a global sensitivity analysis.

The purpose of the present LSA and orthogonality analysis is to expose the effect of regulation in a tractable way, by comparing the LSA of equivalent regulated and unregulated models in Figs. 4 and 5. Clearly, mechanical CV model inputs are more generally influential than regulation factors, with the exception of the system set-point. Equally surprising is the persistence of the pattern in mechanical model sensitivities, as we pass from the unregulated to the regulated system. On the other hand, comparing Figs. 4 and 5, the regulation is seen to suppress relative sensitivity of mechanical input factors - at least for the studied outputs. Notable exceptions include the influence of initial heart period on cardiac output (which might be anticipated) and the influence of arterial compliance on LV pressure.

The qualitative trends and connections from LSA provide tentative verification of Ursino's algorithm, combined with a single ventricle model of the systemic circulation. The response studied here represents an individual at rest, without additional loading applied to the CV system. One expects vagal activity to dominate cardiac output and this is apparent in e.g. the influence of the vagal gain $G_{T,v}$ there. Put another way, the influence of vagal control on heart period, τ_{HR}, and hence cardiac output, is significantly greater than the corresponding sympathetic control.

In Fig. 7, the spread of sensitivity vector orthogonality indices $d_{i,j}$ is more uniform than in the non-regulated equivalent system reported in [21]. It is unlikely that this is due to any shift between mechanical factors, rather the distribution shifts are likely due to the inclusion, in the statistics, of a large number of new regulation input factors.

5 Conclusions and Further Work

We have successfully implemented and tested, within closed-loop operation, the baroreflex regulation model proposed by Ursino [24] *without any algorithmic extensions* e.g. control signal cardiac cycle averaging, reporting LSA and orthogonality results. We are able to conclude, on the basis of our results, that explicit averaging of control signals is not a necessary component of a baroreflex model. Extension of this approach to compare these results with cycle-averaging of control signals, as discussed by other authors is of interest and will be considered in future work, particularly to explore the influence of cycle averaging on (i) system response and stability and (ii) computational time.

The coupled CV mechanical model used in this work is intentionally simple. Extension to include a four chamber cardiac model, the pulmonary circulation and venous tone regulation is tractable using the current approach.

A crucial feature of the physiological baroreflex is the phenomena of neuronal adaptation, which was also also neglected here. The time-varying sensitivity of neurons will be included in future work to examine how this influences system sensitivity values.

LSA, combined with orthogonality analysis, provides the first tranche of information for model personalisation which is a fundamental requirement for all useful digital twins. It is essential, in even moderately complicated models, where a reduction of model inputs (a so-called *model reduction*) is necessary to bring about the decrease of computational cost necessary for a plausible global sensitivity analysis (GSA); put another way LSA is the accepted prelude to the much more costly, variance-based GSA [20], which captures non-linear interactions between inputs, and characterises only the model (in contradistinction to the model *and* its operating point), which we currently have in hand.

Acknowledgments. The publication was created within the project of the Ministry of Science and Higher Education "Support for the activity of Centers of Excellence established in Poland under Horizon 2020" on the basis of the contract number MEiN/2023/DIR/3796. This project has received funding from the European Union's Horizon 2020 research and innovation programme under grant agreement No 857533. This publication is supported by Sano project carried out within the International Research Agendas programme of the Foundation for Polish Science, co-financed by the European Union under the European Regional Development Fund.

Disclosure of Interests. The authors have no competing interests to declare that are relevant to the content of this article.

References

1. Beccia, F., et al.: Personalised Medicine in shaping sustainable healthcare: a Delphi survey within the IC2PerMed project. Eur. J. Public Health **32**(Supplement 3), ckac129.429 (2022). https://doi.org/10.1093/eurpub/ckac129.429
2. Benarroch, E.: The arterial baroreflex functional organization and involvement in neurologic disease. Neurology **71**, 1733–1738 (2008). https://doi.org/10.1212/01. wnl.0000335246.93495.92
3. Bjørdalsbakke, N.L., Sturdy, J.T., Hose, D.R., Hellevik, L.R.: Parameter estimation for closed-loop lumped parameter models of the systemic circulation using synthetic data. Math. Biosci. **343**, 108731 (2022). https://doi.org/10.1016/j.mbs. 2021.108731
4. De Boer, R.: Beat-to-beat blood-pressure fluctuations and heart-rate variability in man: physiological relationships, analysis techniques and a simple model. Ph.D. thesis (1985)
5. Gee, M., Lenhoff, A., Schwaber, J., Ogunnaike, B., Vadigepalli, R.: Closed-loop modeling of central and intrinsic cardiac nervous system circuits underlying cardiovascular control. AIChE J. **69** (2023). https://doi.org/10.1002/aic.18033
6. Harris, D.M.: Regulation of Arterial Pressure, Mohrman and Heller's Cardiovascular Physiology, 10 edn. McGraw Hill, New York (2023). http://accessmedicine. mhmedical.com/content.aspx?aid=1200684113
7. Heldt, T.: Computational Models of Cardiovascular Response to Orthostatic Stress. Ph.D. thesis (2004)
8. Heldt, T., Shim, E., Kamm, R., Mark, R.: Computational modeling of cardiovascular response to orthostatic stress. J. Appl. Physiol. **92**, 1239–1254 (2002). https:// doi.org/10.1152/japplphysiol.00241.2001
9. Jacobs, O.: Introduction to Control Theory. Oxford Science Publ, Oxford University Press (1993). https://books.google.pl/books?id=df8pAQAAMAAJ
10. Kamoi, S., et al.: Continuous stroke volume estimation from aortic pressure using zero dimensional cardiovascular model: proof of concept study from porcine experiments. PloS One **9**, e102476 (2014). https://doi.org/10.1371/journal.pone.0102476
11. Kawel-Boehm, N., et al.: Normal values for cardiovascular magnetic resonance in adults and children. J. Cardiovasc. Magn. Reson. **17**(1), 1–33 (2015)
12. Korakianitis, T., Shi, Y.: A concentrated parameter model for the human cardiovascular system including heart valve dynamics and atrioventricular interaction. Med. Eng. Phys. **28**, 613–628 (2006). https://doi.org/10.1016/j.medengphy.2005. 10.004
13. Mackenzie, J.A.: A 1D model for the pulmonary and coronary circulation accounting for time-varying external pressure. Ph.D. thesis, University of Glasgow (2021)
14. Olsen, C.H., Ottesen, J.T., Smith, R.C., Olufsen, M.S.: Parameter subset selection techniques for problems in mathematical biology. Biol. Cybern. **113**, 121–138 (2019)
15. Otta, M., Halliday, I., Tsui, J., Lim, C., Struzik, Z., Narracott, A.: Sensitivity analysis of a model of lower limb haemodynamics (2022). https://doi.org/10.1007/ 978-3-031-08757-8_7
16. Ottesen, J.T., Mehlsen, J., Olufsen, M.S.: Structural correlation method for model reduction and practical estimation of patient specific parameters illustrated on heart rate regulation. Math. Biosci. **257**, 50–59 (2014). https://doi.org/ 10.1016/j.mbs.2014.07.003. https://www.sciencedirect.com/science/article/pii/ S0025556414001369

17. Ottesen, J., Olufsen, M., Larsen, J.: Applied Mathematical Models in Human Physiology. Mathematical Modeling and Computation. Society for Industrial and Applied Mathematics (SIAM), Philadelphia (2004). https://books.google.pl/books?id=EeNBWyrG-RYC

18. Park, J.H., Gorky, J., Ogunnaike, B., Vadigepalli, R., Schwaber, J.S.: Investigating the effects of brainstem neuronal adaptation on cardiovascular homeostasis. Front. Neurosci. **14**, 470 (2020). https://doi.org/10.3389/fnins.2020.00470

19. Sala, L., Golse, N., Joosten, A., Vibert, E., Vignon-Clementel, I.: Sensitivity analysis of a mathematical model simulating the post-hepatectomy hemodynamics response. Ann. Biomed. Eng. **51**(1), 270–289 (2023)

20. Saltelli, A., et al.: Global Sensitivity Analysis: The Primer. John Wiley & Sons, Hoboken (2008)

21. Saxton, H., Xu, X., Halliday, I., Schenkel, T.: New perspectives on sensitivity and identifiability analysis using the unscented kalman filter (2023)

22. Shi, Y., Lawford, P., Hose, R.: Review of zero-d and 1-d models of blood flow in the cardiovascular system. Biomed. Eng. Online **10**, 33 (2011). https://doi.org/10.1186/1475-925X-10-33

23. Simaan, M.A., Faragallah, G., Wang, Y., Divo, E.: Left ventricular assist devices: engineering design considerations. In: Reyes, G. (ed.) New Aspects of Ventricular Assist Devices, Chap. 2. IntechOpen, Rijeka (2011). https://doi.org/10.5772/24485

24. Ursino, M.: Interaction between carotid baroregulation and the pulsating heart: a mathematical model. Am. J. Physiol. **275**, H1733–H1747 (1998). https://doi.org/10.1152/ajpheart.1998.275.5.H1733

25. Vicente, A., Ballensiefen, W., Jönsson, J.I.: How personalised medicine will transform healthcare by 2030: the icpermed vision. J. Transl. Med. **18** (2020). https://doi.org/10.1186/s12967-020-02316-w

26. Zucker, M., et al.: Changes in mean systemic filling pressure as an estimate of hemodynamic response to anesthesia induction using propofol. BMC Anesthesiol. **22** (2022). https://doi.org/10.1186/s12871-022-01773-8

Healthcare Resilience Evaluation Using Novel Multi-criteria Method

Jarosław Wątróbski[1,2], Aleksandra Bączkiewicz[1(✉)], and Iga Rudawska[3]

[1] Institute of Management, University of Szczecin, ul. Cukrowa 8,
71-004 Szczecin, Poland
{jaroslaw.watrobski,aleksandra.baczkiewicz,iga.rudawska}@usz.edu.pl
[2] National Institute of Telecommunications, ul. Szachowa 1, 04-894 Warsaw, Poland
[3] Institute of Economics and Finance, University of Szczecin, ul. Mickiewicza 64,
71-101 Szczecin, Poland

Abstract. The application of computational science methods and tools in healthcare is growing rapidly. These methods support decision-making and policy development. They are commonly used in decision support systems (DSSs) used in many fields. This paper presents a decision support system based on the newly developed SSP-SPOTIS (Strong Sustainable Paradigm based Stable Preference Ordering Towards Ideal Solution) method. The application of the proposed DSS is demonstrated in the example of assessing healthcare systems of selected countries concerning resilience to pandemic-type crisis phenomena. The developed method considers the strong sustainability paradigm by reducing linear compensation criteria with the possibility of its modeling. The research demonstrated the usefulness, reliability, and broad analytical opportunities of DSS based on SSP-SPOTIS in evaluation procedures focused on sustainability aspects considering a strong sustainability paradigm.

Keywords: Healthcare assessment · Sustainability · Decision support system · Strong sustainability paradigm

1 Introduction

The application of computational science methods and tools such as numerical methods, computational models in healthcare, and smart technologies is growing rapidly. These methods support decision-making, facilitate policy development, and provide computational support in healthcare. One aspect that determines the sustainability of healthcare systems is their resilience to pandemic-type emergencies. As the COVID-19 pandemic highlighted, the vulnerability of healthcare systems contributes to increasing global economic, social, and public health failures caused by health crises [11].

This paper presents a decision support system based on the newly developed SSP-SPOTIS (Strong Sustainable Paradigm based Stable Preference Ordering Towards Ideal Solution) method. The application of the proposed DSS is presented

L. Franco et al. (Eds.): ICCS 2024, LNCS 14835, pp. 188–195, 2024.
https://doi.org/10.1007/978-3-031-63772-8_18

in the example of evaluating healthcare systems of selected countries concerning resilience to pandemic-type crisis phenomena [8]. The developed method considers the strong sustainability paradigm by reducing linear compensation criteria with the possibility of its modeling. SSP-SPOTIS is based on the classic multi-criteria decision analysis (MCDA) method, which is SPOTIS (Stable Preference Ordering Towards Ideal Solution) [3]. SPOTIS method provides a stable ordering of alternatives toward an ideal solution. It is noteworthy that, unlike MCDA methods such as AHP (Analytic Hierarchy Process) [12], ELECTRE (ELimination and Choice Expressing the Reality) [5], and TOPSIS (Technique for Order Preference by Similarity to Ideal Solution) [9], it is resistant to the phenomenon of reversal of rankings. In contrast to methods such as PROMETHEE and ELECTRE [10], SPOTIS has low computational complexity and requires simple information about the decision problem, such as a decision matrix with data performances of alternatives against criteria and specification of bounds values for each criterion. SPOTIS provides a flexible approach that considers decision makers' preferences to determine reference solutions, as opposed to methods such as TOPSIS, VIKOR (VlseKriterijumska Optimizacija I Kompromisno Resenje), and CODAS (COmbinative Distance-based ASsessment) [4].

The rest of the paper is organized as follows. Section 2 demonstrates the methodology of the research. Section 3 presents research results and discusses them. Finally, in Sect. 4, conclusions are outlined, and future work directions are drawn.

2 Methodology

The following steps of the SSP-SPOTIS method are given below. They were developed based on the fundaments of the SPOTIS method, which rules are provided in [3]. The development of the SSP-SPOTIS method is designed to enable the modeling of linear compensation reduction at the Ideal Solution Point (ISP) with respect to which distances of alternatives are computed for the aim of the evaluation. In the classic SPOTIS method ISP is determined considering minimum and maximum bounds of the decision matrix or it is determined individually by decision maker. In the SSP-SPOTIS ISP can be determined automatically by setting s coefficient modeling reducing criteria compensation. Software in Python 3 with the developed method and datasets are available in the GitHub repository at link https://github.com/energyinpython/SSP-SPOTIS.
Step 1. Construct the decision matrix $X = [x_{ij}]_{m \times n}$ as Eq. (1) presents. The decision matrix includes performance values x_{ij} gathered for m alternatives, where $i = 1, 2, \ldots, m$ with respect to n assessment criteria, where $j = 1, 2, \ldots, n$.

$$X = [x_{ij}]_{m \times n} = \begin{bmatrix} x_{11} & x_{12} & \cdots & x_{1n} \\ x_{21} & x_{22} & \cdots & x_{2n} \\ \vdots & \vdots & \vdots & \vdots \\ x_{m1} & x_{m2} & \cdots & x_{mn} \end{bmatrix} \tag{1}$$

Step 2. Compute the Mean Deviation MD for each value x_{ij} from matrix X by subtracting the mean value of each alternative's performance $\overline{x}_j$ for each criterion C_j. Multiply the outcome value by the sustainability coefficient s_j specified for each criterion as a real number from 0 to 1. The s coefficient is determined for each criterion according to the preferences of experts and decision-makers. Equation (2) demonstrates the whole procedure performed in this step.

$$MD_{ij} = (x_{ij} - \overline{x}_j)s_j \tag{2}$$

Step 3. Assign 0 value to these MD values that for profit criteria C_j are lower than 0 (when x_{ij} is less than $\overline{x}_j$) and to these MD values that for cost criteria C_j are higher than 0 (when x_{ij} is higher than $\overline{x}_j$), as Eq. (3) shows,

$$MD_{ij} = 0 \; \forall \; MD_{+ij} < 0 \; \vee \; MD_{-ij} > 0 \tag{3}$$

where MD_{+ij} define MD values for profit criteria and MD_{-ij} represent MD values for cost criteria. This stage protects against unintended improvement of performance values that are outliers from the average toward the worst.

Step 4. Construct the matrix T with reduced compensation to determine minimum and maximum bounds with reduced compensation required for building Ideal Solution Point (ISP) considering reduced compensation. In this aim, subtract MD_{ij} values from performance values x_{ij} contained in decision matrix X as Eq. (4) presents. A compensated decision matrix T is the result of this procedure.

$$t_{ij} = x_{ij} - MD_{ij} \tag{4}$$

Step 5. Define the MCDA problem by specifying minimum and maximum bounds of score values for each criterion included in compensated decision matrix $T = [t_{ij}]_{m \times n}$, as Eq. (5) shows. For each criterion $C_j (j = 1, 2, \ldots, n)$ the minimum and maximum bounds of this criterion are determined respectively by T_j^{min} and T_j^{max}. Size of array with T_{bounds} is $2 \times n$.

$$T_{bounds} = \begin{bmatrix} T_1^{min} & T_j^{min} & \cdots & T_n^{min} \\ T_1^{max} & T_j^{max} & \cdots & T_n^{max} \end{bmatrix} \tag{5}$$

Step 6. Determine the Ideal Solution Point (ISP) defined by $T^\star$ based on T_{bounds}. If for the criterion C_j higher score value is preferred, then the ISP for criterion C_j is $T_j^\star = T_j^{max}$. On the other hand if for the criterion C_j lower score value is preferred, then the ISP for criterion C_j is $T_j^\star = T_j^{min}$. The ideal multi-criteria best solution $T^\star$ is denoted as the point of coordinates $(T_1^\star, T_j^\star, \ldots, T_n^\star)$.

Step 7. Determine the normalized distances d_{ij} from ISP for each alternative A_i using Eq. (6).

$$d_{ij}(A_i, t_j^\star) = \frac{|X_{ij} - T_j^\star|}{|T_j^{max} - T_j^{min}|} \tag{6}$$

Step 8. Calculate the weighted normalized average distance as Eq. (7) presents,

$$d(A_i, t^\star) = \sum_{j=1}^{n} w_j d_{ij}(A_i, t_j^\star) \tag{7}$$

where w_j denotes the weight of jth criterion. Criteria weights were calculated in this research by applying an objective weighting method called CRITIC (Criteria Importance Through Inter-criteria Correlation) [1]. CRITIC determines criteria weights based on a decision matrix with the performances of alternatives.

Step 9. Create alternatives' ranking by sorting $d(A_i, t^\star)$ values in ascending order. The most preferred alternative has the lowest $d(A_i, t^\star)$ value.

The structure model of healthcare systems assessment toward crisis resilience refers to some extension to the conceptual approach introduced by World Health Organization, exposing the significance of how a health system is organized (workforce, physical resources) and financed [2, 13]. The framework was developed with some specific dimensions, including service delivery during the outbreak, absorption of new technologies, and system efficiency and robustness [7]. The proposed model includes 13 criteria (C_1-C_{13}) belonging to seven main groups (G_1-G_7) displayed in Table 1. Profit criteria with the maximization aim are defined by ↑, and cost criteria with the aim of minimization are denoted by ↓.

Table 1. Multi-criteria model for healthcare evaluation.

Criteria group	Proposed indicators
G_1 - System's financial resources [8]	C_1 - Public expenditure on health - share of GDP (↑), C_2 - Public expenditure on health per capita (↑)
G_2 - System's human resources [7]	C_3 - Practicing physicians - density per 1 000 population (↑), C_4 - Practicing nurses - density per 1 000 population (↑)
G_3 - System's infrastructure [2]	C_5 - hospital beds per 1 000 population (↑), C_6 - curative (acute) care beds per 1 000 population (↑)
G_4 - Service delivery [7]	C_7 - Hospitalized patients per million (↓), C_8 - Intensive Care Unit patients per million (↓)
G_5 - Absorption of new medical products and technologies [6]	C_9 - Total tests per thousand population (↑), C_{10} - Vaccination rate (% of population vaccinated) (↑)
G_6 - System's efficiency [7]	C_{11} - Observed case-fatality ratio due to COVID-19 (↓), C_{12} - Deaths per 1.000,000 population due to COVID-19 (↓)
G_7 - System's robustness [6]	C_{13} - Excess deaths (% change from average) (↓)

Data for this research was collected from the following sources: Our World in Data: https://ourworldindata.org/, OECD Health Statistics 2022: https://www.oecd.org/els/health-systems/health-data.htm, The World Bank Data and World Health Organization European Healthcare Information Gateway (Accessed on 4 February 2024). The survey refers to the United States of America (USA) and 25 European countries. The most recent and simultaneously complete data for the countries under consideration against the criteria included in the model was obtained for 2021. Performances of the assessed countries are provided in the GitHub repository.

3   Results

This section provides assessment results of selected countries in relation to criteria of healthcare's resilience to crises using DSS based on the SSP-SPOTIS multi-criteria method. The research was carried out taking into account the modeling of the reduction of the criteria's linear compensation by incrementally increasing the value of the sustainability coefficient. Low values of the s coefficient represent a slight reduction in compensation, while large values of the s coefficient mean that the reduction in compensation is significant. A value of s equal to 0 represents the use of the classic SPOTIS method with full linear compensation of criteria. On the other hand, a value of the s coefficient equal to 1 denotes full compensation reduction. Low values of the s coefficient cause a low reduction in the compensation of weaker values of some criteria by better performances of other criteria. In contrast, the higher the value of the s coefficient, the stronger the prevention of compensating weak values of several criteria by exceptionally favorable performances of other criteria considered in the evaluation. The present work investigates the behavior of rankings under the influence of increasing the value of the s coefficient from 0.0 to 0.5 with a step of 0.05. In the first stage of the research, compensation reduction was modeled by modifying the value of the s coefficient for all model criteria simultaneously.

Table 2 displays a fragment of the matrix, including weighted normalized average distance values of evaluated alternatives obtained by applying increasing s coefficient values in the SSP-SPOTIS method. The complete matrix is provided in the Supplementary material on GitHub.

Table 2. Countries' SSP-SPOTIS scores for all criteria compensation reduction.

Country	0 (SPOTIS)	0.05	0.1	0.15	0.2	0.25	0.3	0.35	0.4	0.45	0.5
United States	0.518	0.517	0.515	0.513	0.512	0.511	0.510	0.509	0.509	0.509	0.509
Austria	0.324	0.313	0.302	0.290	0.277	0.269	0.263	0.260	0.259	0.263	0.269
Belgium	0.377	0.369	0.360	0.350	0.340	0.330	0.319	0.308	0.301	0.294	0.291
Bulgaria	0.661	0.656	0.652	0.647	0.645	0.645	0.648	0.652	0.656	0.660	0.665
...	...	...	...	...	...	...	...	...	...	...	...

Value 0.0 of s coefficient represents using the classic SPOTIS method with no reduced compensation of ISP. Rankings derived from these values sorted in ascending order are visualized in Fig. 1. It can be noticed that Austria is the stable leader of the rankings for the s coefficient in the range from 0 to 0.4. This country dropped to the second place of the ranking only with a reduction in the compensation of the criteria caused by increasing of s coefficient to 0.45 and 0.5.

This means that Austria has balanced performance values for all criteria, and a reduction in compensation does not result in a worse healthcare rating for the country's system. An interesting case is Switzerland, which ranks third in the ranking of the classic SPOTIS method without compensation reduction. However, as compensation reduction increases, it gradually moves up, and finally, with compensation reduction caused by s coefficient set to 0.45, it jumps to the top of the ranking. This proves that the country's healthcare performances are balanced across all the model's criteria. In a similar situation, an increase in compensation reduction not only does not cause a drop from good-ranking positions but actually contributes to advancement, which can be observed in countries such as Belgium, Denmark, France, Ireland, Sweden, the United Kingdom, and Czechia. On the other hand, a drop in ranking with an increase in compensation reduction can be seen for countries such as Norway, Finland, Greece, and the United States. This means that for these countries, good performances for certain criteria are able to compensate for poor values for other criteria. Therefore, when the compensation reduction is increased, the score of these countries decreases.

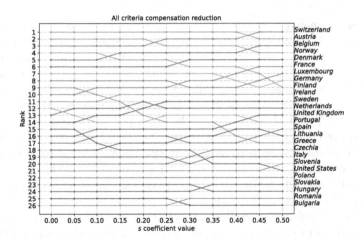

Fig. 1. Rank shifts caused by increasing compensation reduction of all criteria.

In the next step, the impact of an increase in compensation reduction in individual criterion groups was examined. Charts for simulating the compensation reduction of the other G_1-G_7 criteria are provided in the GitHub repository in the material. Shifts in rankings caused by an increase in the s coefficient in the G_2 criterion group show the advancement of Denmark, Germany, Ireland, and Sweden. This means that even if the performances within this group were slightly worse for these countries, as simulated in this analysis, the performances within the rest of the healthcare criteria are stable and balanced enough that these countries not only do not fall in the rankings but still move up relative to the compared countries. In contrast, decreases in ranking with increases in reductions in G_2 criteria compensation were recorded for Belgium, Greece, and the United Kingdom. This suggests that reductions in the performances of the G_2 criteria in these countries are causing the dropping, as the performances for the other criterion groups are not strong and stable enough to prevent this. Noteworthy shifts in the case of G_1 criteria compensation reductions include the decrease of the United States, in the case of G_3 criteria compensation reductions, the advancement of Lithuania and the drop of Germany, in the case of G_4 criteria compensation reductions the promotion of Sweden, and the fall of Greece, in the case of G_5 criteria compensation reductions the decrease of Austria, in the case of G_6 criteria compensation reduction the advance of Belgium, in the case of G_7 criteria compensation reduction the drop of Luxembourg. The analysis demonstrated that Austria, Switzerland, Norway, Belgium, Finland, and Denmark are leaders of the rankings, regardless of the criteria compensation reduction. It confirms the excellent performances and high resilience of these countries' healthcare systems to emergencies according to the adopted model criteria.

4   Conclusions

This paper presented a newly developed methodological framework for DSS for multi-criteria evaluation, considering the modeling of criteria compensation based on the innovative SSP-SPOTIS multi-criteria method. The presented method is compliant with a strong sustainability paradigm, which is essential in sustainability assessment. The possibility to model the reduction of criteria compensation by setting the value of the coefficient s automatizes and facilitates the procedure. The practical application of the developed DSS was demonstrated for a practical example of assessing the resilience of healthcare systems in selected countries in relation to adopted model criteria. The performed research demonstrated the usefulness and reliability of DSS based on SSP-SPOTIS in evaluation procedures focused on sustainability aspects considering a strong sustainability paradigm. The proposed DSS allows the modeling of criterion compensation reduction and conducting simulations, which gives broad analytical opportunities. Future work directions cover developing strong sustainability paradigm based multi-criteria methods based on other MCDA compensation methods and comparative analysis of their results. The proposed approach is planned to be tested further in a benchmarking analysis covering a more comprehensive range of the s coefficient values and including different dimensions of datasets.

Acknowledgments. This research was partially funded by National Science Centre, Poland 2022/45/B/HS4/02960, and Co-financed by the Minister of Science under the "Regional Excellence Initiative" Program.

References

1. Abdel-Basset, M., Mohamed, R.: A novel plithogenic TOPSIS-CRITIC model for sustainable supply chain risk management. J. Clean. Prod. **247**, 119586 (2020). https://doi.org/10.1016/j.jclepro.2019.119586

2. Barasa, E., Mbau, R., Gilson, L.: What is resilience and how can it be nurtured? a systematic review of empirical literature on organizational resilience. Int. J. Health Policy Manag. **7**(6), 491 (2018). https://doi.org/10.15171/ijhpm.2018.06

3. Dezert, J., Tchamova, A., Han, D., Tacnet, J.M.: The SPOTIS rank reversal free method for multi-criteria decision-making support. In: 2020 IEEE 23rd International Conference on Information Fusion (FUSION), pp. 1–8. IEEE (2020). https://doi.org/10.23919/FUSION45008.2020.9190347

4. Do, D.T., Nguyen, N.T., et al.: Investigation of the appropriate data normalization method for combination with preference selection index method in MCDM. Oper. Res. Eng. Sci. Theory Appl. **6**(1) (2023). https://doi.org/10.31181/oresta101122091d

5. Ezbakhe, F., Pérez-Foguet, A.: Decision analysis for sustainable development: the case of renewable energy planning under uncertainty. Eur. J. Oper. Res. **291**(2), 601–613 (2021). https://doi.org/10.1016/j.ejor.2020.02.037

6. Foroughi, Z., Ebrahimi, P., Aryankhesal, A., Maleki, M., Yazdani, S.: Toward a theory-led meta-framework for implementing health system resilience analysis studies: a systematic review and critical interpretive synthesis. BMC Public Health **22**(1), 287 (2022). https://doi.org/10.1186/s12889-022-12496-3

7. Haldane, V., et al.: Health systems resilience in managing the COVID-19 pandemic: lessons from 28 countries. Nat. Med. **27**(6), 964–980 (2021). https://doi.org/10.1038/s41591-021-01381-y

8. Hanefeld, J., et al.: Towards an understanding of resilience: responding to health systems shocks. Health Policy Plan. **33**(3), 355–367 (2018). https://doi.org/10.1093/heapol/czx183

9. Jam, A.S., Mosaffaie, J., Tabatabaei, M.R.: Raster-based landslide susceptibility mapping using compensatory MADM methods. Environ. Model. Softw. **159**, 105567 (2023). https://doi.org/10.1016/j.envsoft.2022.105567

10. Khan, I., Pintelon, L., Martin, H.: The application of multicriteria decision analysis methods in health care: a literature review. Med. Decis. Making **42**(2), 262–274 (2022). https://doi.org/10.1177/0272989X211019040

11. Lee, Y., Kim, S., Oh, J., Lee, S.: An ecological study on the association between International Health Regulations (IHR) core capacity scores and the Universal Health Coverage (UHC) service coverage index. Glob. Health **18**(1), 1–13 (2022). https://doi.org/10.1186/s12992-022-00808-6

12. Mokarram, M., Mokarram, M.J., Gitizadeh, M., Niknam, T., Aghaei, J.: A novel optimal placing of solar farms utilizing multi-criteria decision-making (MCDA) and feature selection. J. Clean. Prod. **261**, 121098 (2020). https://doi.org/10.1016/j.jclepro.2020.121098

13. Stochino, F., Bedon, C., Sagaseta, J., Honfi, D., et al.: Robustness and resilience of structures under extreme loads. Adv. Civil Eng. **2019** (2019). https://doi.org/10.1155/2019/4291703

Plasma-Assisted Air Cleaning Decreases COVID-19 Infections in a Primary School: Modelling and Experimental Data

Tika van Bennekum[1]([✉]) [iD], Marie Colin[2]([✉]) [iD],
Valeria Krzhizhanovskaya[1]([✉]) [iD], and Daniel Bonn[2]([✉]) [iD]

[1] Institute of Informatics, University of Amsterdam, Amsterdam, The Netherlands
tika.v.bennekum@gmail.com, V.Krzhizhanovskaya@uva.nl
[2] Institute of Physics, University of Amsterdam, Amsterdam, The Netherlands
mariecolin103@gmail.com, D.Bonn@uva.nl

Abstract. We present experimental data and modelling results investigating the effects of plasma-assisted air cleaning systems on reducing transmission of SARS-CoV-2 virus among pupils in a primary school in Amsterdam, the Netherlands. We equipped 4 classrooms (120 pupils) with the Novaerus NV800 ICU air cleaning system, and 8 classrooms (240 pupils) had standard ventilation systems. We found a significantly lower number of infections in classrooms with air cleaning systems in the first two weeks after instalment, suggesting that air cleaning decreases aerosol transmission. In the subsequent weeks, however, infection numbers increased in the Netherlands, and the difference between classrooms with and without air cleaning ceased to be significant. We analyzed the experimental results, performed a Kaplan-Meier survival estimation and developed a SIR-based computational model that simulates the results of this experiment. We performed sensitivity analysis, optimised model parameters, and tested several hypotheses. This research gives the potential for implementing improved air quality measures in public spaces, which could result in better air quality regulations in spaces such as schools.

Keywords: SARS-CoV-2 · COVID-19 · aerosol · air cleaning · transmission · prevention · public healthcare · SIR model · sensitivity analysis · Kaplan-Meier survival estimation

1 Introduction

The World Health Organization (WHO) emphasizes the role of aerosols in the transmission of SARS-CoV-2 [9] and states that 'much more research is needed given the possible implications of such route of transmission'. This is particularly relevant for crowded public spaces such as schools, where distancing is not possible and the risk of aerosol transmission of SARS-CoV-2 is high. According to Lu et al., air conditioning allows for the movement of infected droplets, which can

L. Franco et al. (Eds.): ICCS 2024, LNCS 14835, pp. 196–209, 2024.
https://doi.org/10.1007/978-3-031-63772-8_19

infect people if they inhale these droplets [8]. One of WHO's recommendations to decrease aerosol transmission of the Coronavirus is to improve ventilation and use air cleaning, for example by installing air purification systems. Such purification can remove virus particles by means of filtration [7] or inactivate them by non-thermal atmospheric plasma discharge [6,10]. We investigated the effect of a combined HEPA filter and plasma air cleaning on aerosol concentration and persistence time in a primary school in Amsterdam, the Netherlands. This research can help determine whether a serious investment in such air filters is justified and whether advanced air cleaning systems should be incorporated into the policies of public spaces.

To investigate the reduced risk of infection transmission, a computational model can be developed. Various types of Susceptible-Infected-Removed (SIR) disease dynamics models have been used for Covid-19 pandemic modelling [20, 21]. SIR-based models proved to be well-suited for this purpose, and will thus also be employed in this paper. The SIR model classifies the population into three groups: Susceptible, Infected, and Removed. The model predicts the dynamics of these three groups based on the infection and recovery rates characterising the disease.

Our model simulates an experiment conducted in a primary school in Amsterdam, the Netherlands, where the numbers of pupils with Covid infection were recorded daily in 4 classes with air cleaning systems and in 8 classes without these systems (as control group). The model is designed to reflect two distinct situations: school environment during the day (6 h) and outside the school environment the rest of the day.

Model parameters are derived from the existing research on Covid-19 epidemic. Parameter values outside the school environment are obtained from relevant studies on the spread of Covid-19 [12–15]. However, determining appropriate parameter values within the classrooms with air cleaning required extra investigation, as we did not find published data on such experiments. Our results of sensitivity analysis show how much influence each parameter has on the model.

The main research question of this study is: How can the SIR model be modified to simulate the reduced risk of infection in classrooms with a plasma-assisted air filter? The main hypothesis is that the time-varying infection rate can be calibrated to adequately simulate real data. We will show that by adjusting the parameters of the model over time, simulating a low infection rate in the classrooms with a filter during school hours, and adjusting the infection rate outside the school hours depending on the situation in the country the model can be fitted to the data.

2 Methods

2.1 Experiment Description and Experimental Results

The experiment began after the Covid lock-downs on January 10, 2022 and ended on February 16, 2022, after which there was a two-week school holiday. The study population was a group of 360 pupils aged 9 to 12 years old, in a school in Amsterdam, the Netherlands. These pupils were divided into two groups: The first group of 240 individuals was the control group, taking classes in 8 classrooms without plasma-assisted air cleaning system. The second group of 120 individuals studied in 4 classrooms equipped with the plasma air cleaning systems, which reduce the concentration of potentially infectious aerosol particles [5]. The plasma air cleaning system was a Novaerus NV-800 [5], in one classroom a similar air cleaning system was used that was also equipped with a HEPA filter, a Novaerus Defend 400. The graph shows the results of this experiment.

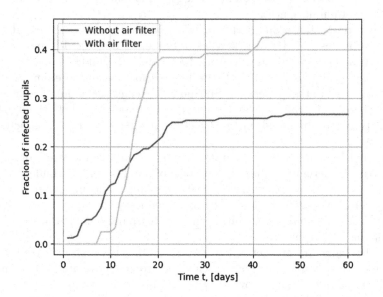

Fig. 1. The number of infected pupils over time in the experiment.

We observe significant differences in the rate with which pupils are infected: after 10 days already more than 8% of all the pupils in the control group have caught Covid-19 (Fig. 3), whereas from the group with the plasma air cleaning system this value is less than 3%. The data also suggest that air cleaning measures only help when the disease pressure is not too high; if too many pupils are infected, the school as a source of new infections is less substantial, and the 'delay' in getting Covid-19 due to the air cleaning rapidly disappears.

Plasma discharge breaks down the aerosols and reduces the concentration of fine dust particles in the air [5]. There were some initial worries at the school that the plasma air cleaning might create ultra-fine dust particles by breaking down the fine dust. We therefore measured the effect of the air cleaning system in a separate test by counting the total number of ultra-fine dust particles (diameters 6–600 nm) in the air of a typical classroom, using a TSI Model 3091 ultra-fine Particle Counter [4]. As shown in Fig. 2, the particle count as a function of time is stable prior to switching on the air cleaning system (green dotted line) and decreases after switching it on (orange dotted line). The graph displays the total number of particles summed over all channels.

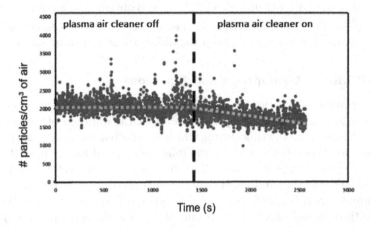

Fig. 2. Concentration of ultra-fine dust particles per cm³ of air. Plasma air cleaner was switched on at 1400 s. Particles with diameters between 6 nm and 600 nm were counted. Dotted lines show linear fits of the two periods (before and after switching the system) as guides to the eye. (Color figure online)

All classes functioned normally without any supplementary measures being taken. The classes consist of generally healthy children with an approximately equal number of males and females. The experiments were done in a period when the Netherlands had an abundance of Covid-tests, and people usually tested at the first symptoms. Parents kept their children at home with either some Covid symptoms or a positive SARS-CoV-2 test, and the school was informed by the parents. The number of reported infections per class was used as the data set for this study.

2.2   Statistical Model for Data Interpretation

The Kaplan-Meier (K-M) survival curve and the log-rank test were used to analyse the experimental results. The Kaplan-Meier analysis [2] estimates the proportion of survivors as a function of time (here, a "survivor" means not infected by Covid-19), describing the conditional probability that the individuals who survived at the beginning of the period, will survive till the end of the period. K-M estimator can be interpreted as (and derived from) non-parametric maximum likelihood estimator. An advantage of the K-M model is that it takes into account the notion of censoring. Indeed, at the end of the study, if the event (here, infection by Covid-19) did not occur, we speak of censorship of the information. For the analysis, the censorship is exact and of type II, i.e. the event declared occurs at the time indicated, and each individual is followed until an event is observed.

We then apply the log-rank test to determine whether the differences between two survival distributions are significant [1]. For each period, the expected number of cases without the difference between the groups (O) and the total number of observed cases (E) is calculated. The statistic is obtained by summing for each group $\frac{(O-E)^2}{E}$. This statistic is then studied according to a chi-square distribution. If a p-value lower than 5%, then the difference is said to be significant.

2.3   SIR Model Assumptions and Equations

Characteristics of Covid-19. For the SIR model, several assumptions need to be taken into account. *Assumption 1* is about the duration of the infectious period, which refers to the time during which an infected individual can transmit the disease to others. For Covid-19, the infectious period ranges from 2.3 to 10 days [12,13]. *Assumption 2* concerns the latent period, which is the time between an individual's infection and when they become infectious to others. For Covid-19, the latent period ranges from 2.2 to 6 days [14,15], which appears to be about the same time period as when symptoms start showing up [11]. *Assumption 3* is about what happens after an individual is recovered from the disease. While it is possible for some individuals to experience severe outcomes, in the case of children, such occurrences are negligible. Children are assumed to develop immunity after recovery. Although the exact duration of this immunity remains uncertain, it is estimated to be between a few months and a few years [16]. Since this experiment runs for a duration of 60 days, it will be assumed that no child contracts the disease more than once. *Assumption 4* is that new individuals are not introduced into the system, because the experiment covers only 2 months and involves a fixed group of school children.

SIR Equations and Modifications. For the group without an air filter, there were no adjustments needed from the basic SIR model described by the following equations for the susceptible group s (1), the infected group i (2) group r (3):

$$\frac{\delta s}{\delta t} = -\beta\, s(t)\, i(t) \tag{1}$$

$$\frac{\delta i}{\delta t} = \beta\, s(t)\, i(t) - \gamma\, i(t) \tag{2}$$

$$\frac{\delta r}{\delta t} = \gamma\, i(t) \tag{3}$$

Where beta is the infection rate and gamma is the recovery rate of the disease. For the group with air cleaning in classrooms, the disease transmission rate coefficient (β) during the 6 h in school is $\beta = \beta_{classroom}$, and during the rest of the day outside of school and on weekends $\beta = \beta_{world}$.

Model Implementation. This SIR system of ordinary differential equations was solved by using the Python package ODEINT. The ODEINT function uses the LSODA algorithm, which is part of the FORTRAN library ODEPACK [17]. The LSODA algorithm dynamically adjusts the step size array to strike a balance between accuracy and efficiency. It achieves this by considering the local error estimate and attempting to maintain it within an acceptable range. If the error estimate falls outside this range, the step size is reduced, and the calculation is recalibrated until the error falls below a user-defined threshold. Once the error estimate is within the acceptable range, the algorithm proceeds to the next time step.

Calibration of Infection Rate and Recovery Rate Parameters. In order to reproduce the observed real data, model parameters are calibrated by minimising the root mean squared error (RMSE) defined as

$$RMSE = \sqrt{\frac{1}{n}\sum_{i=1}^{n}(S_i - D_i)^2}, \tag{4}$$

where D_i is the observed data point, S_i is the simulation result, and n is the number of observations.

To overcome the problem of a slight mismatch of the varying time steps in simulations (which leads to the simulation results output at different moments not matching exactly the experimental time stamps of 1 day), we used linear interpolation for simulated results [18].

For model parameter calibration, Grid Search technique is employed in this study, systematically testing possible combinations of parameter values to identify the optimal combination. This is used to optimise the infection rate and recovery rate parameters.

2.4   Sensitivity Analysis on SIR Model

Sensitivity Analysis is a valuable method for investigating the influence of parameter variations on a model. It examines how the outputs of a system are connected to and impacted by its inputs [19]. It is a way to understand how changes in independent variables affect the model while considering specific assumptions. It does so by exploring the different sources of uncertainty within a mathematical model and quantifying their impact. For this paper, SALib (Sensitivity Analysis Library) is used. This is a Python library that provides tools for conducting sensitivity analysis. One of the methods it supports is the computation of first-order Sobol indices. The computed Sobol indices provide a measure of the relative importance of each input parameter. Higher indices indicate that a parameter has a larger influence on the output variability, while lower indices suggest a lesser impact.

3   Results

3.1   Statistical Modelling Results

Figure 3 shows the estimated 'survival probability', i.e. the probability of remaining in the uninfected group without Covid-19 during the period of the test. We can see that during the first 15 days, the survival probability in Group 1 (with the air cleaning system) is significantly higher than in Group 2 (without the air cleaning). We can therefore make a preliminary conclusion that the plasma air cleaning system reduces the risk of infection of Covid-19. However, we see that after Day 15, the difference between the groups becomes insignificant (falls within the 95% confidence interval).

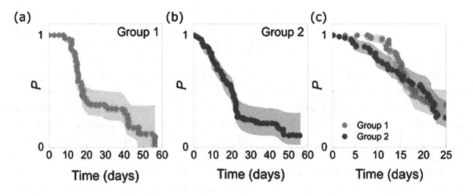

Fig. 3. Kaplan-Meier plot of the probability of not catching Covid-19 as a function of time. (a) in classrooms with plasma air cleaning system (Group 1); (b) in classrooms without the air cleaning (Group 2). The shaded area shows the 95% confidence interval; (c) both groups during the first 25 days.

To explain the slightly lower survival probability in Group 1 after Day 15, we looked at the total number of daily registered Covid cases in the Netherlands [3] and found an exponential growth during that period, with the number of cases sky-rocketing from 31,321 cases in Day 9 to 104,549 in Day 22 (see Fig. 6). This means that after the first 2 weeks of the experiment, the disease pressure in the country becomes so large that the probability of getting Covid-19 outside the school completely dominates, and air cleaning during the 6 h in school can no longer reduce the infection spread in the population.

For further analysis of the statistical results we used the log-rank test to see whether the observed difference between the classrooms with and without air cleaners is significant. The calculated p-value is smaller than 5%, therefore we conclude that the difference is significant.

3.2 SIR Modelling Results

We experimented with several SIR model modifications to simulate the reduced risk of infection in classrooms with an air filter and to reproduce the observed dynamics. Below we present the simulation results.

For the Grid Search algorithm, we defined the bounds for the parameter values based on experimental studies. For instance, in a study on the spread of Covid-19 in Canada [22], a time period of 320 days was divided into three time periods and assigned respective disease transmission rates of 0.18, 0.30 and 0.13. We therefore set the range of the disease transmission rate β values from 0.1 to 0.7.

Our experimental data (Fig. 1) showed that the group with an air filter had a higher percentage of the Covid-19 infections at the end of the experiment. This is counter-intuitive and can only be explained by a different disease transmission rate β_{world} outside the classroom, higher than in the group without air filter. The unexpectedly high number of Covid-19 cases in the group with the air filter could be caused by a birthday party in the group or by some pupils visiting regions with a very high prevalence of Covid-19 cases.

There is no research available yet to get an estimate for the disease transmission rate β specifically for the air-filtered classrooms. Thus, this was first estimated and further adjusted in numerical experiments. For the disease transmission rate inside the classrooms β_{class} we use the range from 0 to 0.7.

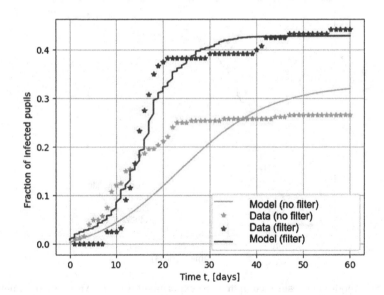

Fig. 4. SIR modelling results compared to experimental data. Calibrated $\beta_{world\ filter} = 0.4$, $\beta_{class} = 0.66$ and $\gamma_{filter} = 0.129$ for the model with the filter. The model without the filter the values are $\beta_{no\ filter} = 0.6$ and $\gamma_{no\ filter} = 0.5$. (Color figure online)

The model with the air filter (blue curve) produces significantly similar results to the data, having a quick rise in infected pupils after which the infections slow down. The main difference being that the rise starts earlier then is shown by the data. As we see, the modelled number of infected pupils with no filter (red curve) grows slower than in the observed data. Further analysis of experimental data shows that the initial fraction of infected people in this group is higher than in the group with air filter, we should therefore adjust the I_0 parameter. With more initially infected pupils, the number of infected people will grow faster. The bounds for I_0 will be from 1 to 10 infected people (Table 1).

Table 1. Optimal model parameters.

Parameter	Minimum	Maximum	Optimum filter	optimum no filter
β_{class}	0.1	0.7	0.56	–
β_{world}	0.1	0.7	0.4	0.5
γ	0.083	1	0.129	0.5
I_0	1	10	1	4

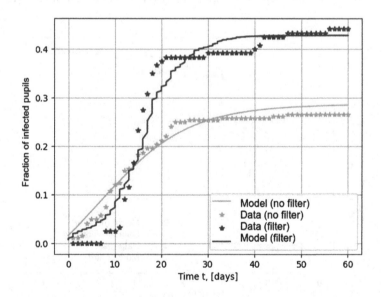

Fig. 5. SIR modelling results compared to experimental data. An extra parameter I_0 is calibrated indicating the initial number of pupils infected. All other parameters are fixed as in Fig. 4.

National Infection Cases. The infection rates observed in the experiment are undoubtedly affected by the situation in the Netherlands. Therefore, the effect of the national infection statistics (Fig. 6) on the model is tested. The hypothesis is that incorporating the daily count of new infections in the Netherlands will improve the model accuracy.

The data on the daily infection cases in the Netherlands is used to calculate the daily infection rate β_{world}. The pre-calibrated infection rate parameter is multiplied by the number of cases at the current day divided by the mean of the daily new infections. This ensures that days with a relatively high number of infections in the Netherlands will have a higher infection rate in the simulation. This modification only applies to the infection rates outside the classroom.

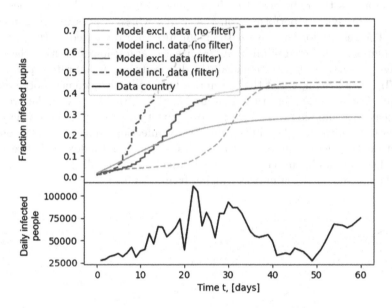

Fig. 6. SIR modelling results including and excluding the national data. For visual clarity, the number of daily infections in The Netherlands is plotted below.

Simulation results of this adjusted model are plotted against the model without the national data input in Fig. 6. We can see that the models including the national infection data yield very different results. For the model with air filter, incorporating the data results in a substantial increase in the number of infected pupils. Conversely, in the model without an air filter, the total number of infected pupils remains unchanged, but the increase in infections is delayed in the simulation. The model without an air filter was already closely aligned with the experimental data, and the inclusion of national infection statistics only reduced its accuracy. Similarly, the model without an air filter was already quite accurate, but required a slight delay in the number of infections. This delay is evident in the model with the incorporated national data, but the overall outcome significantly deviates from the observed data.

Incorporating the national infection numbers did not improve the accuracy of the models, perhaps because the data is generalised over the whole country, while the local numbers where the pupils resided could vary significantly. To investigate that, further analysis and advanced model adjustments should be performed in the future.

4   Uncertainty Quantification

A sensitivity analysis has been conducted on the SIR model presented in this paper to assess the influence of each parameter on the model's outcomes. The sensitivity analysis is performed on the model whose results are presented in Fig. 5. The results for the case without air filter are presented in (Fig. 7). The initial number of infected pupils (Fig. 7a) has the largest impact during the starting days. In agreement with this idea, the disease transmission rate β parameter (Fig. 7b) has a low impact at the beginning, but gradually becomes more influential over time. The recovery rate γ parameter (Fig. 7b) follows a similar trend to the disease transmission rate, having a low impact at the start and growing influence as time advances. The initial number of infections thus has the strongest impact on the model outcomes, with a first-order Sobol index S_i between 0.5 and 1. The beta and gamma parameters have a lower but still significant impact, with the first-order Sobol indices ranging from 0 to 0.25.

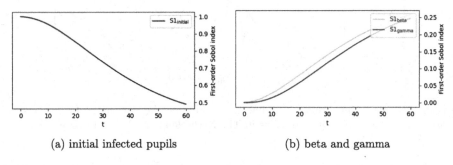

(a) initial infected pupils   (b) beta and gamma

Fig. 7. The firs-order Sobol indexes of the model parameters for the model without air filter over 60 days.

The results for the case with air filter are presented in Fig. 8. The initial number of infected pupils (Fig. 8a) again has the highest influence in the beginning of the model, then quickly decreases, but stays relatively high (first-order Sobol index $S_i = 0.92$ for the rest of the period. This indicates that the initial number of infected pupils has a greater impact on the model with an air filter compared to the model without one.

The disease transmission rate of the world β_{world} parameter (Fig. 8b) starts with S_i zero at the beginning of the simulation, but rapidly climbs to an index of 0.0125. Compared to the model without an air filter, the overall impact of β_{world} is lower. The disease transmission rate in the classroom β_{class} follows a similar trend as β_{world} but with a delay. The overall impact of the classroom parameter is very low, with a first-order Sobol index ranging between 0 and 0.0005. Although the disease transmission rate in the classroom has a low direct influence on the model, it plays a crucial role in influencing the impacts of other parameters. This is because the influences of other parameters are significantly different in the model with an air filter compared to the model without one.

The recovery rate γ (Fig. 7b) follows a similar trend as the disease transmission rates. It initially has a low impact on the model, but its influence gradually increases over time. The first-order Sobol index for the recovery rate ranges from zero to 0.03, this means it has a higher impact on the model with an air filter than the model without one.

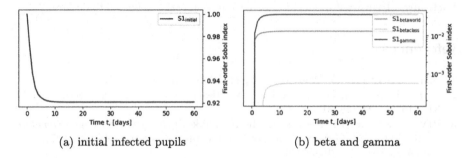

(a) initial infected pupils   (b) beta and gamma

Fig. 8. The first-order Sobol indexes of the model parameters for the model with air filter over 60 days.

5   Conclusion

Adding the plasma air cleaning system on top of the normal ventilation significantly reduces the aerosol concentration and persistence during (school) activity [5]. This has important implications: In addition to preventive measures as wearing facemasks and space ventilation, active air-clearing of aerosols by a dedicated system such as the plasma system tested here, further reduces the SARS-CoV-2 transmission risk. In this study we found that a significant reduction in Covid-19 infections can be realized by adding a plasma air cleaning system to the existing ventilation in a primary school in at least the first two weeks of installment.

The Kaplan-Meier results reveal a significant reduction in Covid-19 infection risk in classrooms equipped with a plasma air cleaning system during the initial 15 days of the experiment. After this period, this difference becomes statistically insignificant, falling within the 95% confidence interval.

We showed that a SIR model with some adjustments and model calibration is capable of reproducing the experimental data. The most optimal model was found by using a time-varying infection rate. This means that the pupils with air filters in their classrooms had different infection rate values based on the time of day and the day in the week. The sensitivity analysis showed that the number of initially infected pupils had the highest impact on the model, decreasing its influence over time. The infection rate and recovery rate parameters have a lower influence in the first days of the simulation, increasing as the simulation progressed. This analysis confirmed the conclusions of the healthcare authorities that it is critically important to monitor the early onset of the highly contagious deceases like the SARS-CoV-2.

These results show the possibility to implement better air quality regulation in public spaces such as elementary schools, by incorporating advanced air cleaning systems. In future research it would be interesting to determine the number of such systems needed for different spaces and different number of people.

Disclosure of Interests. The authors have no competing interests to declare.

References

1. Bland, J., Altman, D.: The Logrank test. BMJ **328**, 1073 (2004)
2. Bland, J., Altman, D.: Survival probabilities (the Kaplan-Meier method). BMJ **317**, 1572–1580 (1998)
3. RIVM, I&V & EPI: Covid-19 besmettelijke personen per dag. *Rivmdata* (2021). https://data.rivm.nl/meta/srv/dut/catalog.search#/metadata/097155aa-75eb-4caa-8ed3-4c6edb80467e
4. Wensing, M., Schripp, T., Uhde, E., Salthammer, T.: Ultra-fine particles release from hardcopy devices: sources, real-room measurements and efficiency of filter accessories. Sci. Total Environ. **407**, 418–427 (2008)
5. Somsen, G., Bonn, D.: Infection control unit suppresses airborne aerosols during cardiac stress testing in an outpatient cardiology clinic. J. Clin. Exp. Cardiolog. **12**, 692 (2021)
6. Smith, S., et al.: Probability of aerosol transmission of SARS-CoV-2. *MedRxiv*, pp. 2020-07 (2020)
7. Elias, B., Bar-Yam, Y.: Could air filtration reduce COVID-19 severity and spread. New Engl. Compl. Syst. Inst. **9**, 1–2 (2020)
8. Lu, J., et al.: COVID-19 outbreak associated with air conditioning in restaurant, Guangzhou, China, 2020. Emerg. Infect. Dis. **26**, 1628 (2020)
9. Organization, W.: Transmission of SARS-CoV-2: implications for infection prevention precautions: scientific Brief, 09 July 2020
10. Fennelly, M., et al.: Effectiveness of a plasma treatment device on microbial air quality in a hospital ward, monitored by culture. J. Hosp. Infect. **108**, 109–112 (2021)
11. Xiang, Y., Jia, Y., Chen, L., Guo, L., Shu, B., Long, E.: COVID-19 epidemic prediction and the impact of public health interventions: a review of COVID-19 epidemic models. Infect. Dis. Model. **6**, 324–342 (2021)

12. Wang, C., et al.: Evolving Epidemiology and Impact of Non-pharmaceutical Interventions on the Outbreak of Coronavirus Disease 2019 in Wuhan, China, MedRxiv (2020)

13. Zhao, S., et al.: Imitation dynamics in the mitigation of the novel coronavirus disease (COVID-19) outbreak in Wuhan, China from 2019 to 2020. Ann. Transl. Med. **8** (2020)

14. Chinazzi, M., et al.: The effect of travel restrictions on the spread of the 2019 novel coronavirus (COVID-19) outbreak. Science **368**, 395–400 (2020)

15. Kucharski, A., et al.: Early dynamics of transmission and control of COVID-19: a mathematical modelling study. Lancet. Infect. Dis **20**, 553–558 (2020)

16. Baraniuk, C.: How long does Covid-19 immunity last? BMJ **373** (2021)

17. Petzold, L.: Automatic selection of methods for solving stiff and nonstiff systems of ordinary differential equations. SIAM J. Sci. Stat. Comput. **4**, 136–148 (1983)

18. Bayen, A., Siauw, T.: Chapter 14 - interpolation. In: An Introduction to MATLAB®Programming and Numerical Methods For Engineers, pp. 211–223 (2015)

19. Razavi, S., et al.: The Future of Sensitivity Analysis: an essential discipline for systems modeling and policy support. Environ. Modell. Softw. **137**, 104954 (2021). https://www.sciencedirect.com/science/article/pii/S1364815220310112

20. Davidsson, R.: Modelling Covid-19 Interventions with Machine Learning and SIR Models (2021). https://scripties.uba.uva.nl/search?id=record_28773

21. Carcione, J., Santos, J., Bagaini, C., Ba, J.: A simulation of a COVID-19 epidemic based on a deterministic SEIR model. Front. Public Health. **8**, 230 (2020)

22. Jayatilaka, R., et al.: A mathematical model of COVID-19 transmission. Mater. Today Proc. **54**, 101–112 (2022). International Conferences & Exhibition on Nanotechnologies, Organic Electronics & Nanomedicine - NANOTEXNOLOGY 2020

Modelling Information Perceiving Within Clinical Decision Support Using Inverse Reinforcement Learning

Ashish T. S. Ireddy$^{(\boxtimes)}$ (ID) and Sergey V. Kovalchuk (ID)

ITMO University, Saint Petersburg, Russia
{ireddy,kovalchuk}@itmo.ru

Abstract. Decision support systems in the medical domain is budding field that aims to improve healthcare and overall recovery for patients. While treatment remains specific to individual symptoms, the diagnosis of patients is fairly general. Interpreting the diagnosis and assigning the appropriate care treatment is a crucial part undertaken by medical professionals, however, in critical scenarios, having access to recommendations from a clinical decision support system may prove life-saving. We present a real-world application of inverse reinforcement learning (IRL) to assess the implicit cognitive state of doctors when evaluating decision support data on a patient's risk of acquiring Type 2 Diabetes mellitus (T2DM). We show the underlying process of modelling a Markov Decision Process (MDP) using real-world clinical data and experiment with various policies extracted from sampled trajectories. The results provide insights into the approach to modelling real-world data into interpretable solutions via IRL.

Keywords: Inverse Reinforcement Learning · Markov Decision processes · Decision Support systems · Clinical decision making · Diabetes mellitus

1 Introduction

The medical sector has forever been one of the most vital industries in the world with a fragile margin of error. From mild cases to severe, diagnosis and selection of appropriate treatments are key to the patient's recovery. While accuracy and precision are key characteristics that healthcare professionals strive to possess, in many cases, they face crucial situations where a diagnosis or opinion falls short causing unforeseen circumstances. Most often a group of doctors consult together and decide on an appropriate course of action but this cycle eventually gets modulated over time (i.e. due to data scarcity, lack of experience etc). Decision support systems (DSS) have taken up the role of providing a layer of confidence and trust for decision-makers to take necessary actions. Despite being moderately used in the medical domain and under an experimental status, the use of Clinical decision support systems (CDSS) has widely increased to incorporate it into general practice. The current state of CDSSs allows professionals to feed data (present and historical) and evaluate the best possible

© The Author(s), under exclusive license to Springer Nature Switzerland AG 2024
L. Franco et al. (Eds.): ICCS 2024, LNCS 14835, pp. 210–223, 2024.
https://doi.org/10.1007/978-3-031-63772-8_20

scenarios, optimal solutions, risk conditions and many such parameters that can be fine-tuned to provide the matching recommendations respective to individual patients. However, one of the main features that it lacks is the ability to assess situations as humans do. The underlying impact such as social leverage over fear of treatment can be overruled via CDSS. Understanding the way a human views a situation is complex and varies widely based on each activity [4]. Replicating the cognitive state during an activity via traditional reinforcement learning (RL) methods [14] is tedious and extremely demanding. Hence, learning by observation proves to be a more efficient way of modelling the human mind's cognitive state. Within our work, we provide such a case of modelling the human cognitive state where medical professionals interpret true patient information and prediction data from a diabetes prediction model, to assess if a patient will have Type 2 Diabetes mellitus (T2DM). This data is further evaluated by doctors according to their level of perception, understandability, agreement and usability.

Using Inverse reinforcement learning (IRL), we have modelled and extracted the underlying reward functions based on optimal policies that describe the cognitive state of the doctor and the strategy during the evaluation of patient and prediction model data. We demonstrate the use of Linear IRL using Markov decision processes (MDPs) and provide a basic outlook of apprentice learning using cyclic MDP environments. Our results and investigation provide insight and foundational results to approach IRL and MDPs using expert trajectories from real-world data. Further, we provide early results of complex models for IRL policy extraction using expert trajectories and possible frameworks to interpret reward functions and policies to scenarios.

The remaining part of the paper is structured as follows: Sect. 2 describes the background and related works on IRL and MDPs. Section 3 is the methodology where we elaborate on modelling our MDP and the IRL setup. Section 4 is the results section where the generated simulations are investigated and its features elaborated. Section 5 is the discussion and Sect. 6 is the conclusion.

2 Related Works

Learning a task is a process of discovering the outcome after taking a sequence of actions to fulfil an objective. There can be instances when certain actions lead to swift fulfilment of the objective while other approaches might not be conclusive at all. We can view a majority of the day-to-day activities in the real world as a reinforcement problem, where an environment with states, actions and outcomes are defined to reach a certain goal. These activities have already been discovered and documented where the stages ahead involve optimization and improvement of the process itself. A simple example of tuning a guitar shows the effect of over-tuning that leads the strings to snap while under-tuning leads to noise instead of notes. By tuning the guitar to a certain scale, we obtain ideal sound notes and can therefore produce music. However, in open-ended processes, the same cannot be said all the time. Training an autonomous car to mimic the behaviour of a person is one of the most widely used examples to describe the complexity of teaching the penalty, reward and justification for taking specific actions given a situation [11]. Other examples include flying aircraft, learning to play table tennis [9] etc., the direct strategy to fulfil the goal is not explicitly defined yet, and the outline of the problem and its rules are stipulated via observable behaviour of the ideal scenario that is discovered [7]. IRL is aimed at resolving problems where the complete do's and don'ts are not provided by the user and all that remains is a set of expert demonstrations of performing the task from which the ideal behaviour, actions, ethics and rules can be defined by relatively modelling the implicit strategy allowing replication of the behaviour to produce similar results and therefore learning the process itself [2]. Further, from the data standpoint, RL and supervised learning demand explicitly defined pathways and boundary conditions to perform a task on par with humans, IRL stands out by learning over demonstrations and therefore minimizes the initial effort of conditioning the data and preparing the model itself. Depending on certain activities, IRL algorithms can learn existing scenarios and collectively map newer pathways which can prove to be a lucrative incentive in many medical decision making situations where implicit and explicit factors play an interdependent role in decision making. A simple example is the discharge of patient from a hospital due to early recovery impacted by improving weather conditions, traditional approaches may categorize such an occurrence as an outlier or a "by-chance occurrence" without investigating the underlying reasoning. IRL models may be able to derive such relations, thereby, recommending actions often observed in doctors behaviour.

In the community, one of the most common applications of IRL is the gird world and shortest path problem. During our search we found only one such work that provides insight into using IRL to assess human risk-taking characteristics using real-world data [8]. Apprentice modelling [1] is an extension of full-fledged IRL that can take advantage of the full extent of using expert trajectories to find the best reward function and its matching policy relative to the actual process over data. These results can then be interpreted and understood as the plausible cognitive state space or process of taking actions by a human based on the incentive they gain by reaching the final state. Therefore learning the process and its etiquette while also trying to find an undiscovered pathway that may have been overlooked by the human subject [17]. The prospects of using imitation learning are massive yet we believe the problem lies with modelling the data into appropriate state-feature spaces and MDPs. There may be instances where datasets may not directly be modelled into an MDP without additional modifications or modulations of state spaces. Within our work, we have shown such an example of modelling data into an MDP without the need for modulation. Further, there is also the conundrum of using RL to solve an IRL problem being less efficient. An answer to this has been provided by the authors in [14] where they aim to reset the learning trajectory of the IRL model such that follows the start state of the expert's demonstration and avoids exploring all possible combinations in the state space.

3   Modelling Perceiving in Clinical Decision Process with CDSS Interaction

In this section we define our approach to modelling our decision data for IRL using MDPs as a baseline interpretation. We define notations that will be used throughout the paper here forth.

A set of n (finite) **expert trajectories** $E_T = \{\tau_1; \tau_2; ....\tau_n\}$ constitute to a combination of state action combinations, defined via the following terms:

- $S = \{s_1, s_2, ...s_n\}$ is a finite set of all possible **states** the agent can take in E_T
- $A = \{a_1, a_2, ...a_n\}$ is a set of **actions** an agent can taken in E_T
- $T_{PA}(.)$ are the state **transition probabilities** of moving to state s' from s upon taking action a, i.e. $T(s, a, s')$, extracted from E_T
- $\gamma \in [0, 1)$ is the **discount factor** that dictates the weightage for long-term-short-term reward strategy
- π is the **policy** function that defines the action to be taken in each state i.e. $(\pi : S \rightarrow A)$
- π^* is the **optimal policy** that defines the optimal actions to take in each state such that the generated reward is maximum
- $\tau = \{(s_0, a_1, s_1); (s_1, a_2, s_2); ....(s_{n-1}, a_n, s_n)\}$ is a **trajectory** describing one complete iteration of the agent in the MDP (i.e. until it concludes or reaches an end state)

Traditional RL involves finding the optimal policy of a problem using a defined reward function. IRL is defined as the opposite where the reward function is sought using the perceived optimal policy from a set of expert demonstrations. When taking into account real-world scenarios of IRL i.e. using expert trajectories, we assume that the experts' actions in given trajectories are the optimal behaviour to be followed. With this in mind, given a set of expert trajectories E_T, we assume there are n policies $\{\pi_1, \pi_2, ...\pi_n\}$ that can generate maximal rewards. We use the linear programming approach for IRL as described in [10] to find the maximal reward function for assumed policy π^*. On finding the maximum reward, we iterate through the various policies possible from the data to get a complete overview of reward functions existing within the expert trajectories. Simply put, we run RL inside IRL to find the reward function respective to our assumed policies from the expert trajectories. Figure 1 provides an overview of our setup for the simulation.

3.1   CDSS Data

The dataset used here is from [6] where the authors have performed an experiment to analyse the effect of having decision makers (e.g. doctors, physicians) supported by information from a prediction model, a FINDRISK measure and case-based explanation to assess the concurrent perceptional state through subjective metrics. Through a survey, physicians were asked to assess the chances

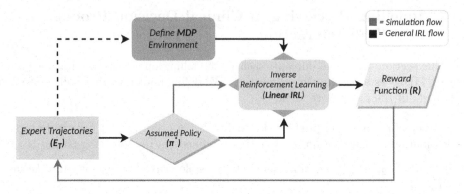

Fig. 1. A schematic overview of our IRL simulation setup where E_T is the real world experts trajectories from which the MDP environment and the policies π are extracted, The MDP tuple is fed to the linear IRL algorithm which assumes optimal policy π^* to be extracted from E_T after which the reward function R is formulated. Our simulation flow involves repeating the process of initialising the MDP respective to each subset of data and all possible policies existing within E_T

of a patient encountering T2DM in the future based on the amount of data presented in three alternative settings:

- **Setting 1:** From prediction model only
- **Setting 2:** From a FINDRISK scale and prediction model
- **Setting 3:** From an explanation, FINDRISK scale and prediction model

The physicians were provided with the patient's basic information (age, height, weight, BMI (body mass index), gender, blood, physical activity level, blood sugar, heredity, arterial pressure) and one of three prediction settings. Physicians were then asked to assess each setting with three subjective perception measures via a Likert scale from 1 (strongly disagree) to 5 (strongly agree), *Understandability* denoting the level of interpretation of the information, *Agreement* corresponding to the acceptance of the model's prediction as per the data and *Usability* reflecting the subsequent usage of the data in further diagnosis. A total of 541 cases of patient assessment data were found to be usable for our experiment.

3.2 Initializing MDP for CDSS Data

From the CDSS data we model our MDP as a system of:

- S : 5 **states** = $\{End,\ Understandability,\ Agreement,\ Usability,\ Completion\}$
- A : 2 **actions** = $\{Continue,\ Terminate\}$
- T_P: Transition probabilities of moving from state s to s' extracted from E_T using the following strategy:

$$T_P(s, a, s') = \frac{\# \ of \ times \ (s \ \to \ s') \ occurs \ in \ T_E}{Total \ \# \ of \ occurences \ in \ T_E} \tag{1}$$

Given a trajectory τ of a patient's data evaluated by the physician, the MDP is initialized in the state of understandability and takes an action to either continue or terminate the MDP based on the scored evaluation from the physician respective to each state and metric. The decision to terminate or continue is deterministically relative to the doctors' assessment via a threshold M_T. We have introduced three thresholds for each measure to assess the impact of having strict vs moderate evaluation criteria. Relative to the 1–5 Likert scale we have selected thresholds of $M_T = [2, 3, 4]$ to evaluate the changes in reward function per policy over various settings of data. For a complete assessment, the MDP should traverse through the states of understandability, agreement, and usability to finally reach the completion state by taking action *continue* when the given data for each measure is greater than the metric threshold M_T. Likewise, if the condition is not satisfied, by taking action *terminate* the MDP reaches the state of End. Figure 1 describes the MDP and its feature space (Fig. 2).

$$\pi(S) \to A = \begin{cases} Continue, & if[Understandability \ or \ Agreement \ or \ Usability] > M_T. \\ Terminate, & otherwise. \end{cases} \tag{2}$$

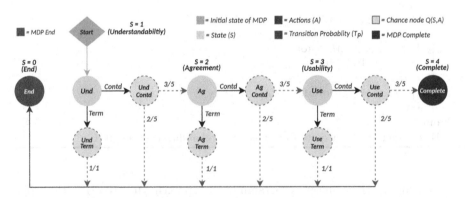

Fig. 2. A visualization of our MDP adapted to CDSS data space. The MDP is initialised in the state of understandability, within each state two actions *Continue* or *Terminate* can be performed, $Q(S, A)$ denotes the chance node and the probability of arriving in state S' on taking action A from which the transition probabilities T_P is extracted

4   Case Study: T2DM Risk Prediction Perceiving

4.1   Simulating Risk Prediciton Using MDP

To model the risk perception from CDSS data, we first divided the collected responses of physicians into 4 groups. The three settings as initialized in the previous section and a combined version that holds trajectories of settings 1,2 and 3 together. Next, we set up the MDP as mentioned in the prior section to simulate the flow of the feature space. The transition probabilities T_P were generated using a self-created script 1 where, S and A correspond to the number of states and actions in the MDP, $\tau_n = [Und, Ag, Use]$ is a tuple of metric scores for a set of trajectories, M_T is a metric threshold that is used to generate the transition probability using Eq. (1).

Algorithm 1: Calculation of transition probabilities

Data: S, A, set of n trajectories $\tau_n = [Und, Ag, Use]$, $M_T = [2, 3, 4]$
Result: A matrix with size $[S, A, S]$ with probabilities $\in [0, 1]$
Feed individual settings data [*setting 1, setting 2, setting 3, combined settings*];
for n *trajectories in* τ_n **do**

> iterate through each patient trajectory;
> **if** $\tau_n[Und] > M_T$ **then**
> > $\# \; of \; A(Continue)_{\tau_n[Und]} += 1$;
>
> **else**
> > $\# \; of \; A(Terminate)_{\tau_n[Und]} += 1$;
>
> **end**
> **if** $\tau_n[Ag] > M_T$ **then**
> > $\# \; of \; A(Continue)_{\tau_n[Ag]} += 1$;
>
> **else**
> > $\# \; of \; A(Terminate)_{\tau_n[Ag]} += 1$;
>
> **end**
> **if** $\tau_n[Use] > M_T$ **then**
> > $\# \; of \; A(Continue)_{\tau_n[Use]} += 1$;
>
> **else**
> > $\# \; of \; A(Terminate)_{\tau_n[Use]} += 1$;
>
> **end**

end
Using equation (1) acquire the transition probability between $[0, 1]$

We used the linear programming approach to resolve the IRL function as implemented in [3]. We fed the IRL algorithm a tuple $(S, A, \tau_P, [\pi], R_{max}, \gamma, L1)$ the number of states S, actions A, transition probabilities T_P, a set of policies $\pi = [\pi_1, ...\pi_8]$, a discount factor, maximum reward R_{max}, a discount factor γ and an $L1 \in [0, 1]$ regularization value. The set of policies from our data was defined based on the possible combinations of the MDP reaching the states [*End, Und, Ag, Use, Completion*] over the threshold M_T. This results in eight policies as shown in Table 1.

The resultant of the IRL algorithm is a reward function of specific weights for each state respective to the policy. We chose our γ value to be 0.9, a long-term strategy, relative to predicting the patients' assessment in the future and not in the present state. When a smaller γ value was selected we observed that reward was only awarded in a policy where all three states had a *continue* action and the

Table 1. All combinations of policy π possible within our MDP space extracted from our CDSS data

Case	Policy	Understandability	Agreement	Usability
1	[0, 0, 0, 0, 0]	Terminate	Terminate	Terminate
2	[0, 0, 0, 1, 0]	Terminate	Terminate	Continue
3	[0, 0, 1, 1, 0]	Terminate	Continue	Continue
4	[0, 1, 1, 1, 0]	Continue	Continue	Continue
5	[0, 1, 1, 0, 0]	Continue	Continue	Terminate
6	[0, 1, 0, 0, 0]	Continue	Terminate	Terminate
7	[0, 0, 1, 0, 0]	Terminate	Continue	Terminate
8	[0, 1, 0, 1, 0]	Continue	Terminate	Continue

state of *Complete* was reached, for other policies the reward was 0. This can be attributed to having a short-term strategy of γ. Similarly, our $L1$ regularization value was also set to 0.9 as smaller values produced no plausible reward as output (Fig. 3).

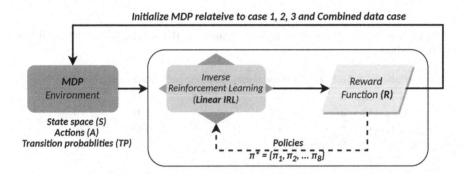

Fig. 3. Overview of the simulation process to extract reward function R for all possible policies in a given data setting. The MDP is generated for each data setting (including the transition probabilities T_P) after which the IRL algorithm is fed with MDP tuple and a set of assumed optimal policies $[\pi^* = \pi^1, ...\pi^8]$ over which the maximum reward is extracted. The same process is carried out for other cases of data.

4.2   Inferring Reward Functions for Trajectories

We initialized the MDP for settings 1, 2, 3 and a combined version. The transition probabilities T_P reflecting the physicians' assessment for each setting varied as the level of data fed was different. This was also observed in the metric ratings provided by physicians when filtered by threshold M_T. It was observed that *setting 1* consisting only of a model prediction data had the highest metric ratings of all cases. *Setting 2* adds one more level of information i.e. FINDRISK score, notably the agreement metric for *setting 2* was lower than the latter metrics, this may be attributed to a steep FINDRISK score. *Setting 3* consists of the most information among all cases where the model prediction, FINDRISK score and an explanation are provided to the physician making it more interpretable. Collectively, *setting 3* has a higher amount of data fed to the physician with specificity in details this provides a better score across all metrics over *setting 2* yet not better than *setting 1*. We assume that due to the broad level of details in *setting 1*, the chances of assigning a higher score is more likely, however having greater details as in *setting 3* may have caused minor disagreements with the results which therefore have been reflected in the scores. The behaviour of the metric scores across levels of data can be seen in Fig. 4.

From our simulations, we acquired 96 reward functions for 8 policy combinations over 4 settings of data and 3 metric thresholds. We describe the behaviour and attributes of the reward function as per the metric threshold M_T. When $M_T = 2$, across all 4 settings, we saw the same reward function produced. For policies where two or more Continue actions were performed, the reward was awarded to the concurrent state reached. However, for all other policies and states, the reward acquired was zero. Table 2 describes the reward function acquired for the policies under the threshold.

Table 2. Reward functions of all policies π under a threshold $M_T = 2$ for a all four settings of data

Policy	End	Understandability	Agreement	Usability	Complete
[0, 0, 1, 1, 0]	0	0	0	0	10
[0, 1, 1, 1, 0]	0	0	0	10	10
[0, 1, 1, 0, 0]	0	0	0	10	0
[0, 1, 0, 1, 0]	0	0	0	0	10

With a threshold of $M_T = 3$, across all settings, we observed the reward function to follow the behaviour as in the previous threshold scenario indicated in Table 2. However, for settings 2 and 3 we observed an exception in the reward produced for reaching *Understandability* state in the policies where there are 3 continued actions performed. The observed reward function is shown in Table 3

Having $M_T = 4$, unlike in previous threshold scenarios, we observe the reward function to emphasize penalties to reaching a state over rewards. In all settings, 1, 2, 3 and combined where the policy has the action *continue* occurring in states of Agreement and Usability, a negative reward was observed when reaching the End state and a small negative reward when reaching the state of agreement with usability action being *continue*. Table 4 shows the reward function observed.

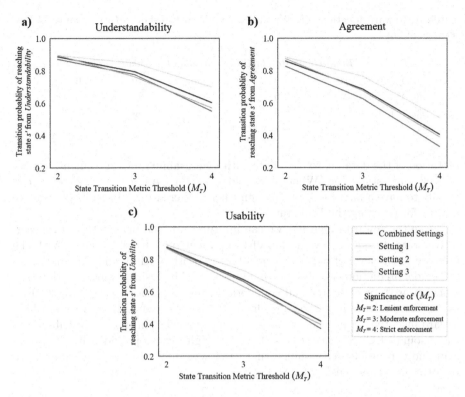

Fig. 4. Observed transition probabilities of trajectories of settings 1, 2, 3 and combination when traversing between the states with metric threshold set to lenient, moderate and strict enforcement of the implicit policy; (a) Transition probability of *Understandability* over thresholds M_T; (b) Transition probability of *Agreement* over thresholds M_T (c) Transition probability of *Usability* over thresholds M_T

Table 3. Reward functions of all policies π under a threshold $M_T = 3$ for a all four settings of data

Policy	End	Understandability	Agreement	Usability	Complete
[0, 0, 1, 1, 0]	0	0	0	0	10
[0, 1, 1, 1, 0]	0	-10	0	10	10
[0, 1, 1, 0, 0]	0	0	0	10	0
[0, 1, 0, 1, 0]	0	0	0	0	10

Overall, we observe that with a low threshold $M_T = 2$, the reward function is lenient and allows for a broader range of policies to have higher rewards this is due to many trajectories easily overcoming the threshold and therefore having a higher transition rate. With the threshold $M_T = 3$, the reward function can be termed as locally balanced by adding a penalty term to the understandability state when having a complete policy case. This may be attributed to the inter-

Table 4. Reward functions of all policies π under a threshold $M_T = 4$ for a all four settings of data

Policy	End	Understandability	Agreement	Usability	Complete
[0, 0, 0, 0, 0]	10	0	0	0	0
[0, 0, 1, 1, 0]	-10	0	-2.1	0	0
[0, 1, 1, 1, 0]	-10	0	0	0	0

pretation of the data i.e. to have data with high usability and agreement, it must be well understandable. When threshold $M_T = 4$, only a few trajectories can qualify to have a high rate of transition, therefore creating a baseline of negative reward for reaching the End state.

From the reward functions, we can interpret a broad perspective of the cognitive state during the evaluation of the patient data by physicians. With the increase in the level of data (i.e. the addition of FINDRISK and explanation to the model prediction) we observe an improved and inclusive reward function that alots rewards for close and full completion while penalties for reaching an End state. The increase in metric threshold M_T is observed to have a diminishing reward incentive with penalties being more prevalent. This may be attributed to the amount of CDSS trajectories that have high ratings provided by the physician and can be further assumed that cases with high scores are more likely to pass through all stages of the MDP.

5   Discussion

In this section we address issues and open topics that we came across during our work.

Perceptional State of AI and Human for Collaboration: Modelling the cognitive state of a decision maker (human) can be considered as achieving partial progress towards active human-artificial intelligence (AI) collaboration. The rest of this process lies in crafting the environment, validation metrics and the AI model itself. One of the ways this can be achieved with great accuracy is by trying to interpret the perceptional state of the human mind and the AI model from either perspective [5]. On a general basis, the human decision maker views the AI model (DSS) as an accessory tool providing suggestions, rather than considering it as a capable collaborator in the decision-making process. This can be attributed to the nature of AI models being tediously sensitive or providing recommendations and solutions that are broadly accurate but not specific enough. There is also the factor of trust in AI as it is relatively new to fully formal usage in the real world [15]. Decision-makers often see their recommendations with skepticism and prefer to re-validate solutions to ensure there are no errors, this implicitly brings about another layer of work and scrutiny which decision-makers have to take on, thereby reducing the trust and dependence on AI. Parallelly, from the AI's perspective, there is a bigger dimension of unanswered questions that are not explicitly demanded yet should be investigated to aid the AI model's evolution to provide more aligning solutions to the human decision-maker [12] e.g. if a physician decision maker selects to discharge a patient 2 days earlier than scheduled yet the AI models suggestion was the latter, the reasoning behind such a decision is left explicitly unanswered to the AI model but implicitly answered to the physician. This may cause a loss of understanding from the AI's perspective considering the action taken. Hence, we believe that by using metrics to evaluate decisions in the form of subjective and descriptive measures, while also training the AI model to perceive the possible cognitive state of a human in a situation, the recommendations and alignment of AI models to that of the human decision-maker will greatly improve, therefore positively impacting trust and reliability in AI models.

Modelling the Data a Greater Issue than IRL: During our experiment, we encountered numerous instances where modelling the real-world CDSS data into an MDP was a tedious and grueling task. Initially, we assumed the possibility of using all patient characteristics, metrics evaluations and settings together in a single MDP, yet the development of such an MDP was unsuccessful as it led to multi-dimensional state spaces with non-comprehensible solutions. As IRL aims to learn the complete policy and behaviour of experts within the trajectories, we can model the MDP features space with nested dimensions or with branched recurrences thereby reducing the computation complexity by a significant level. E.g. within our CDSS data, we tried to model a 125-state space MDP where each state corresponds to an iteratively expanding variable with 5 actions that correspond to the metric evaluation from the physician, yet the layer of patient data here was not included in the state space as it wasn't rational in its solution. In one of our approaches to model a multi-dimensional state space with patient data included, we created a theoretically working state-action feature space MDP, however, when trying to feed the CDSS data into the designed architecture the complexity of data extraction and processing was high. Further, when trying to feed the data to the IRL algorithm, the dimensionality of input tuples was unable to be resolved without splitting and transforming the data into simpler formats (i.e. a four-dimensional state space had to be transformed into a single dimension vector to fit into the tuple format and it may cause it to lose its accurate weightage).

6   Conclusion and Future Work

The results of our work provide a first-stage implementation of IRL to assess the cognitive state of the human mind using real-world data from a clinical decision support system and its evaluation performed by physicians. We uncover the underlying policies and reward functions using linear programming IRL that explains the cognitive state of the physician during the analysis of patient data and decision support data to predict the chances of the patient acquiring T2DM in the future. We demonstrate our construction of MDP, our approach to performing IRL using RL and investigate the reward functions over a set of policies under various levels of information provided to the physician and the effect of having data thresholds that modulate the reward function.

In the future, we are planning to extend the study into a deeper investigation of physicians' professional behaviour and decision-making. The ultimate goal of the study is to improve human-AI interaction in the CDSS context with better interaction and collaboration between decision-makers and AI agents. One of the ways to achieve this goal is to attain a better understanding of human intentions and decisions via IRL, technology acceptance models, etc. An important future direction is the incorporation of model and human correctness, agreement, and final decision performance metrics into the model. Also, we aim to include more patient data and create a denser MDP feature space to improve the specificity of the reward function obtained along with using diverse IRL algorithms such as deep learning and Bayesian approach [13, 16]. We also plan to include newer real-world datasets for various diseases and clinical decisions that can provide more unexplored dimensions of data therefore providing a dynamic representation of the cognitive state during decision-making scenarios.

Acknowledgement. This research is financially supported by The Russian Science Foundation, Agreement #24-11-00272.

References

1. Abbeel, P., Ng, A.Y.: Apprenticeship learning via inverse reinforcement learning. In: Proceedings of the Twenty-first International Conference on Machine Learning, p. 1 (2004)
2. Adams, S., Cody, T., Beling, P.A.: A survey of inverse reinforcement learning. Artif. Intell. Rev. **55**(6), 4307–4346 (2022)
3. Alger, M.: Inverse reinforcement learning (2017). https://doi.org/10.5281/zenodo.555999
4. Damacharla, P., Javaid, A.Y., Gallimore, J.J., Devabhaktuni, V.K.: Common metrics to benchmark human-machine teams (HMT): a review. IEEE Access **6**, 38637–38655 (2018)
5. Howes, A., Jokinen, J.P., Oulasvirta, A.: Towards machines that understand people. AI Mag. **44**(3), 312–327 (2023)
6. Kovalchuk, S.V., Kopanitsa, G.D., Derevitskii, I.V., Matveev, G.A., Savitskaya, D.A.: Three-stage intelligent support of clinical decision making for higher trust, validity, and explainability. J. Biomed. Inform. **127**, 104013 (2022)
7. Lee, K., Rucker, M., Scherer, W.T., Beling, P.A., Gerber, M.S., Kang, H.: Agent-based model construction using inverse reinforcement learning. In: 2017 Winter Simulation Conference (WSC), pp. 1264–1275. IEEE (2017)
8. Liu, Q., Wu, H., Liu, A.: Modeling and interpreting real-world human risk decision making with inverse reinforcement learning. arXiv preprint arXiv:1906.05803 (2019)
9. Muelling, K., Boularias, A., Mohler, B., Schölkopf, B., Peters, J.: Learning strategies in table tennis using inverse reinforcement learning. Biol. Cybern. **108**, 603–619 (2014)
10. Ng, A.Y., Russell, S., et al.: Algorithms for inverse reinforcement learning. In: ICML, vol. 1, p. 2 (2000)

11. Phan-Minh, T., et al.: DriveIRL: drive in real life with inverse reinforcement learning. In: 2023 IEEE International Conference on Robotics and Automation (ICRA), pp. 1544–1550 (2023). https://doi.org/10.1109/ICRA48891.2023.10160449
12. Pinski, M., Benlian, A.: AI literacy-towards measuring human competency in artificial intelligence (2023)
13. Ramachandran, D., Amir, E.: Bayesian inverse reinforcement learning. IJCAI **7**, 2586–2591 (2007)
14. Swamy, G., Wu, D., Choudhury, S., Bagnell, D., Wu, S.: Inverse reinforcement learning without reinforcement learning. In: International Conference on Machine Learning, pp. 33299–33318. PMLR (2023)
15. Ueno, T., Sawa, Y., Kim, Y., Urakami, J., Oura, H., Seaborn, K.: Trust in human-AI interaction: scoping out models, measures, and methods. In: CHI Conference on Human Factors in Computing Systems Extended Abstracts, pp. 1–7 (2022)
16. Wulfmeier, M., Ondruska, P., Posner, I.: Deep inverse reinforcement learning. CoRR, abs/1507.04888 (2015)
17. Ziebart, B.D., Maas, A.L., Bagnell, J.A., Dey, A.K., et al.: Maximum entropy inverse reinforcement learning. In: AAAI, Chicago, IL, USA, vol. 8, pp. 1433–1438 (2008)

Modelling of Practice Sharing in Complex Distributed Healthcare System

Chao Li[1]([🖂]) [ID], Olga Petruchik[2] [ID], Elizaveta Grishanina[1], and Sergey Kovalchuk[1] [ID]

[1] ITMO University, St Petersburg, Russia
{316325,eogrishanina}@niuitmo.ru, kovalchuk@itmo.ru
[2] Taganrog City Clinical Emergency Hospital, Taganrog, Russia

Abstract. This research investigates how collectives of doctors influence their diagnostic method preferences within small-world network social structures through participation in diverse types of medical practice-sharing activities across different scales. We propose an approach based on vectorization of the preferences for various diagnostic methods among physicians, quantifying their openness to these methods using the Shannon diversity index. Utilizing theoretical foundations from threshold models, influence models, and the Hegselmann-Krause model, we designed simulation experiments for teaching activities and seminars to explore the dynamic changes in preference vectors and Shannon diversity indices among these doctors in a small-world network. We evaluated our approach with a real-world data set on vertigo treatment by several clinical specialists of different specialty (neurologists, otolaryngologist). Building on real data from this initial group, we then simulated data for a large number of doctors from various medical communities to examine phenomena in larger-scale systems . Hierarchical networks featuring small-world properties were developed to simulate "local" within-community and "global" across-community seminars, reflecting different intra- and inter-community scenarios. The experiments show different patterns of practice converging during simulation in various scales and scenarios. The findings of this study provide significant insights for further research into practice-based knowledge sharing among healthcare professionals, highlighting the nuanced interplay between social network structures and professional consensus formation.

Keywords: community behavior · complex networks · practice sharing · complex systems · diagnostic method preferences · hierarchical network

1 Introduction

Recognizing that competencies are distributed within a healthcare setting (i.e., collective competence) is vital [1]. The healthcare sector is not only largely distributed and fragmented but it also exhibits a high degree of diversity with strong local autonomy [2, 3]. In the healthcare sector, various medical tasks face diverse, multi-level, large-scale, and complex challenges that are intrinsically linked with the concept of "distributed" systems [4, 5].

© The Author(s), under exclusive license to Springer Nature Switzerland AG 2024
L. Franco et al. (Eds.): ICCS 2024, LNCS 14835, pp. 224–238, 2024.
https://doi.org/10.1007/978-3-031-63772-8_21

Healthcare is usually considered as highly regulated systems with large number of norms such as clinical recommendations, protocols, internal and external hospital rules, etc. Still, due to high complexity of disease, multiple decisions are made by doctors in accordance to their experience or known "best practices" within certain degree of freedom under regulation. Here we consider "practice" as a significant pattern of clinical decision making. Such patterns are optimized and refined through continuous dynamic interaction, thereby facilitating the sharing and transfer of knowledge between different but related tasks [6]. In the healthcare domain, practice specifically includes diagnostic practice [7], emergency response practice [7], disease management practice [8], preventive medical practice [9], and rehabilitation practice [9]. Therefore, it is necessary to investigate the process of practice sharing within distributed healthcare environments with multiple doctors with different communication channels.

In medical practice, vertigo is a common symptom but poses diagnostic and treatment challenges due to its diverse etiologies and complex clinical presentations. When faced with patients exhibiting symptoms of vertigo, different doctors might make varying clinical decisions even in the face of similar cases. This variability partly stems from the lack of unified clinical guidelines and recommendations, and partly from each physician's preferences for clinical examination and diagnostic tests. Significant differences in the adoption rates of specific diagnostic tests by doctors have been observed, even for similar symptoms [10, 11]. Some doctors may prefer to use balance tests like the Romberg's test, while others might rely more on hearing tests, such as Weber's test. Furthermore, for the assessment of vertigo, some physicians might frequently use the Headshake test, while others might more commonly employ gait analysis tests and tests related to respiratory responses, such as the hypercapnic response.

Building upon this foundation, our research motivation is to deeply understand how groups of physicians influence and shape their diagnostic method preferences through the sharing of "practice" within distributed healthcare environments. Our research aims to identify the key factors influencing the formation of physicians' diagnostic method preferences and to understand how these preferences evolve in a distributed medical environment through patterns of practice sharing[12].

This study not only offers a new perspective for understanding the sharing of "practice" in medical decision-making but also provides a theoretical foundation for further exploration of practice-based knowledge sharing among healthcare professionals. Through this research, we aim to offer insights into the mechanisms of knowledge sharing and social interaction in medical decision-making and practice, especially in addressing diagnostic challenges. We seek to demonstrate how improving knowledge sharing and social interactions can enhance the quality and efficiency of medical services.

2   Modelling Practice Sharing in Complex Healthcare System

Here, by "practice" in healthcare we consider patterns in clinical decision making which appears multiple times in treatment of similar patients. The most important role is played by practices in situation where official regulation (by clinical recommendations, protocols, etc.) give certain degree of freedom to a doctor, or in situations where there is no strict protocols (e.g. in appearance of new disease, or in complex diseases). In turn, "practice sharing" refers to the dissemination of medical knowledge and information related to practices within a distributed healthcare environment where good practices may be shared or recommended by experienced specialists (e.g. explicitly during dedicated meetings or implicitly during common information sharing).

Several studies have touched upon the concept of practice sharing. However, these investigations often did not formally define or scientifically model the concept, leaving room for more rigorous research and analysis in this area. Prasidh Chhabria et al. [6] and Kyunghoon Hur et al. [13] actually explored practice sharing between different healthcare tasks. Wei Gong et al. discussed the practice sharing of various smart intensive care units [14]. Corinna Maier et al. discussed the "practice sharing" of continuous precise drug dosage use across hospitals or research centers [15].

We introduce a constructive modeling approach and process for the nascent field of distributed medical practice sharing. Our modeling approach studies physicians, hospitals, and different types of medical sharing events. The model structure is shown in Fig. 1.

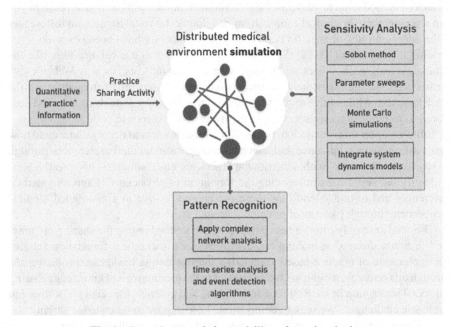

Fig. 1. General approach for modelling of practice sharing

2.1 Quantitative Medical Practice

The content of medical data sets is diverse [2, 16], medical practice information exists in datasets in various implicit forms. We suggest mining this information from the perspective of medical practice preferences. Specifically, we use medical Practice Preference Vectors and Physician Shannon Diversity Index to measure medical practices.

Given a medical dataset D and a binary vector representation for each physician i based on X unique diagnostic items, the real Practice Preference Vector P_i for physician i is defined as a vector of proportions $[p_{i1}, p_{i2}, \ldots, p_{ix}]$, where each element p_{ij} represents the proportion of times diagnostic item j was employed by physician i relative to their total usage of all X diagnostic items. This formulation refines the initial binary encoding to quantify the diagnostic preferences of physicians, capturing the relative frequency of each diagnostic item's use. Formally, P_i is obtained by:

$$P_i = \left[\frac{n_{i1}}{N_i}, \frac{n_{i2}}{N_i}, \ldots, \frac{n_{iX}}{N_i} \right] (1)$$

where n_{ij} is the number of times physician i used diagnostic item j, N_i is the total number of diagnostic items used by physician i, X is the total number of unique diagnostic items in dataset D.

Physicians' practice preferences are typically formed based on long-term experience and are not prone to significant changes in the short term, making longitudinal analyses more relevant and meaningful. The preference for specific practices is distinct from the dynamics and adaptability of practices themselves. Short-term analyses or assessments at a specific point in time fail to capture meaningful patterns or trends in physicians' practice preferences. Therefore, we employ the Shannon diversity index for a macro-level quantification of physicians' overall preferences for different diagnostic methods over extended periods. This approach not only reflects the overall distribution of physicians' preferences but also effectively reveals the diversity and breadth of diagnostic practices on a larger scale. Such a macro-analytical method provides a valuable perspective for understanding physicians' diagnostic preferences, aligning well with the practical requirements of medical practice data analysis. The physician's Shannon diversity index H can be calculated using the following equation:

$$H = -\sum_{i=1}^{R} p_i \log(p_i) (2)$$

where, R is the total number of diagnostic methods, p_i is the probability of the i_{th} diagnostic method (i.e., the frequency of use of this method relative to the total number of uses).

2.2 Physician Practice Sharing Activity

In the actual medical environment, the sharing of practices in their implicit forms among physicians is realized through a variety of sharing activities with different types and natures. These activities are divided into two main categories: educational activities with a presenter and seminars among physicians. The preferences in practices among physicians may shift after participating in these different activities. Within our modeling approach, it is essential to simulate both categories of activities to explore how these practice-sharing endeavors in their implicit forms affect the quantified information on medical practices we have established. We provide original simulation algorithms based on different theoretical models.

Teaching Activities. We employ a hybrid application of Influence Models [17] and Threshold Models [18] as the theoretical basis for the implementation of "teaching activities." Teaching activities are hosted in rotation by doctors, with the presenting doctor exerting an influence (ϕ) on their "direct neighbors" within the network structure. Due to the potential difference in magnitudes between a doctor's Shannon diversity index and the total usage of diagnoses by the doctor, we need to balance their impacts. We employ min-max normalization: converting the values of diagnostic usage to a range between 0 and 1. After normalization, we can calculate the weighted average. The influence is calculated using the Weighted Average Method, with the formula as follows:

$$\phi = w_s \times S_{speaker} + w_u \times U_{norm} \quad (3)$$

where, ϕ represents the influence of the activity. $S_{speaker}$ is the Shannon diversity index of the speaker. U_{norm} is the normalized total usage of diagnoses. w_s and w_u are the weights assigned to the speaker's Shannon diversity index and the total usage of diagnoses, respectively.

After determining the influence (ϕ), we update the Shannon diversity index for direct neighbors of the speaker doctor, simulating knowledge exchange and adaptation dynamics. The update utilizes a linear adjustment, based on the relative Shannon diversity indices ($S_{speaker}$ and $S_{neighbor}$), encapsulating learning responses. The unified formula for both positive and negative adjustments is:

$$S'_{neighbor} = S_{neighbor} + \alpha \times \left(\text{sign}\left(S_{speaker} - S_{neighbor} \right) \right) \times \left| \phi - S_{neighbor} \right| \quad (4)$$

For preference vectors, the update is governed by a linear adjustment towards the presenting doctor's preferences, encapsulated as:

$$P'_n = P_n + \alpha \times (P_s - P_n) \quad (5)$$

where P'_n and P_n represent the updated and current preference vectors of a neighbor doctor, respectively, P_s is the preference vector of the presenting doctor, and α is the learning rate. This formula ensures that each neighbor's preferences incrementally align with those of the presenter, reflecting the adaptive learning process within professional networks.

Seminars for Practice Sharing. Given the nature of seminars as spaces for professional dialogue and learning, the HK model [19] is particularly suitable for simulating these events. Based on the HK model, the Shannon index update algorithm is as follows.

$$S_{i\prime} = \frac{1}{|N_i| + 1} \times \left(S_i + \sum_{j \in N_i, \lceil S_j - S_i \rceil \leq \tau} S_j \right) \quad (6)$$

where, $S_{i\prime}$ represents the updated Shannon diversity index of doctor i. S_i is the current Shannon diversity index of doctor i. N_i denotes the set of neighbors of doctor i whose Shannon indices differ from S_i by no more than a threshold τ (opinion acceptance threshold). $|N_i|$ is the count of such neighbors. τ is defined as the boundary of the confidence interval, wherein the influence between nodes is considered for adjustment only if the difference in their Shannon diversity indices falls within this range. This implies that an individual considers the opinions (or, by analogy, diagnostic practices) of their peers to be sufficiently credible or relevant only when the disparity in their Shannon diversity indices does not exceed τ.

Extending the Hegselmann-Krause model to preference vector updates in seminars involves calculating the updated preference vectors P'_i of doctors by averaging the preferences of neighbors within a specific Euclidean distance (δ), then adjusting towards this average with a strength (α). The core update formula simplifies to:

$$P'_i = \alpha \times \left(\frac{\sum_{j \in N_i, d(P_i, P_j) \leq \delta} P_j}{|N_i|} \right) + (1 - \alpha) \times P_i \qquad (7)$$

P_i and P'_i are the current and updated preference vectors of doctor i, $d(P_i, P_j)$ is the Euclidean distance between the preferences of doctors i and j, δ is the distance threshold for considering neighbors' influences, N_i is the set of neighbors within δ of i, α controls the update intensity, blending the average neighbor preference with $i's$ current preference.

2.3 Distributed Medical Network Structure.

The doctors in the hospital are a closely connected professional community for communication. Reflecting this real-world characteristic, we adopt the "small-world network" model to simulate the complex communication and influence propagation paths among doctors [20]. We construct the small-world network here using three key parameters: the number of doctors, the rewiring probability, and the mean degree (number of connections per node).

We consider each doctor as a node within the network, with nodes featuring three attributes: the doctor's name, practice preference vector, and Shannon diversity index. Edges represent the practice-sharing relationships among doctors. Not all doctors have actual sharing relations in reality. Thus, we simulate this aspect by adjusting the mean degree and the rewiring probability. For instance, a higher mean degree reflects close communication and cooperation relationships among doctors within a community. The rewiring probability models the opportunities for communication within the community, even among doctors who are geographically distant or have slightly different professional orientations.

To visually represent the Shannon diversity index of doctors, we utilize the color of the nodes in the network diagram, where the similarity in colors indicates the closeness of the Shannon diversity indexes.

2.4 Simulation

The purpose of the simulation experiment is to investigate the specific impacts of two types of sharing activities on the practices of physicians within hospitals, with a primary focus on two aspects.

1.Simulation Time. Depending on the experimental scenarios, datasets, and re-search objectives, it is necessary to set varying simulation durations to capture the dynamics of the system. This may include simulations based on multiple specific time points, long-term simulations, and repetitive cyclical simulations.

2. Simulation Scale. The size and complexity of the simulation must be tailored to accommodate the scope of the experimental framework and the granularity of the analysis desired. This involves determining the number of agents or entities, the extent of the networked environment, and the volume of data to be processed. Choices range from small-scale simulations focusing on detailed interactions within a confined setting, to large-scale simulations that aim to replicate broader system-wide dynamics across multiple interconnected scenarios.

2.5 Evaluation Analysis Methods

The final step in the model is the evaluation and analysis of data derived from the simulation experiments. The main methods can be divided into 2 categories.

1. Sensitivity Analysis: Implement global sensitivity analysis, such as the Sobol method, to quantitatively assess the impact of varying input parameters on simulation outcomes; Utilize parameter sweeps and Monte Carlo simulations to evaluate the robustness of the model against parameter variations and identify key parameters; Integrate system dynamics models to assess system behavior under parameter changes and use this information to optimize the model.

2. Pattern Recognition: Apply complex network analysis to identify collective behaviors and diffusion patterns in medical practice, such as using community detection algorithms to discover group structures within practice sharing;Employ time series analysis and event detection algorithms to track and recognize the temporal dynamics and trends of practice sharing.

3 Practice Sharing in Vertigo Treatment

In this section, we conduct a study on diagnostic practice sharing using a dataset from a 2016–2020 vertigo clinic in Rostov-on-Don. This exemplifies our practice sharing research model, illustrating its application in studying how physician groups shape diagnostic preferences through small-world networked activities.

3.1 Data Set and Processes

Vertigo, particularly Benign Paroxysmal Positional Vertigo (BPPV), is a complex condition characterized by a multitude of etiologies and the involvement of various medical specialists, including neurologists and otolaryngologists, among others. The diagnosis and treatment of vertigo and BPPV involve a range of methods, from specific diagnostic tests to repositioning maneuvers, such as the Dix-Hallpike test and Epley maneuver, underscoring the multifaceted approach required to manage this condition effectively [21].

   The original data is composed of 10 structured.xlsx files with 40 fixed headers.These headers have five main classes: patient basic information, diagnostic and treatment information, medical history and status records, treatment and recommendations, and patient background information. Our dataset's mixed-format data was processed to abstract "practice" information via tokenization, keyword extraction, and subsequent lemmatization, followed by categorization to compute proportions and formulate both practice information data and doctor preference vectors.

During pre-processing phase, we structure the practice information, which contains detailed medical diagnostic information aggregated by unique appointment card numbers. Each unique appointment card number represents an individual patient and is associated with 125 different diagnostic items. For each diagnostic item, the file meticulously records the specific di-agnostic outcomes, provided in text format.

We use the algorithm in Sect 2.1.to obtain the preference vector of each doctor under the current data set. The table below, Table 1, describes the structure of the Doctor Preference Vectors.

Table 1. Structure of the Doctor Preference Vectors

Doctor Neurologist	Total usage	Romberg's test	Hallpike test	123 items remaining
Doctor A	538	0.3086	0.2472	***
Doctor B	18024	0.1353	0.1263	***
Doctor C	585	0.4154	0.3487	***
Doctor D	1078	0.3878	0.2653	***
Omit 6 doctors' names	***	***	***	***

3.2 Model Identification, Validation, and Sensitivity Analysis Based on Actual Data

First Experiment and Parameter Sensitivity Analysis Based on Actual Data.
The first experiment was conducted based on additional information provided by the dataset creators. From this first experiment, we obtained parameters that fit the dataset information.

From 2016 to 2020, there were nine "teaching activity" events and eight "seminar" events conducted. During this period, from 2017 to 2020, each year featured two "teaching activities" and two "seminars," with 2016 hosting only one "teaching activity."

Initially, we assigned default values to the parameters in Formulas 2 to 6 for simulating 17 events within our model on actual timelines. We algorithmically extracted the real preference vectors of ten doctors before December 31st each year and evaluated their fit with the simulation by calculating the Euclidean distance to the actual vectors. Through parameter sweeps and Monte Carlo simulations, we fine-tuned the simulation to closely match the real data, setting rewiring_prob = 0.3, mean_degree = 4, α =0.05, w_s=0.5, w_u=0.5, τ =0.2, and δ =0.5 after iterative adjustments.

With the given parameters, we measured yearly Euclidean distances between simulated and actual preference vectors for 10 doctors, summarizing with average, median, and range (max and min) to evaluate our model's accuracy, as Table 2 illustrates.

Table 2. Preference vector comparison results.

Year	Average Distance	Median Distance	Max Distance	Min Distance
2016	0.64	0.5	2	0
2017	1.21	1	2	0
2018	1.38	1.41	2	1
2019	0.88	1	2	0
2020	1.11	1	2.24	0

We compared two preference vectors, each composed of 125 binary values corresponding to 125 diagnostic items. The average value calculated in column 2 of Table 2 is 0.88, indicating that the annual difference between simulated and actual doctor preference vectors is less than one out of 125 items. In other words, the preference changes in the 125 simulated diagnostic items closely match those observed in the actual dataset.

3.3 Simulation Scenario Expansion: Long-Term, Large-Scale, and Variant Studies

In the experimental results based on the above parameters and actual time nodes, although the final preference vectors of the 10 doctors are consistent with the data in the data set, there was no consensus among the doctors at the end of 2020.

So based on parameters obtained from the first experiment, we extended the simulation period from January 1, 2016, to December 31, 2026. "Teaching activities" were scheduled monthly, while "seminars" occurred weekly.

By increasing the frequency of "practice sharing" events, we aimed to observe the system's dynamics over an extended simulation. The experiment revealed that after 117 months, the Shannon diversity indices of all ten doctors converged to a single value, resulting in identical preference vectors for each doctor (see Fig. 2). In Fig. 2, A to J represent the 10 doctors in the original data set.

Subsequently, leveraging information extracted from a real dataset comprising ten doctors, we employed methods such as non-uniform probability selection, random non-repetitive sampling, normalized random allocation, and an overarching simulation framework to generate preference vector files for 1000 doctors from ten different medical sharing communities.

We then refined the seminar model based on the Hegselmann-Krause (HK) model to simulate "local seminars" within communities and "global seminars" across communities. Different reconnection probabilities and average degrees were used to mimic the network structures within and between communities, constructing a hierarchical small-world network across ten communities. Higher average degrees and reconnection probabilities were used in intra-community networks, reflecting the higher frequency of interaction and closer collaboration among doctors within the same community (understood as the same hospital, city, or group of doctors with similar professional backgrounds). For inter-community network configurations, lower average degrees were adopted to represent the greater challenges and fewer direct contacts in interactions between doctors from different communities (or different specialties and geographical locations).

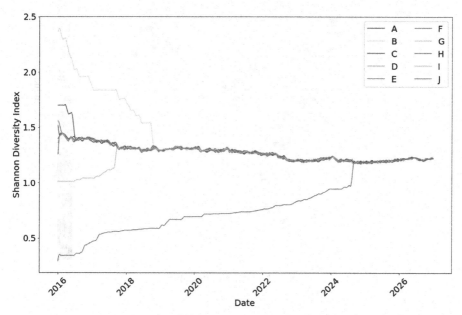

Fig. 2. Doctor Shannon index change chart

In our extended simulation running from 2016 to 2026, designed to reflect large-scale conditions, we convened local seminars quarterly within communities and global seminars annually across communities.Over the course of 120 months, it was found that consensus on Shannon indices was reached for only 64.3% of the doctors, from an initial pool of 1000, signifying a divergence in practice sharing.

These doctors coalesced into 18 distinct consensus groups based on their Shannon indices, with the largest group encompassing 643 physicians. The remaining 17 groups, although smaller in size, each achieved consensus internally. Doctors whose Shannon index differences were less than 0.15 by the end of the simulation were categorized into the same group and visually represented through color coding within the hierarchical network structure, as depicted in Fig. 3. This visualization illustrates not only the majority consensus but also the presence of multiple subgroups persisting with distinct practice-sharing patterns.

In our large-scale system analysis, we captured the number of doctors in each group and the varying times taken for different groups to converge. Groups with fewer than ten doctors generally converged within one month; hence, we have presented only those groups with more than ten doctors in Fig. 4.

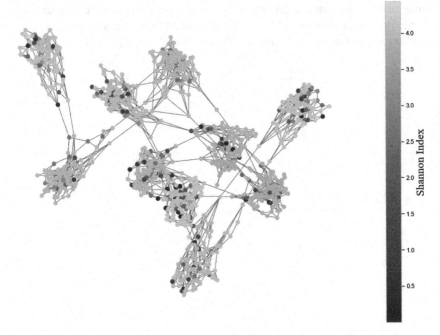

Fig. 3. Large-scale simulation

The Fig. 4 indicates that smaller groups (e.g., 10, 13, or 14 members) exhibit a faster median convergence time, suggesting that streamlined communication within tighter cohorts facilitates swifter consensus. Conversely, as group sizes expand (notably to 81, 98, or 643 members), both the median and interquartile range of convergence times increase, reflecting the broader range of opinions and the complexity involved in harmonizing these views within larger collectives. In the largest group, particularly, the significant spread and outliers in convergence times underscore the distinct challenges some doctors may encounter in aligning their practices with group consensus.Through experiments, we found that the convergence speed in large-scale systems has no obvious dependence on the number of communities and the number of doctors in the community.

In exploring the impact of seminar frequency on the convergence time of practice sharing among physicians(in the 18 final automatically formed groups), we applied a grid search methodology, considering only scenarios where the global seminar frequency does not exceed that of the local seminars to ensure the practical viability of the experimental setup.

The boxplots reveal a slight decrease in the median convergence time with an increasing frequency of local seminars, particularly when the global seminar frequency is set to every three months. Moreover, the shortest convergence times are observed at a global frequency of three months, indicating that frequent global interactions facilitate quicker consensus building. The reduction in interquartile range and decreased variability reflect an increased data concentration, suggesting that a tight seminar schedule positively influences convergence. Overall, the charts underscore the significance of increasing seminar frequency in shortening the convergence time between physicians. This trend suggests that frequent interaction through seminars may play a critical role in harmonizing practices among medical professionals. Furthermore, the data implies that strategic planning of educational activities could be pivotal in fostering a unified approach to healthcare within the community (see Fig. 5.).

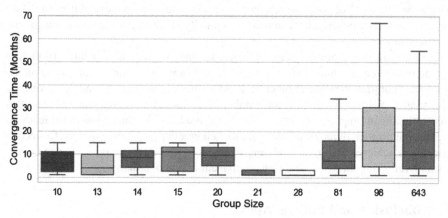

Fig. 4. Group convergence dynamics by Size

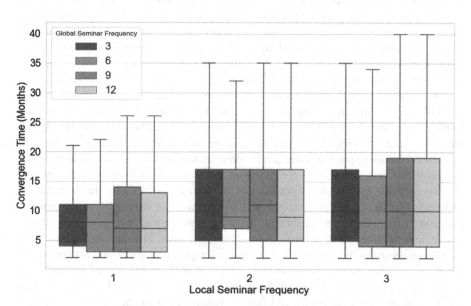

Fig. 5. Activity frequency versus convergence time

4 Discussion

In the preliminary small-scale experiment involving ten doctors, we simulated "Practice Sharing" by incorporating real activity timing information from the dataset. The simulated preference vectors closely aligned with the actual preference vectors, demonstrating the fundamental scientific validity of our modeling approach for "Practice Sharing" within the "Vertigo" dataset. Subsequently, we simulated the scenario of these ten doctors over a decade, where all participants eventually reached a consensus in diagnostic practice choices.

In the third phase of our study, we expanded the experiment to include 1000 doctors across ten communities. We examined the dependency of convergence time on the number of communities, the number of doctors within communities, and the frequency of activities. We found that both the number of doctors and the frequency of activities have a significant impact on convergence. In the model, individual physicians' learning is conceptualized as dynamic adjustments to their practice preference vectors. Collective learning is achieved through simulating interactions and information exchange among doctors, reflecting the social learning component in practice sharing.

5 Conclusion and Future Work

In this study, we have investigated the effectiveness of our model and approach in simulating the "Practice Sharing" process among doctors. Our model successfully replicated the evolution of doctors' preference vectors across networks of varying sizes, revealing the potential for achieving consensus within small groups and the complexities encountered in broader communities. This underscores the value of our approach in understanding and facilitating knowledge sharing and consensus formation in medical practice.

In our future work, building on the extension of practice preference studies, we are particularly interested in delving into the micro-level fusion and switching of practices. In this regard, introducing a research paradigm based on Distributed Constraint Markov Decision Processes (DEC-MDP) through reinforcement learning is intriguing. Regarding practice sharing, our focus extends to incorporating greater real-world complexity. We posit that interdisciplinary collaboration and information exchange can enhance the diversity and innovation of practices, while also acknowledging the potential for differentiation in practice preferences. Medical policies can shape physicians' practice preferences through the establishment of standardized procedures, promotion of specific treatment methods, or restrictions on certain practices. The recognition by authoritative bodies plays a significant role in the widespread adoption and acceptance of practices, with historical data suggesting significant shifts in physicians' practice preferences before and after the release of clinical guidelines. Additionally, factors such as medical culture, local ambitions, and funding reflect varying dependencies on the real world.

Acknowledgments. This research is financially supported by The Russian Science Foundation, Agreement #24–11-00272.

References

1. Schoenherr, J.R.: Trust and explainability in A/IS-mediated healthcare: Operationalizing the therapeutic alliance in a distributed system. In: 2021 IEEE International Symposium on Technology and Society (ISTAS). pp. 1–8. IEEE, Waterloo, ON, Canada (2021). https://doi.org/10.1109/ISTAS52410.2021.9629118
2. Rajpurkar, P., Chen, E., Banerjee, O., Topol, E.J.: AI in health and medicine. Nat. Med. **28**, 31–38 (2022). https://doi.org/10.1038/s41591-021-01614-0
3. Guleva, V., Shikov, E., Bochenina, K., Kovalchuk, S., Alodjants, A., Boukhanovsky, A.: Emerging Complexity in Distributed Intelligent Systems. Entropy **22**, 1437 (2020). https://doi.org/10.3390/e22121437
4. Grimson, J., Stephens, G., Jung, B., Grimson, W., Berry, D., Pardon, S.: Sharing health-care records over the Internet. IEEE Internet Comput. **5**, 49–58 (2001). https://doi.org/10.1109/4236.935177
5. Funkner, A.A., Egorov, M.P., Fokin, S.A., Orlov, G.M., Kovalchuk, S.V.: Citywide quality of health information system through text mining of electronic health records. Appl Netw Sci. **6**, 53 (2021). https://doi.org/10.1007/s41109-021-00395-2
6. Chhabria, P.H.: A Bayesian Nonparametric Approach to Multi-Task Learning for Contextual Bandits in Mobile Health. (2022)
7. Félix, N.M., Gommeren, K., Boysen, S.: Translational medicine between human and veterinary emergency and critical care medicine: a story meant to have a happy ending. Crit. Care **23**, 361 (2019). https://doi.org/10.1186/s13054-019-2659-3
8. Text="CE:[CROSSREF]Score: 0.00crsid: H55BcHtype: bibarticleAuthors: Francesca Pistone, Davide Testa, Irene Martelli, Gabriella Sciascia, Grazia Maria Luisa Rizzelli, Alessandra Tripoli, Salvatore De Marco, Paola Migliorini, Antonio Gaetano TavoniYear: 2023JournalTitle: Clinical and Experimental RheumatologyArticleTitle: Mucocutaneous and gastrointestinal involvement of MIS-A in a 67-year-old man: a case reportBibArticleDOI: 10.55563/clinexprheumatol/5ox5gj" Pistone, F., Testa, D., Martelli, I., Sciascia, G., Rizzelli, G.M.L., Tripoli, A., De Marco, S., Migliorini, P., Tavoni, A.G.: Mucocutaneous and gastrointestinal involvement of MIS-A in a 67-year-old man: a case report. Clinical and Experimental Rheumatology. (2023). https://doi.org/10.55563/clinexprheumatol/5ox5gj
9. Ren, C., Jiang, B., Lu, N.: Task Adaptation Meta Learning for Few-Shot Fault Diagnosis under Multiple Working Conditions. In: 2023 6th International Symposium on Autonomous Systems (ISAS). pp. 1–5. IEEE, Nanjing, China (2023). https://doi.org/10.1109/ISAS59543.2023.10164461
10. Strupp, M., Dlugaiczyk, J., Ertl-Wagner, B.B., Rujescu, D., Westhofen, M., Dieterich, M.: Vestibular Disorders. Deutsches Ärzteblatt international. (2020). https://doi.org/10.3238/arztebl.2020.0300
11. Figtree, W.V.C., Menant, J.C., Chau, A.T., Hübner, P.P., Lord, S.R., Migliaccio, A.A.: Prevalence of Vestibular Disorders in Independent People Over 50 That Experience Dizziness. Front. Neurol. **12**, 658053 (2021). https://doi.org/10.3389/fneur.2021.658053
12. Kovalchuk, S.V., et al.: A Conceptual Approach to Complex Model Management with Generalized Modelling Patterns and Evolutionary Identification. Complexity **2018**, 1–15 (2018). https://doi.org/10.1155/2018/5870987
13. Hur, K., et al.: GenHPF: General Healthcare Predictive Framework for Multi-Task Multi-Source Learning. IEEE J. Biomed. Health Inform. **28**, 502–513 (2024). https://doi.org/10.1109/JBHI.2023.3327951
14. Gong, W., Cao, L., Zhu, Y., Zuo, F., He, X., Zhou, H.: Federated Inverse Reinforcement Learning for Smart ICUs With Differential Privacy. IEEE Internet Things J. **10**, 19117–19124 (2023). https://doi.org/10.1109/JIOT.2023.3281347

15. Maier, C., De Wiljes, J., Hartung, N., Kloft, C., Huisinga, W.: A continued learning approach for model-informed precision dosing: Updating models in clinical practice. CPT Pharmacom & Syst Pharma. **11**, 185–198 (2022). https://doi.org/10.1002/psp4.12745

16. Distributed Analytics on Sensitive Medical Data: Beyan, O., Choudhury, A., Van Soest, J., Kohlbacher, O., Zimmermann, L., Stenzhorn, H., Karim, Md.R., Dumontier, M., Decker, S., Da Silva Santos, L.O.B., Dekker, A. The Personal Health Train. Data Intellegence. **2**, 96–107 (2020). https://doi.org/10.1162/dint_a_00032

17. Gold, D., Katz, E., Lazarsfeld, P.F., Roper, E.: Personal Influence: The Part Played by People in the Flow of Mass Communications. Am. Sociol. Rev. **21**, 792 (1956). https://doi.org/10.2307/2088435

18. Granovetter, M.: Threshold Models of Collective Behavior. Am. J. Sociol. **83**, 1420–1443 (1978). https://doi.org/10.1086/226707

19. Perrier, R., Schawe, H., Hernández, L.: Phase coexistence in the fully heterogeneous Hegselmann-Krause opinion dynamics model. Sci. Rep. **14**, 241 (2024). https://doi.org/10.1038/s41598-023-50463-z

20. Watts, D.J., Strogatz, S.H.: Collective dynamics of 'small-world' networks. Nature **393**, 440–442 (1998). https://doi.org/10.1038/30918

21. Petruchik, O.V., Funkner, A.A., Efremov, R.V., Kovalchuk, S.V., Kubryak, O.V.: Identification of Elements of the Formal Language in Medical Descrip-tions of Vertigo in the Context of Diagnosis. (2021). https://doi.org/10.13140/RG.2.2.30845.69605/1. (in Russian)

Simulation and Detection of Healthcare Fraud in German Inpatient Claims Data

Bernhard Schrupp[1,2]($\boxtimes$), Kai Klede[1], René Raab[1], and Björn Eskofier[1]

[1] Machine Learning and Data Analytics Lab, Department Artificial Intelligence in Biomedical Engineering (AIBE), Friedrich-Alexander-Universität Erlangen-Nürnberg (FAU), Erlangen, Germany
{kai.klede,rene.raab,bjoern.eskofier}@fau.de

[2] AOK Bayern - Die Gesundheitskasse, Munich, Germany
bernhard.schrupp@by.aok.de

Abstract. The German Federal Criminal Police Office (BKA) reported damages of 72.6 million euros due to billing fraud in the German healthcare system in 2022, an increase of 25% from the previous year. However, existing literature on automated healthcare fraud detection focuses on US, Taiwanese, or private data, and detection approaches based on individual claims are virtually nonexistent. In this work, we develop machine learning methods that detect fraud in German hospital billing data.

The lack of publicly available and labeled datasets limits the development of such methods. Therefore, we simulated inpatient treatments based on publicly available statistics on main and secondary diagnoses, operations and demographic information. We injected different types of fraud that were identified from the literature. This is the first complete simulator for inpatient care data, enabling further research in inpatient care.

We trained and compared several Machine Learning models on the simulated dataset. Gradient Boosting and Random Forest achieved the best results with a weighted F1 measure of approximately 80%. An in-depth analysis of the presented methods shows they excel at detecting compensation-related fraud, such as DRG upcoding. An impact analysis on private inpatient claims data of a big German health insurance company revealed that up to 12% of all treatments were identified as potentially fraudulent.

Keywords: Healthcare · Inpatient Claims · Healthcare Fraud · Fraud Detection · Data Generation · Inpatient Claims Simulation

1 Introduction

Public health insurance companies in Germany spend roughly one-third of their budget each year on hospital treatments [1]. However, only about 10% of detected fraudulent billing claims are accounted for hospital treatments, with most cases uncovered due to tips from insiders or patients [2]. This suggests a significant number of potentially fraudulent cases within the inpatient care system.

The international research on fraud detection in healthcare focuses on identifying fraudulently acting participants, while in Germany, proving fraud for each claim separately is necessary. Limited accessibility to German inpatient claims data due to legal constraints poses a challenge for research. Addressing these issues, this study aims to identify potentially fraudulent claims in German inpatient billing data and generate simulated inpatient claims data to facilitate Machine Learning.

The project code is available at https://github.com/mad-lab-fau/inpatient-claims-simulator.

2   Related Work

Fraud in inpatient claims can be performed in various ways. Jürges and Köberlein [3] noted a significant decrease in reported newborn weights after the introduction of Germany's inpatient treatment billing system. Hospitals may manipulate weight following thresholds defined in the Diagnosis-Related Group (DRG) catalog to receive higher compensation [4]. Similar manipulation occurs with ventilation hours, where higher numbers lead to increased financial compensation [5].

Changes in the order of diagnoses are a more elaborate way to manipulate the billing of services. Shifting a profitable secondary condition (secondary = not the cause of the current hospitalization) to a primary diagnosis (primary = the cause of the current hospitalization) can be a rewarding fraudulent practice [6], often requiring access to medical documents not readily available to paying organizations.

Discharging a patient a few days earlier than medically necessary is another form of fraud, as these additional days would not increase the financial compensation, apart from the additional reimbursement for care [7].

Performing healthcare without medical indication is also considered fraudulent. A vivid example of this pattern of unnecessarily incurred hospital costs is the increase in cesarean deliveries compared with vaginal deliveries. Performing a cesarean section can be highly advantageous for hospitals as it almost doubles revenue while requiring fewer resources than complicated births [7]. A second example, also performed on newborns, is identifying the necessity of parental care throughout a hospital stay, meaning the child cannot stay there without one parent being nearby [5].

These fraud patterns will be injected in the simulated dataset and used to train fraud detection models (see Table 1).

3   Data Generation

As public inpatient claims billing data on an individual claims level is unavailable, a data simulation approach based on publicly available information is presented to facilitate Machine Learning development. The model operates under

Table 1. After simulating regular inpatient claims, we inject the following fraud patterns derived from the literature for 3.07% of all records.

Fraud Pattern	Source	Share in Dataset
Changing order of ICD codes	[6]	1.73%
Reducing duration of stay	[6]	0.85%
Adding need for personal care when treating newborns	[5]	0.34%
Changing a vaginal birth to a cesarean section	[7]	0.05%
Increasing number of ventilation hours to reach next threshold	[5]	0.06%
Decreasing reported weight of newborns to reach next threshold	[3]	0.05%

assumptions such as treatments solely occurring in the main department, uninterrupted treatments, and patients not being transferred between hospitals or departments within the same hospital. Admissions are based solely on referral, and discharges are always medically justified.

3.1 Inpatient Claims Modeling

The simulation process was initialized with patients and hospitals, defined by unique IDs and locations (zip code). The zip code was sampled by following the German population density and hospital addresses [11]. Patient attributes include age and gender, randomly sampled to reflect observed distributions in primary ICD statistics. Hospitals and patients are matched based on distance, and each treatment is defined by primary and secondary ICD codes (diagnosis codes; version ICD-10-GM-2021) and OPS codes.

The treatment defining parameters primary ICD, a list of up to 20 secondary ICD codes [8,9], and up to 20 OPS codes (treatments carried out, including imaging, surgeries and medication) [10] were generated by sampling according to publicly available statistics and demographics. When simulating treatment codes, constraints in form of relationships of OPS codes with diagnoses were formed, excluding combinations causing errors in a DRG Grouper [12].

Next, length of stay was determined using thresholds from relevant DRGs and a Gaussian distribution. The duration was added to a randomly selected admission date (between 01.01.2021–31.12.2021) to calculate the discharge date.

To complete the simulation, for cases with OPS codes referencing to ventilation we sampled the value according to a power-law distribution. If the patient is a newborn (age < 1 year), it is necessary to determine the weight, which was sampled according to WHO information [13,14] using a Gaussian distribution.

Fraudulent behavior was injected by randomly selecting 20% of all claims for adjustments. Where possible, the values for ventilation hours and weight at birth were altered accordingly. If a birth is coded in the billing information, it was changed to a cesarean section. In case a newborn is hospitalized, the need for assistance with personal care was added to the list of secondary diagnoses. To achieve the pattern of exchanged primary ICDs, the combination of diagnoses with the highest relative weight was chosen (ignoring combinations of primary

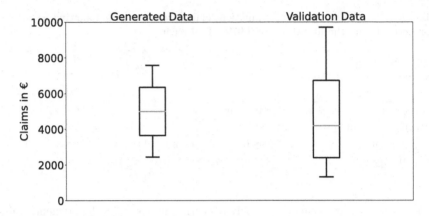

Fig. 1. Comparing the generated inpatient claims with the validation data, the median (highlighted in both plots) is 5,005 Euros in the generated data and 4,199 Euros in the validation data. Although the box plot clearly shows the lack of outliers in both directions (whiskers indicate the 10th and 90th percentiles), this dataset is the first and only simulation of inpatient treatments, and future research could likely improve the model.

ICDs starting with 'Z'). If the duration exceeds the lower threshold defined [4], it is reduced by 1/4, but no less than 1 and no more than five days. After the fraud creation and injection process, the final claim was calculated.

3.2   Evaluation of the Simulation Results

Simulation results are evaluated against publicly available information and proprietary data from AOK Bayern. That dataset comprises all 2021 inpatient treatments, including diagnoses, treatments, admission and discharge data, claim amounts, DRG codes, the hospital's ID, patient information, location, and over 1 million individual claims.

To evaluate the claims data generated, some preliminary assumptions were made to achieve comparability between the validation data and the generated data. Adding additional surcharge for extra charges, inflation and additional charges for particularly complex diagnoses (such as severe burns or hemophiliacs), the average claim rises to 4,620.65 Euro, still neglecting the inflation from 2018 to 2019. In 2021, the average claim for inpatient treatments observed in the validation data was 5,537.26 Euros, with a median absolute deviation (MAD) of 1,328.05 Euro. The MAD in the generated dataset is 1,187.86 Euro. While this indicates that simulation and validation data resemble each other, Fig. 1 implies, that large outliers are skewing standard deviation and interquartile range.

While the average length of stay in the simulated data is 0.39 days or 5.4% shorter than in the statistical data (6.81 days in comparison to 7.2 days) [15], the ratio of short stays (within three days) is two percentage points smaller than the statistically observed numbers in 2021 (38% compared to 40%). Comparing the generated data with information from the validation data, these effects increase further. The median, which is less prone to outliers, is 5.0 days per inpatient stay in the generated data, while it is 6.86 days in the validation set, with a MAD of 3.0 days in both cases. These effects indicate either a slight underestimation of stays within three days or the missing representation of a few very long hospital stays. The longest hospital stay in the validation set is 730 days, four times longer than the longest generated case.

Finally, 15 cases were randomly selected for an expert review by a public health insurance's hospital controller to validate the hospital claims in their respective context. Consistently, the problem of too many codes for imaging (OPS Chap. 3), and too few actual surgeries (OPS Chap. 5) was observed. This pattern was also visible in statistical comparisons (Chap. 3 occurred 16% too often, while Chap. 5 occurred 32% too rarely).

4   Fraud Detection

The primary objective of this paper is to identify instances of fraud in individual cases by employing Machine Learning (ML) models. To train these, the previously simulated inpatient claims were used. Initially, we adjusted the features to avoid hospital-based over-fitting by replacing the hospital ID with six individual attributes calculated for each hospital. These attributes include average claim per treatment, average number of ventilation hours, average rate of cesarean section deliveries, average weight of newborns, number of inpatient treatment cases, and average distance between the hospital and its patients.

For fraud detection using ML models, standard algorithms such as Decision Tree (DT), Random Forest (RF), Gradient Boosting (GB), Multi-Layer Perceptron (MLP), and Logistic Regression (LR) from scikit-learn [16] were trained. Li et al. [17] noted that most research on healthcare fraud detection focuses on MLPs and DTs, suggesting their potential performance superiority after hyperparameter tuning.

4.1   Results

All five models perform well at classifying claims with no prevalent fraud, achieving a precision of over 98%. However, slight differences exist. While RF, GB, and MLP have a recall of over 99%, DT and LR labeled less than 99% of all non-fraudulent cases correctly (see also Table 2).

Regarding the precision of models in predicting fraudulent claims, RF performs the best, correctly identifying 98.4%, closely followed by GB with a precision of 91.2%. Conversely, LR and DT (precision of 15.0% and 4.6%) perform the poorest. However, in terms of recall, LR outperforms others, detecting 67.4% of all fraudulent cases detected, tailed closely by DT at 62.9%. GB detects every second fraudulent case, while MLP and RF (recall of 42.4% and 39.4%) perform slightly worse. Overall, RF and GB achieve the best results.

Table 2. Performance comparison of Decision Tree (DT), Random Forest (RF), Gradient Boosting (GB), Multi-Layer Perceptron (MLP), and Logistic Regression (LR) on the synthetic fraud detection dataset. As expected, all models detect the majority class (No Fraud) with high precision; however, the performance differs significantly for the minority class (Fraud). Random Forest and Gradient boosting achieve the highest weighted F1 score and are therefore selected for further studies.

Model	AUC	Weighted F1 score	Class 0 (No Fraud)		Class 1 (Fraud)	
			Precision	Recall	Precision	Recall
DT	0.615	0.416	0.981	0.600	0.046	0.629
RF	0.697	0.777	0.982	**0.999**	**0.984**	0.394
GB	0.753	**0.824**	0.985	0.998	0.912	0.513
MLP	0.709	0.759	0.982	0.995	0.706	0.424
LR	**0.778**	0.589	**0.989**	0.882	0.150	**0.674**

When analyzing recall and precision on the six included fraudulent patterns, differences between models and patterns are evident across all models and settings. RF and MLP fail to predict increases in ventilation hours and changes to cesarean sections. The fraud pattern best detected by all classifiers is changes in the order of ICD codes.

RF and GB, the top-performing models on the generated dataset, were applied to AOK's data. Among 899,610 cases of interest, RF classified 1% of all claims as fraudulent, while GB did so at 12%. Although these values vary widely and slightly exceed the internationally observed healthcare fraud rate of up to 10% [18]. However, further validation is required.

4.2 Discussion

Implementing these models in practical applications can improve processes if the limitations are taken into account. RF and GB exhibit high precision scores, resulting in a lower false positive rate. However, this comes at the cost of a relatively low number of detected fraud cases in the dataset, with only 39% (RF) and 51% (GB). Despite this, they remain favorable compared to DT, MLP, and LR for practical applications.

As all models struggle with fraud patterns where only small changes occur, additional approaches are necessary. The structure of the data implies that outlier detection algorithms could be suitable. This is especially promising for parameters such as the number of ventilation hours. In our data an unexpectedly high number of occurrences can be observed at threshold values, indicating potentially fraudulent behavior.

The employment of RF and GB on the real-world dataset shows the potential of applying models only trained with simulated data. As the results lie within the expected range, the applicability of the proposed approach is supported. Nonetheless, the ML models' dependency on the quality of the inpatient claims simulation is conclusive. With improvements in this regard, an increased performance of the prediction algorithms seems to be assured. These results demonstrate the possibility of inpatient fraud detection based on previously generated data. Even though the method is not yet refined, sufficient results have been achieved, both on simulated as well as on real data.

5   Conclusion

We presented the first known approach to simulate inpatient billing data and developed machine learning methods to detect inpatient claims fraud based on it. We leverage publicly available statistics and inject known fraudulent behaviors into the dataset to enable the supervised training of fraud detection methods. While most inpatient care cases are simulated accurately, outliers are underrepresented in the simulation.

Nonetheless, our method achieved sufficient results for fraud detection in the next step. Based on the simulated data, Gradient Boosting and Random Forest were the most convincing models, with weighted F1 scores of around 80%. To detect small changes in only one parameter, alternative approaches should be considered to improve performance. Outlier Detection methods are suspected to be promising. When applying this approach to uncover inpatient claims fraud on German health insurance data, the observed rate is up to 12%. While this claim is backed by literature, further validations are necessary.

The proposed approach for data simulation may inspire further research in the inpatient care domain, specifically in fraud detection, hospital planning, and healthcare resource allocation.

References

1. Bundesministerium für Gesundheit: Vorläufige Finanzergebnisse der GKV für das Jahr 2021. https://www.bundesgesundheitsministerium.de/presse/pressemitteilungen/vorlaeufige-finanzergebnisse-gkv-2021.html. Accessed 23 Dec 2023
2. AOK Bundesverband GbR: Fehlverhalten im Gesundheitswesen. Bericht über die Arbeit und die Ergebnisse der Stellen zur Bekämpfung von Fehlverhalten im Gesundheitswesen (2021). https://aok-bv.de/imperia/md/aokbv/presse/pressemitteilungen/archiv/taetigkeitsbericht_fv_im_gesundheitswesen_2018-2019.pdf. Accessed 23 Dec 2023
3. Jürges, H., Köberlein, J.: First do no harm. Then do not cheat: DRG upcoding in German neonatology. DIW Discussion Papers (2013)
4. Institut für das Entgeltsystem im Krankenhaus: Fallpauschalen-Katalog gem. §17b Abs. 1 S. 4 KHG Katalog ergänzender Zusatzentgelte gem. §17b Abs. 1 S. 7 KHG Pflegeerlöskatalog gem. §17b Abs. 4 S. 5 KHG. https://www.g-drg.de/ag-drg-system-2021/fallpauschalen-katalog/fallpauschalen-katalog-2021. Accessed 26 Dec 2023
5. Busse, R., Geissler, A., Aaviksoo, A.: Diagnosis related groups in Europe: moving towards transparency, efficiency, and quality in hospitals? BMJ (Clin. Res. Ed.) (2013). https://doi.org/10.1136/bmj.f3197

6. van Herwaarden, S., Wallenburg, I., Messelink, J.: Opening the black box of diagnosis-related groups (DRGs): unpacking the technical remuneration structure of the Dutch DRG system. Health Econ. Policy Law (2020). https://doi.org/10. 1017/S1744133118000324

7. Sievert, J.: Möglichkeiten der Abrechnungsmanipulation im Krankenhaus. Logos, Berlin (2011)

8. Statistisches Bundesamt: 23131-0003: Krankenhauspatienten: Deutschland, Jahre, Geschlecht, Altersgruppen, Wohnort des Patienten, Hauptdiagnose ICD-10 (1-3-Steller Hierarchie) (2022). https://www-genesis.destatis.de/genesis/downloads/ 00/tables/23131-0003_00.csv. Accessed 26 Dec 2023

9. Statistisches Bundesamt: 23141-0003: Nebendiagnosen der vollstationären Patienten: Deutschland, Jahre, Geschlecht, Altersgruppen, Wohnort des Patienten, Nebendiagnosen ICD-10 (1-3-Steller Hierarchie) (2022). https://www-genesis. destatis.de/genesis//online?operation=table&code=23141-0003. Accessed 26 Dec 2023

10. Statistisches Bundesamt: 23141-0111: Operationen und Prozeduren an vollstationären Patienten: Bundesländer, Jahre, Geschlecht, Altersgruppen, Operationen und Prozeduren (1-4-Steller Hierarchie) (2022). https://www-genesis.destatis.de/ genesis//online?operation=table&code=23141-0111. Accessed 26 Dec 2023

11. Statistisches Bundesamt: Neues Krankenhausverzeichnis (2021). https://www. destatis.de/DE/Themen/Gesellschaft-Umwelt/Gesundheit/Krankenhaeuser/ krankenhausverzeichnis.html. Accessed 26 Dec 2023

12. IMC clinicon: IMC Navigator https://www.imc-clinicon.de/tools/imc-navigator/ index_ger.html. Accessed 26 Dec 2023

13. World Health Organization: Weight-for-age BOYS. https://cdn.who.int/media/ docs/default-source/child-growth/child-growth-standards/indicators/weight-for-age/wfa-boys-0-13-zscores.pdf. Accessed 26 Dec 2023

14. World Health Organization: Weight-for-age GIRLS. https://cdn.who.int/media/ docs/default-source/child-growth/child-growth-standards/indicators/weight-for-age/wfa-girls-0-13-zscores.pdf. Accessed 26 Dec 2023

15. Statistisches Bundesamt: Grunddaten der Krankenhäuser. https://www. destatis.de/DE/Themen/Gesellschaft-Umwelt/Gesundheit/Krankenhaeuser/ Publikationen/Downloads-Krankenhaeuser/grunddaten-krankenhaeuser-2120611217004.pdf. Accessed 26 Dec 2023

16. Pedregosa, F., Varoquaux, G., Gramfort, A.: Scikit-learn: machine learning in Python. J. Mach. Learn. Res. (2011)

17. Li, J., Huang, K.-Y., Jin, J.: A survey on statistical methods for health care fraud detection. Health Care Manag. Sci. (2008). https://doi.org/10.1007/s10729-007-9045-4

18. Gee, J., Button, M., Brooks, G.: The financial cost of healthcare fraud. University of Portsmouth and MacIntyre Hudson LLP (2010). https://pure.port.ac.uk/ ws/portalfiles/portal/1925942/The-Financial-Cost-of-Healthcare-Fraud---Final-%282%29.pdf. Accessed 26 Dec 2023

The Past Helps the Future: Coupling Differential Equations with Machine Learning Methods to Model Epidemic Outbreaks

Yulia Abramova and Vasiliy Leonenko$^{(\boxtimes)}$

ITMO University, 49 Kronverksky Pr., St. Petersburg 197101, Russia
vnleonenko@itmo.com

Abstract. The aim of the research is to assess the applicability of methods of artificial intelligence to the analysis and prediction of infectious disease dynamics, with an aim to increase the speed of obtaining predictions along with enhancing quality of the results. To ensure the compliance of the forecasts with the natural laws governing the epidemic transmission, we employ Physics-Informed Neural Networks (PINN) as our main tool for the forecasting experiments. With the help of numerical experiments, we show the applicability of the approach to infectious disease modeling based on coupling classic approaches, namely, SIR models, and the cutting-edge research related to machine learning techniques. We compare the accuracy of different implementations of PINN along with the statistical models in the task of forecasting COVID incidence in Saint Petersburg, thus choosing the best modeling approach for this challenge. The results of the research could be incorporated into surveillance systems monitoring the advance of COVID and influenza incidence in Russian cities.

Keywords: Mathematical epidemiology · SIRD models · Physics-informed neural networks · Forecasting · Python

1 Introduction

Simple mathematical models, such as SIR models, based on differential equations, are known from the beginning of XX century and are still in wide use for epidemic outbreak prediction, having a great ability to generate easily interpretable results. In fact, a large part of the decision-making frameworks developed around the world to analyze and combat COVID-19 were based on SIR models (e.g., [1, 2]). However, they are often unable to capture peculiarities of disease dynamics related to stochastic effects, as well as to consider the uncertainty in input data. The differential model output is 100% determined by the values of the input parameters and is represented by a smooth incidence curve, although the real incidence usually has fluctuations around the trend. Also, despite the fact that SIR models belong to the simplest explanatory models available for simulating disease dynamics, they still require implementation of the calibration algorithm and might suffer from the local optima problem, giving incoherent results. In

L. Franco et al. (Eds.): ICCS 2024, LNCS 14835, pp. 247–254, 2024.
https://doi.org/10.1007/978-3-031-63772-8_23

this aspect SIR models are more complicated to handle than the statistical models, especially to the domain specialists not directly connected with modeling and differential equations, so they are far from being "out-of-the-box" solution for decision making in epidemiology. At the same time, the statistical approaches, although suitable for long-term forecasting of seasonal illnesses [3], are hardly applicable for predicting peaks and outbreak longevity because of the complicated physical laws governing outbreak incidence dynamics.

To provide a more effective prediction tool which is also easy to use, machine learning approaches can be employed [4]. While more sophisticated and often more accurate than simple statistical models, they also have their disadvantages. Particularly, LSTM networks which are commonly used for forecasting purposes in mathematical epidemiology [5, 6], require massive datasets for training, which are often not available, especially in cases of infections caused by novel or mutated viruses.

One of the ways to overcome the problem of the ML-based predictors, which do not "know" the laws of incidence dynamics due to scarce amount of training data, is essentially to feed them those laws directly, thus obtaining a hybrid approach incorporating ML techniques and SIR-type models. Thus, there exists an opportunity of enhancing the quality of cutting-edge methods with the help of age-old [7] differential equations. The corresponding approach is called PINN (physics-informed neural networks) [8]. Although PINNs have already received appreciation in the domains like physics, their application to epidemiological problems is still not widespread. In most of the corresponding papers known to the authors, PINNs are used to predict cumulative dynamics of epidemics in big territories, of the scale of separate countries, thus the regarded incidence time series are rather smooth and could be easily handled by neural networks. The efficiency of PINN applied to noisy city-level incidence data is yet to be assessed.

In this research, we use PINN for the aim of predicting COVID-19 incidence in Saint Petersburg and comparing its efficiency depending on modifications of PINN. The research question addressed is whether PINN can be considered an efficient "out-of-the-box" solution for the problem of disease forecasting. The answer on that question will clarify whether PINN can become a core component of decision-making frameworks used by domain specialists and policy makers to plan control measures on the city level.

2   Methods

2.1   SIRD Model

The SIRD model is one of the primary models for studying and modeling the spread of epidemics. It is based on a simpler SIR model [15] but includes an additional category, thus dividing the population into four categories: Susceptible (S), Infected (I), Recovered (R), and Dead individuals (D). In this system, each class of individuals is assigned to a specific compartment, and transitions between compartments represent the movement of individuals between different states. When "Susceptibles" (S) come into contact with infected individuals, they can also become infected and transition to the "Infected" category (I). "Infected" individuals (I) either recover and are moved to the "Recovered" category (R) or, they may die and move to the Dead category (D). All the rates, i.e. of infection transmission α, of recovery β, and of mortality γ, are considered constant.

We assume that individuals who recover from the virus gain complete immunity against future infection, so our model is limited to the simulation of a single COVID-19 wave. The total population $N = S + I + R + D$ is considered to be constant. The corresponding differential equations have the following form:

$$\frac{dS}{dt} = -\beta SI, \frac{dI}{dt} = \beta SI - \gamma I - \delta I, \frac{dR}{dt} = \gamma I, \frac{dD}{dt} = \delta I.$$

2.2 Physics-Informed Neural Network

Physics-informed neural networks (PINNs) are specialized neural networks designed for addressing supervised learning tasks while incorporating the principles of physics, especially those associated with intricate nonlinear partial differential equations and ordinary-differential-equations (ODEs) [8]. Models based on PINNs adhere to physical laws by incorporating a loss function that includes the residuals from physics equations and boundary conditions. These models utilize automatic differentiation to compute the derivatives of the neural network output concerning its inputs (spatial and temporal coordinates, and model parameters) [9]. By reducing the loss function, the network can accurately approximate solutions [10, 11]. By leveraging these differential equations, PINNs enhance the learning process, enabling the algorithm to converge toward the correct solution, even when the available training data is limited.

In a specific application, PINNs were utilized for data approximation and the identification of unknown parameters of the SIRD model. Following an in-depth analysis of the system, the model can calculate the coefficients within the system of differential equations within a predetermined range, facilitating interpretation of the results [12].

Typically, a PINN architecture consists of multiple fully connected layers with numerous neurons and incorporates non-linear activation functions between them. The input for a PINN comprises batches of time steps, while the output represents tensors that convey the network's estimations of the compartments within the SIRD model at each time step [13]. These estimations are constrained by the conditions derived from the SIRD system:

$$f_1 = \frac{dS}{dt} - (-\beta SI),$$

$$f_2 = \frac{dI}{dt} - (\beta SI - \gamma I - \delta I),$$

$$f_3 = \frac{dR}{dt} - \gamma I, f_4 = \frac{dD}{dt} - \delta I.$$

Per training iteration, we compute the usual data error defined as

$$loss_U = mean(S - S')^2 + mean(I - I')^2 + mean(R - R')^2 + mean(D - D')^2$$

Here, X is the actual data that the model was provided, X' is the prediction the model computed. We obtain the physics-informed part of the loss function, the residual error, per training iteration as $loss_F = mean(f_1)^2 + mean(f_2)^2 + mean(f_3)^2 + mean(f_4)^2$.

The neural network's parameters can be acquired through the process of minimizing the mean squared error loss. This loss function incorporates a weighting factor denoted as μ, which falls within the range of $\mu \in [0,1]$ and helps balance the importance of accurately reproducing the data and conforming to the differential equations: $loss = \mu loss_U + (1 - \mu) loss_F$.

2.3 Data

In our research, we work with daily incidence cases and cumulative incidence data covering a period of 359 days from July 5, 2020, to June 28, 2021 [14]. To enhance the computational speed, we reduced the dataset using every tenth record of current and cumulative incidence. Although our experiments concentrated exclusively on one particular epidemic season, the PINN described here can be easily adapted for the diverse epidemic seasons.

3 Results

3.1 Forecasting of Post-peak Incidence

The prediction quality of the models was evaluated based on the root mean square error (RMSE), which measures the average squared difference between the predicted and actual values. We examined various architectures of our fully connected neural network, which forms the basis of PINNs calibrated to two types of input data: daily registered cases and cumulative incidence data (Fig. 1). After the first layer, the activation function was Rectified Linear Unit (ReLU) in both cases. We established that for cumulative data, it is more advantageous to consider ReLU as the activation function, whereas for daily registered cases the hyperbolic tangent (tanh) is more suitable. We also compared our results with ARIMA predictions, with parameters (2,2,3).

It is worth noting that the resulting incidence forecast made by ARIMA outperforms both versions of PINN: namely, RMSE for the prediction by ARIMA is 2541.22, whereas RMSE for PINN (ReLU) is 15786.79 and RMSE for PINN (tanh) is 6516.33. Also, one can notice that the incomplete data for the experiments were taken in such a way that the peak incidence was included in the dataset, i.e., we predicted the second half-wave of declining incidence. It is obvious that the ARIMA forecasting made on incomplete data before the peak will be unable to replicate the falling of incidence and will show a growth to infinity instead. At the same time, our experiments showed that in such conditions PINNs behave in similar fashion, despite the laws of disease incidence incorporated into the loss function. In fact, the form of the prediction curve in PINN rather replicates the activation function than the typical incidence curve. Since predicting the peak incidence and the day of maximal number of the infected is of crucial importance for healthcare specialists, the next series of experiments were dedicated to the modification of PINN algorithm to make it useful for peak forecasting.

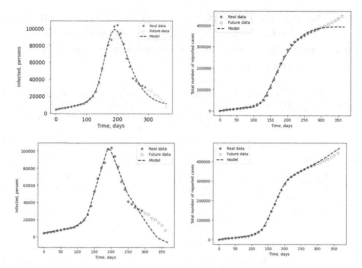

Fig. 1. Upper row: PINN's (tanh) prediction of incidence (left) and cumulative infected (right). Lower row: PINN's (ReLU) prediction of incidence (left) and cumulative infected (right).

3.2 Peak Prediction

Let us consider the case when we impose an additional condition at the boundary by explicitly assuming a particular duration of the outbreak t*, i.e. the value of modeled incidence which corresponds to t* should be close to zero. Since, we don't have sufficient historical data to derive the typical epidemic duration (in case of COVID-19), or, alternatively, the variation in this value is rather big (in case of seasonal influenza), we iterated the value of the desired epidemic duration over a certain range and assessed the forecasting error in each of the iterations. The modified formula for the loss function is $loss = \mu loss_U + (1 - \mu)loss_F + loss_T$, where $loss_T$ is the additional boundary condition. Using the synthetic dataset, we compared the performance of PINNs depending on presence or absence of the boundary condition in the loss function. The inclusion of the additional boundary condition greatly enhanced the prediction quality. As a result, we use this modified error function for addressing prediction tasks on actual data, which we do below, assuming that 160 initial incidence points are used for training. We considered a range of values from 260 to 380 to search for peak predictions. We selected several durations for consideration (Fig. 2, left), for which we executed 10 iterations of the PINN with different initial seeds to account for the output variation due to the stochasticity of the algorithm. This allowed us to generate a total of 10 peak predictions for each hypothetical length of the epidemic waves. The training process was being halted once the error function fell below a predetermined threshold of 0.0001. The accuracy of fitting to the known incidence points (calibration error) and the difference between the expected and the actual incidence trajectory (prediction error) are given in Table 1. We show the quality of the peak prediction made according to these trajectories in Fig. 2, right. The maximal incidence reached during an outbreak and the day of its occurrence is a crucial epidemic indicator for the healthcare specialists, since it largely defines the maximal workload of the healthcare system in terms of the resources spent and hospital beds allocated. The points on the graph represent the prediction bias for the peak day (dt) and the ratio between the modeled and actual outbreak peak heights (dh). The ideal peak prediction scenario corresponds to the intersection of dt = 0 (vertical dashed line) and dh = 1 (horizontal dashed line).

Table 1. Median RMSE depending on the assumed outbreak duration. The smallest values of RMSE are marked in bold

t*, days	260	280	310	330	360	380
Calibration	565.9	392.9	527.2	**349.2**	457.0	437.9
Prediction	26186	20073	13651	**12153**	20123	18961

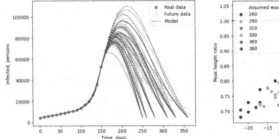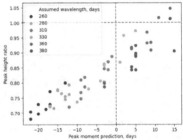

Fig. 2. Left: PINN's prediction represented by a color-coded gradient illustrating possible scenarios for each sample value across ten iterations. Right: Biases of the peak prediction for the daily incidence data depending on the assumed wavelength

In the ideal case, we would expect from calibration error (Table 1) to be convex as a function of assumed disease duration, and, moreover, to have the local minimum at the same value of outbreak duration t* as the function of predicted error, with both giving us the duration of the real outbreak (approx. 360 days). This result could be indeed reached on synthetic data generated by SIRD model, with the smooth and concave incidence curve, however, it is not the case for the real outbreak. As we can see from the tables, values of median RMSE fluctuate with changing t*. The optimal values of RMSE are reached for t* = 330 and this value also delivers us the optimal forecast quality, but it is not equal to 360. As it is demonstrated in Fig. 2 (right), even for the assumed wavelength corresponding to the lowest median RMSE (t* = 330, marked with orange dots) the bias both in peak height and peak day might be dramatic, depending on the simulation run. Also, it is interesting that almost all of the predictions underestimated the peak height. We consider it a peculiarity related to the regarded incidence dataset which cannot be attributed to the forecasting method itself.

4   Discussion

In this research, we described the modeling techniques for the retrospective analysis and forecasting of COVID-19 incidence in Saint Petersburg based on physics-informed neural networks. We showed that PINN has several advantages compared with SIR models and classical statistical approaches. Apart from SIR models, PINN can reproduce variation in incidence data contrary to the smooth forecasting curves provided by solving differential equations (which is however not very visible in our graphs due to data sampling and big wavelength in general). When used as a part of the forecasting framework with data assimilation, PINN calculates consecutive forecasts much faster than SIR. In general, PINN does not require complex calibration procedures, like SIR models, which is especially beneficial when the data reanalysis is performed on weekly basis.

Compared to standard ML models, PINN can produce an adequate forecast without requiring too much data for training. This is essential because the field of epidemiology generally does not provide big datasets for training.

Apart from simple statistical approaches, like ARIMA, PINN can foresee and thus to reproduce a typical form of the disease incidence curve thanks to the physical laws incorporated into its loss function. This gives PINN an advantage of accurately predicting the incidence before the peak of the outbreak, which cannot be done using ARIMA and similar methods. However, for that purpose PINN requires a modification with adding a boundary condition, which was described in detail in the previous section. Also, when predicting the disease incidence after reaching the peak, PINN loses its benefits and can be easily replaced by ARIMA.

Adding assumed length of the outbreak into the loss function helped obtain more reasonable peak predictions and limit the calibration time by discarding options with unreasonable outbreak lengths. At the same time, in small training datasets different forecasts with dramatically different assumed wavelengths can demonstrate similar RMSE. However, it cannot be considered a problem of the PINN, because it is a known issue of forecasting on small incidence data when a plausible forecast could be only obtained when approaching the incidence peak [16].

Despite being based on ODEs, PINN does not guarantee that the conservation laws governing the epidemic will be complied in a model. As an example, in many cases PINN shows a negative number of infective persons at the second part of the outbreak and thus the trajectory should be artificially cut at zero. This should be taken into account when using PINN to derive disease indicators different from the predicted incidence (for instance, the number of the recovered individuals).

PINN cannot deliver plausible parameter values for the ODE on which it is based. While this could be true on smooth simulated data, in general case it does not work, as our experiments showed. Thus, apart from ODEs, PINN cannot be used to assess the disease parameter values. If there is a need to solve this task using ML techniques, there are better solutions for that, for instance, NeuroODE [17], since it does not add a stochastic component to the obtained solution.

The conclusions made during the experiments might be somewhat limited because only one wave of COVID-19 was used for testing. In future studies, we plan to generalize the results further by justifying them on additional incidence data.

To sum up, based on our research, we can conclude that the technique of PINN is indeed capable of addressing the challenges which are more often tackled by SIR models, but it also has its drawbacks, and cannot be named a universal out-of-the-box approach. The selection of the proper tool for epidemic surveillance largely depends on the modeling aim and it is up to the researcher to make an ultimate choice of the most suitable technique.

Acknowledgments. This research was supported by The Russian Science Foundation, Agreement #22-71-10067.

Disclosure of Interests. The authors declare no competing interests.

References

1. Read, J., et al.: Novel coronavirus 2019-nCoV (COVID-19): early estimation of epidemiological parameters and epidemic size estimates. Philos. Trans. R. Soc. B **376**, 20200265 (2021)
2. Maier, B., Brockmann, D.: Effective containment explains subexponential growth in recent confirmed COVID-19 cases in China. Science **368**, 742–746 (2020)
3. Kondratyev, M., Tsybalova, L.: Long-term forecasting of influenza-like illnesses in Russia. Int. J. Pure Appl. Math. **89**(4), 619–641 (2013)
4. Shahid, F., Zameer, A., Muneeb, M.: Predictions for COVID-19 with deep learning models of LSTM, GRU and Bi-LSTM. Chaos,Solit Fract **140**, 110212 (2020)
5. Yang, Z., et al.: Modified SEIR and AI prediction of the epidemics trend of covid-19 in china under public health interventions. J. Thoracic Dis. **12**(3), 165–174 (2020)
6. Leonenko, V.N., Bochenina, K.O., Kesarev, S.A.: Influenza peaks forecasting in Russia: assessing the applicability of statistical methods. Procedia Comput. Sci. **108**, 2363–2367 (2017)
7. Kermack, W.O., McKendrick, A.G.: A contribution to the mathematical theory of epidemics. Proc. R. Soc. Lond. Ser. A Contain. Pap. Math. Phys. Character **115**(772), 700–721 (1927)
8. Raissi, M., Perdikaris, P., Karniadakis, G.E.: Physics-informed neural networks: a deep learning framework for solving forward and inverse problems involving nonlinear partial differential equations. J. Comput. Phys. **378**, 686–707 (2019). ISSN: 0021-9991
9. Baydin, A.G., Pearlmutter, B.A., Radul, A.A., Siskind, J.M.: Automatic differentiation in machine learning: a survey (2015). arXiv:1502.05767
10. Rao, C., Hao, S., Yang, L.: Physics-informed deep learning for incompressible laminar flows. Theor. Appl. Mech. Lett. **10**, 207–212 (2020)
11. Faroughi, S.A., Datta, P., Mahjour, S.K., Faroughi, S.: Physics-informed neural networks with periodic activation functions for solute transport in heterogeneous porous media (2022). arXiv preprint arXiv:2212.08965
12. Cuomo, S., Di Cola, V.S., Giampaolo, F., Rozza, G., Raissi, M., Piccialli, F.: Scientific machine learning through physics—informed neural networks: where we are and what's next. J. Sci. Comput. **92**(88) (2022)
13. Shaier, S., Raissi, M., Seshaiyer, P.: Data-driven approaches for predicting spread of infectious diseases through DINNS: disease informed neural networks. Lett. Biomath. **9**(1), 71–105 (2022)
14. Kouprianov, A.: Monitoring COVID-19 epidemic in St. Petersburg, Russia: data and scripts (2021). https://github.com/alexei-kouprianov/COVID-19.SPb.monitoring
15. Huaman, I., Leonenko, V.: Does Complex mean accurate: comparing COVID-19 propagation models with different structural complexity. In: Computational Science. ICCS 2023: Computational Science—ICCS 2023, pp. 270–277
16. Leonenko, V.N., Ivanov, S.V.: Prediction of influenza peaks in Russian cities: comparing the accuracy of two SEIR models. Math. Biosci. Eng. **15**(1), 209–232 (2018)
17. Chen, R.T., Rubanova, Y., Bettencourt, J., Duvenaud, D.K.: Neural ordinary differential equations. Adv. Neural Inf. Process. Syst. **31** (2018)

Combining Convolution and Involution for the Early Prediction of Chronic Kidney Disease

Hadrien Salem[1]([✉]), Sarah Ben Othman[1], Marc Broucqsault[2],
and Slim Hammadi[1]([ID])

[1] CRIStAL CNRS UMR 9189, Villeneuve d'Ascq, Nord, France
{hadrien.salem,sara.ben-othman,slim.hammadi}@centralelille.fr
[2] Altao, Lille, Nord, France
mbroucqsault@altao.com

Abstract. Chronic Kidney Disease (CKD) is a common disease with high incidence and high risk for the patients' health when it degrades to its most advanced stages. When detected early, it is possible to slow down the progression of the disease, leading to an increased survival rate and lighter treatment. As a consequence, many prediction models have emerged for the prediction of CKD. However, few of them manage to efficiently predict the onset of the disease months to years prior. In this paper, we propose an artificial neural network combining the strengths of convolution and involution layers in order to predict the degradation of CKD to its later stages, based on a set of 25 common laboratory analyses as well as the age and gender of the patient. Using a dataset from a French medical laboratory containing more than 400 000 patients, we show that our model achieves better performance than state-of-the-art models, with a recall of 83%, F1-score of 76%, and 97% overall accuracy. The proposed method is flexible and easily applicable to other diseases, offering encouraging perspectives in the field of early disease prediction, as well as the use of involution layers for deep learning with time series.

Keywords: Machine Learning · Chronic Kidney Disease · Disease prediction · Data processing · Big Data Analytics · Artificial Neural Networks · Convolutional Neural Networks · Involutional Neural Networks

1 Introduction

Chronic Kidney Disease (CKD) is a long-term condition corresponding to a malfunction of the kidneys. The kidneys play a crucial role in filtering waste products and excess fluids from the blood to form urine. In CKD, the kidneys gradually lose their ability to function effectively over time. This can cause a wide range of complications, such as anemia or an increased risk of cardiovascular diseases. In the most severe cases, CKD can lead to kidney failure, making heavy treatments such as dialysis or kidney transplantation necessary.

© The Author(s), under exclusive license to Springer Nature Switzerland AG 2024
L. Franco et al. (Eds.): ICCS 2024, LNCS 14835, pp. 255–269, 2024.
https://doi.org/10.1007/978-3-031-63772-8_24

When detected early, it is generally possible to slow down the progression of the disease with lifestyle changes and medication. While it is normal for kidney function to decrease with age, certain patients can experience an accelerated degradation and quickly evolve to advanced stages of the disease. Therefore, those patients are likely to not be properly monitored and risk more severe health consequences.

However, it is difficult to predict the speed at which one's renal function might decrease. There lies the crux of the challenge : using seemingly-normal biological data to predict an abnormal evolution of the patients' renal function. Risk factor calculations exist, such as the Kidney Failure Risk Equation (KFRE) [26]. However, establishing the formulae for these calculations can be difficult, and depends on the targeted population. In that regard, an alternative based on machine learning could be beneficial, as they can easily be retrained on new data. In particular, this would allow every medical laboratory to maintain their own model, catered to their own patient base.

In this paper, we propose a neural network model based on convolution and involution layers for the early prediction of a degradation in kidney function, thus helping with the monitoring of CKD for patients at risk. Our model is based solely on common laboratory tests, and its predictions could therefore be integrated seamlessly into a biological report, without requiring any additional information on the patient.

2   Related Work

Disease prediction and detection has received a lot of attention in the literature, especially in recent years. The type of data and the techniques used to analyze it depend on the disease that is studied. Diabetes, for example, has been extensively researched, both for risk prognosis [16,28] and blood glucose prediction [17]. Other diseases, such as Alzheimer's disease [1,8] or colorectal cancer [12] have been studied in a similar manner with various types of data: not only laboratory tests, but also imaging, and even environmental and lifestyle data [4].

In addition to "classic" machine learning approaches, such as tree-based methods, support vector machines, or artificial neural networks, other methods such as survival analysis [21,22,29] and knowledge-based approaches [19,27] have been used to establish prediction or risk-stratification models, as well as discovering semantic relationships in medical data.

Chronic Kidney Disease (CKD) in particular is one of the conditions that has received the most attention in the literature, because of its high incidence. Machine learning tasks often focus on the detection of CKD using Electronic Health Records (EHRs) and laboratory results like creatinine or eGFR (estimated Glomerular Filtration Rate) [11,23], using various methods such as Support Vector Machine, Random Forest, XGBoost, Logistic Regression, Neural networks, AdaBoost, etc. These models usually achieve very good accuracy (sometimes over 99%), but it could be argued that this classification task is easy, because CKD itself is diagnosed using eGFR (refer to Sect. 3.1 for a more detailed explanation).

In other words, if eGFR is part of the learning features, the use of machine learning for the diagnosis of CKD at one point in time can be questionable.

In order to prevent complications in CKD, it is much more interesting to be able to predict the onset or the progression of the disease months or years in advance. A lot less research can be found in the literature on the subject. Bernardini et al. [3] propose a semi supervised learning for short term prediction of ckd stage using 2.5 years of history, but is quite restrictive on the patients that the model can apply to, since all fields must have been observed in the patients' history. Zhao et al. [30] use a combination of genetic and non genetic features to study risk factors in CKD for patients with a 5-year followup, but their dataset contains only a small sample of patients with similar characteristics. Chuah et al. [6] predict the progression of patients to end-stage CKD within 2 years and compare their results against those obtained by clinicians. While they obtain, 93.9% global accuracy, they have a high false negative rate leading to 60% recall. Finally, Razavian et al. [24] propose a prediction model for 133 diseases, including end-stage CKD, based on a set of 18 common laboratory tests. Using ensemble algorithms based on convolutional neural networks (CNN) and long short-term memory networks (LSTM), they manage to predict the onset of these conditions up to 15 months prior. They obtain 92% AUC for end-stage CKD, but do not present other performance metrics. Besides, if their model shows good AUC on certain diseases, simultaneously predicting such a high number of diseases leads to poorer performance on many of them.

While recurrent networks (such as Gated Recurrent Units [7] or Long Short-Term Memory networks [9]) have been shown to provide good results for time series classification, including in the domain of disease prediction [24], we would like to explore the use of new types of layers and their ability to efficiently extract meaningful features in a patient's biological history. In particular, to the best of our knowledge, no works in the literature use involution layers [14] to perform disease prediction based on biomarkers. This type of layer, designed as an "inversion" of convolutional layers, has given good results in the literature on other tasks, in particular for image recognition. However, just like CNNs, they can be used to perform predictions using time series. Liang et al. [15] even show that combining involution and convolution can provide better results than using one or the other separately.

In this paper, our focus will be the prediction of the progression of CKD using a model combining convolutional and involutional layers. In a population of more than 400 000 patients, we will detect those who move to a critical stage of the disease within the next year.

Our contribution is as follows:

- We propose a combined convolutional and involutional neural network model to perform the early prediction of the progression of CKD. Involutional networks in particular have not, to the best of our knowledge, been used for disease prediction in the literature. This type of architecture is novel in this field, and opens up new perspectives to build neural networks that are better able to capture meaningful features when studying longitudinal biological data.

- We use a set of features consisting exclusively of the patient's age and gender, and 25 commonly-measured laboratory analyses. These markers do not only include biomarkers related to renal functions, such as eGFR or creatinine, but also extends to other groups, including the lipid panel and blood counts. This is an alternative to the most common approach consisting in finding targeted biomarkers, which is difficult and disease-dependent. Using common biomarkers makes our framework more flexible and easily applicable than most in the literature.
- We analyze the performances of our model on a real dataset originating from a French medical laboratory, with over 400 000 patients. Our model provides good results despite large amounts of missing data, which matches real-life use cases. We compare our results with those obtained by state-of-the-art models, and show that the combination of convolution and involution provides better results in terms of accuracy, precision and F1-score, than either of the networks used individually.

3   Methods

In this section, we will define precisely the machine learning problem we are trying to solve, before describing in detail the model.

3.1   Problem Definition

In this study, our goal is the early prediction (months to years prior) of the progression of Chronic Kidney Disease (CKD) to an advanced stage of the disease.

In order to classify the gravity of this disease, physicians use the estimated Glomerular Filtration Rate (eGFR), which quantifies the filtering capacity of the kidneys. It is expressed in $mL/min/1.73\,m^2$ and computed with the Chronic Kidney Disease Epidemiology Collaboration (CKD-EPI) equation [13]:

$$eGFR = 141 \times \min(\frac{Cr}{\kappa}, 1)^\alpha \times \max(\frac{Cr}{\kappa}, 1)^{-1.209} \times 0.993^{Age} \tag{1}$$

In the above equation, $\kappa = 0.9$ for males and $\kappa = 0.7$ for females, $\alpha = -0.411$ for males and $\alpha = -0.329$ for females, and Cr is the level of serum creatinine expressed in $\mu mol/L$.

Following the guidelines provided by the Kidney Disease - Improving Global Outcomes (KDIGO) organization in their 2022 clinical practice guidelines for diabetes management in CKD [25], we consider the five following stages of CKD:

- **Stage 1** (normal or high kidney function): eGFR $\geq 90\,mL/min/1.73\,m^2$
- **Stage 2** (mildly decreased kidney function): $60 \leq$ eGFR $< 90\,mL/min/ 1.73\,m^2$
- **Stage 3** (mildly to moderately decreased kidney function): $30 \leq$ eGFR $< 60\,mL/min/1.73\,m^2$

- **Stage 4** (severely decreased kidney function): $15 \leq$ eGFR $< 30\,\mathrm{mL/min/}$ $1.73\,\mathrm{m}^2$
- **Stage 5** (kidney failure): eGFR $< 15\,\mathrm{mL/min/1.73\,m}^2$

In the above classification, stage 1 corresponds to the lowest stage of progression of the disease, while stage 5 corresponds to end-stage CKD and usually either requires dialysis or kidney transplant for the patient. Stages 2 and 3 are intermediate stages: while stage 2 is almost never concerning, stage 3 can sometimes indicate a more severely decreased kidney function depending on other parameters, such as age or albuminuria levels.

In this study, we will consider stages 4 and 5 to be the "positive" cases. This differs from other works in the literature, where only kidney failure (stage 5) is predicted [2,23]. Indeed, from a medical point of view, it is interesting to predict earlier stages of the disease to start monitoring and treating the patient [25].

3.2   Model

With these stages established, we can define the machine learning task as follows: we select a cohort of patients, and try to determine whether they will progress to stage 4 or 5 during a certain prediction window, using a set of laboratory analyses (biomarkers) over time. We are thus treating a binary classification problem.

The difficulty of modeling the patients' history is that there is variability for both dimensions of the input: the length of the patients' history can vary because they do not necessarily have the same amount of records, and the analyses they take at each given point in time can also vary. To address this issue, similarly to the works of Razavian et al. [24], we use a sliding window framework: each patient has a variable-length, continuously-valued history of laboratory analyses X. For each of those matrices, at each time point t, we select the B months before t to form the input $X_{t-B:t}$. Data is marked as missing for the months where no analyses were taken.

We then look at the value of eGFR in a prediction window P after a "gap" G in order to label the input matrix. The gap ensures that clinical tests realized just before the prediction window are not used by the model and guarantees an "early" prediction. Data is labeled using the different stages defined in Sect. 3.1. For a label to be attributed, two measurements separated by at least 90 days must be observed within one of the stages' ranges. This precaution, recommended by KDIGO2022 [25], ensures that the renal disease is indeed chronic, and not the result of a temporary phenomenon. It also helps reduce noise in the dataset.

In Fig. 1, we show an example. Each square on the timeline represents a month, and the white square corresponds to the observation date t. The parameters in this example are $B = 6$, $G = 2$ and $P = 5$. This means that using six months of biological history (in light blue), the model would predict whether this patient reaches stage 4 or stage 5 CKD within a 5-month prediction window (in dark blue) after $t + G$. The gap (in dark grey) is arbitrarily set to two months in this illustration.

While it is theoretically possible to set any value for B, G and P, all the results in this article were obtained with the parameters listed in 4.1.

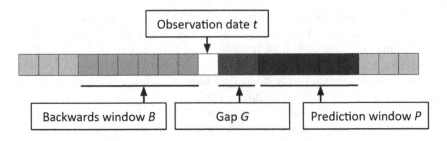

Fig. 1. Time model of the patients' biological history

3.3   Cohort Selection

Figure 2 shows our patient selection process. Out of 888 961 patients with at least one eGFR measurement in our dataset, we select those who have at least three eGFR measurements, since our approach relies on studying the history of biomarkers. For each of the 468 933 patients we have left, we are able to create several matrices using the sliding window framework described in Sect. 3.2. Each matrix is labeled using the criteria defined in 3.1. We recall that stages 4 and 5 constitute the positive case (labeled 1).

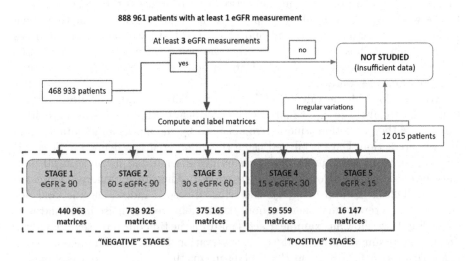

Fig. 2. Patient selection flowchart

3.4 Feature Engineering

The abundance of missing data is one of the main challenges in dealing with medical data - especially so with laboratory data, because different patients are prescribed different analyses (or sets of analyses). However, there are some biomarkers that are very commonly measured compared to others, and form a "baseline" of analyses upon which other biomarkers can be added in specific cases. Targeting these baseline biomarkers is fundamental to our study, since we aim to detect abnormalities in common laboratory data. As a consequence, we select all features that are present for a significant amount of patients, in this case 25%. Out of the 76 available biomarkers available in our dataset, we are thus left with 25 commonly-measured laboratory analyses, and two demographic features (age and gender). It is important to note that those markers are not all traditionally associated with the renal function, as we use for example the lipid panel and blood counts. This means that our model can be used routinely, without necessarily suspecting CKD and prescribing specific tests for it, which is a strong advantage in the context of disease screening.

The full list of features is provided in Table 1. Preliminary analyses have shown that the results given with this subset of analyses are equivalent to those obtained using all biomarkers, and are being computed faster thanks to the reduced dimensionality.

3.5 Pre-processing

Management of Missing Data. By using a set of common biomarkers, we have already limited the amount of missing data. However, due to the way we create matrices as presented in Sect. 3.2, there remains a lot of gaps in the matrices, because biomarkers are almost never measured every month. As a consequence, we need a strategy for the management of this missing data. We impute missing values for a feature based on the mean value for that column within the patient's history. We decide not to compute the mean based on the whole dataset, so that the imputed values remain close to the patient's, therefore being more meaningful.

Management of Class Imbalance. The classes we have defined are heavily imbalanced in the dataset, since patients at stages 4 and 5 a minority in the dataset (as seen in Fig. 2). As such, positive samples represent about 5% of the total number of samples.

Several strategies have been employed in the literature to deal with class imbalance, such as over-sampling the minority class, under-sampling the majority class, Synthetic Minority Over-sampling Technique (SMOTE) [5] and its variants, or the definition of imbalance-aware cost functions such as Balanced Binary Cross-Entropy (BBCE) [18].

In order to avoid discarding the majority of the dataset, we decide to use an oversampling technique to address class imbalance. In particular, we use random oversampling, which provides the best results in our preliminary tests.

3.6   Machine Learning Model

Neural Network Model. Convolution layers are popular in artificial neural networks, notably in the fields of image recognition, image segmentation, and computer vision. They rely on the learning of filters, or kernels, which are able to capture abstract patterns in data. When dealing with images, these patterns can be local edges or textures. However, convolution layers and convolutional neural networks (CNN) have been shown to also be able to tackle time series, including in disease prediction [24]. Figure 3a shows a simple example of convolution, where a 2×2 kernel was learned by the network and applied to the data. The output is obtained by convolution product between the data matrix and the kernel, and represented on the right of the figure.

Involution layers were introduced by Li et al. [14] as an "inversion" of convolution, in that they are spatial-specific and channel-agnostic. In particular, it means that instead of learning a filter and applying it to the entire data matrix, an involution layer learns a unique function that will create different filters for each element of the matrix (pixels for an image, biomarker values in our case). This process has a twofold advantage: involutional networks require less parameters in general, since only one function needs to be learned to generate the filters; and they are by nature better at capturing local phenomena. Figure 3b shows an example of involution. The network learned a kernel generation function ϕ, which is used to generate a 3×3 kernel from the centermost element of the data matrix. This kernel is then applied to the matrix, which results in the output depicted on the right.

Liang et al. [15] show that combining the respective strengths of convolution and involution lead to increased performance on image classification. Similarly, we decide to create a neural network architecture combining convolution and involution for the prediction of CKD progression. Both types of layers can be adapted for use with time series by considering a two-dimensional input, where each column corresponds to a biomarker, and each row corresponds to a specific date. In other words, each row represents a biological record for the patient at a given date.

The architecture of our neural network model is presented in Fig. 4. We use a convolutional layer with twelve 1×3 kernels. ReLU activation is used for non-linearity, and followed by a batch normalization layer [10]. After this, we append two blocks constituted of an involution layer with 3×3 kernels and ReLU activation, and a MaxPooling layer. The output of that layer is flattened and fully connected to 64 neural units, before the output layer provides the binary prediction. Binary cross-entropy is used to compute the loss, and the whole model is implemented using the Keras Python library. Hyperparameters have been fine-tuned by using a sample of the data.

In order to evaluate the performance of this model, which we will call **combined model**, we define two other models:

– In the first one, we replace each involution block by another convolution block, similar to the one defined above (convolution layer, batchnorm and

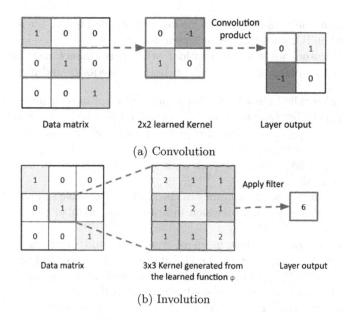

(a) Convolution

(b) Involution

Fig. 3. Schematic illustrations for convolution and involution

ReLU activation). This model will be referred to as the **convolution-only network**.

– In the second one, we remove the convolution block, leaving only the involution blocks. This model will be referred to as the **involution-only network**.

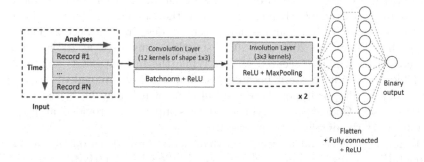

Fig. 4. Architecture of our neural network

Model Evaluation. Accuracy, which is defined as the proportion samples classified correctly, is not enough to assess the performance of a model with high class imbalance. If the proportion of samples in the majority class is significantly higher than in the minority class, then good accuracy can be achieved by only predicting the majority class. This must be avoided in real-life applications, especially when it comes to disease screening.

For this reason, we will also consider other performance metrics that take into account the quality of prediction on both the majority and minority class:

- **Recall** (also called sensitivity or true positive rate) measures the proportion of actual positive samples that are correctly identified by a classifier. It is computed as $\frac{TP}{TP+FN}$, where TP is the number of true positives (positive samples classified correctly) and FN is the number of false negatives (positive samples classified incorrectly). A high recall is desirable in our context of disease screening.
- **Precision** measures the accuracy of the model on positive samples. It is computed as $\frac{TP}{TP+FP}$. A high precision indicates a low number of false positives, which is generally preferred.
- **F1-score** is the harmonic mean of sensitivity and precision. As such, it gives a balanced measure of the classifier's performance. It is computed as $\frac{2 \times Precision \times Recall}{Precision + Recall}$. The F1-score gives equal weight to precision and recall, making it a popular metric to study imbalanced datasets.
- **Area under ROC curve (AUC).** The Receiver Operating Characteristic curve plots the true positive rate against the false positive rate at different classification thresholds. A higher area under this curve indicates better performance.

4   Results and Discussion

4.1   Experimental Setup

The experiments were realized on a dataset containing nearly 900 000 patients that had at least one eGFR measurement in their biological history, which we use to create our training and test datasets as described in Sect. 2. The data was gathered between 2010 and 2022, strictly anonymized at its source and was hosted on a highly secured server. While the full dataset is not publicly available, we decide to use it because it contains a large amount of day-to-day laboratory results, as opposed to hospital data, which may be biased towards positive cases. Code and anonymized data samples are provided at https:// github.com/CristalOSL/ckd-prediction.

Regarding the technical setup, all treatments are realized on the Windows 10 operating system with an Intel Xeon Gold 6126 CPU (2.60 GHz) and 128 GB RAM. We use Python 3.9 as our programming language. Data is manipulated with the Dask library (2023.2.0) for parallel computing, and Pandas (1.5.3) for lighter treatments. Data imbalance is managed using Imbalanced-learn (0.10.1), and the neural networks are implemented using Keras (2.14.0).

The set of features we have used for this study is detailed in Table 1. For all biomarkers, we have also provided the associated Logical Observation Identifiers Names and Codes (LOINC) standardized code [20].

Table 1. Features used in the study

Feature	LOINC
Age	–
Gender	–
Creatinine	2160-0
Estimated Glomerular Filtration Rate	62238-1
Mean Corpuscular Hemoglobin Concentration	28540-3
Red blood cell distribution width	30385-9
Platelets in plasma	26516-5
Leukocytes	26464-8
Hemoglobin	718-7
Neutrophils	30448-5
Lymphocytes	26474-7
Monocytes	26484-6
Hematocrit	20570-8
Erythrocytes	26453-1
Mean Corpuscular Hemoglobin	28539-5
Mean Corpuscular Volume	30428-7
Cholesterol	2093-3
Triglyceride	2571-8
Cholesterol in HDL	2085-9
Cholesterol non HDL	43396-1
Glucose	2345-7
Potassium	2823-3
Alanine aminotransferase	1742-6
Aspartate aminotransferase	1920-8
Sodium	2951-2
Chloride	2075-0
Prostate Specific Antigen	2857-1

In the following simulations, the model described in Sect. 3.2 is parameterized as follows: we use a backwards window of $B = 12$, a prediction window of $P = 12$ and a gap of $G = 3$. In other words, we perform the prediction of the evolution of CKD towards stage 4 or 5 within the year after a three-month gap, using a year of biological history.

4.2   Experimental Results

The first part of Table 2 shows the results obtained in our prediction task, comparing the three models described in 3.6: the involution-only model, the convolution-only model, and the model combining both convolution and involution layers. The performance metrics are those that were presented in Sect. 3.6: accuracy (Acc.), area under ROC curve (AUC), precision (Prec.), recall (Rec.) and F1-score (F1). A threshold of 0.5 was used to compute the latter. In order to compare these results with those obtained in the literature, the lower part of Table 2 displays these performance metrics for different models. Unavailable figures are indicated by a dash. Since all models do not perform the exact same prediction task, the "Positive" column indicates which stages are considered to be the positive samples for each individual model. Besides, all of these studies have been performed on different datasets, making the comparison difficult. Because of these differences, Table 2 should be interpreted as an overview of state-of-the-art models and their respective performance.

Overall, we observe that our models achieve good performance compared to the literature: we obtain the best overall accuracy, recall and F1-score, as well as competitive AUC and precision. As explained in Sect. 3.6, obtaining a high recall is particularly important in our context of disease screening, since it indicates that few positive cases are missed. Our high F1-score also indicates a good balance between the detection of positive cases and the accuracy of those predictions.

If we compare the combined model to the involution- and convolution-only models, we notice that it achieves the best performance in terms of overall accuracy, precision and f1-score, while remaining comparable in terms of AUC and recall. Precision in particular is significantly higher, being 20% above the involution-only model in that regard. We conclude that this combined model strikes a good balance between the performance metrics (as indicated by the high F1-score), and manages to detect most positive cases while keeping a relatively low proportion of false positives. This shows that the combination of convolution and involution layers is able to better capture patterns within the patients' biological history in order to predict the onset of CKD.

Table 2. Experimental results and overview of results obtained for CKD prediction in state-of-the-art models

Model	Positive	Acc.	AUC	Prec.	Rec.	F1
Involution-only model	Stages 4 & 5	0.967	0.911	0.610	0.850	0.710
Convolution-only model	Stages 4 & 5	0.952	0.908	0.500	0.860	0.630
Combined model	Stages 4 & 5	0.976	0.905	0.710	0.830	0.760
Razavian et al. ensemble model [24]	Stage 5	–	0.920	–	–	–
Razavian et al. ensemble model [24]	Stage 3	–	0.864	–	–	–
Chuah et al. XGBoost [6]	Stage 5	0.939	–	0.750	0.600	0.670
Bernardini et al. SS-MTL model [3]	Stages 2 to 5	0.746	0.811	0.657	0.731	0.665
Zhao et al. logistic regression model [30]	Stage 5	–	0.894	–	0.827	–

5   Conclusion

In this paper, we have proposed a combined convolutional and involutional neural network for the prediction of stage 4 and stage 5 CKD a year in advance, using a year of biological history. Our model includes 25 very commonly-measured biomarkers, as well as age and gender, thus being adapted for practical use by medical laboratories. By testing our model on a large dataset containing laboratory analyses for over 400 000 patients, we have demonstrated that its performance was an improvement over state-of-the-art models in terms of accuracy, recall and F1-score, and offers more balanced results than models using only involution or convolution. More generally the use of such networks with time series should be further explored.

A future avenue of research is the generalization of this model to other diseases, as it only uses common laboratory results and makes no assumptions on the patients, meaning it could be easily transferable to conditions like diabetes or even prostate cancer. However, the generalization to other diseases poses new challenges, such as the selection of an optimal set of analyses allowing for maximum performance, and the selection of the best model parameters with regards to the backwards window and prediction window, since different diseases may evolve at different speeds. One could also explore the use of automated machine learning to adapt the model's hyperparameters to the various diseases that would be studied, and the possibility to perform transfer learning between them.

Disclosure of Interests. The authors have no competing interests to declare that are relevant to the content of this article.

References

1. Abuhmed, T., El-Sappagh, S., Alonso, J.M.: Robust hybrid deep learning models for alzheimer's progression detection. Knowl.-Based Syst. **213**, 106688 (2021)
2. Nagaraj, S.B., Pena, M.J., Ju, W., Heerspink, H.L., Consortium, B.D.: Machine-learning-based early prediction of end-stage renal disease in patients with diabetic kidney disease using clinical trials data. Diabetes Obes. Metab. **22**(12), 2479–2486 (2020)
3. Bernardini, M., Romeo, L., Frontoni, E., Amini, M.R.: A semi-supervised multi-task learning approach for predicting short-term kidney disease evolution. IEEE J. Biomed. Health Inf. **25**(10), 3983–3994 (2021)
4. Chang, V., Ganatra, M.A., Hall, K., Golightly, L., Xu, Q.A.: An assessment of machine learning models and algorithms for early prediction and diagnosis of diabetes using health indicators. Healthcare Anal. **2**, 100118 (2022)
5. Chawla, N.V., Bowyer, K.W., Hall, L.O., Kegelmeyer, W.P.: Smote: synthetic minority over-sampling technique. J. Artif. Intell. Res. **16**, 321–357 (2002)
6. Chuah, A., et al.: Machine learning improves upon clinicians' prediction of end stage kidney disease. Front. Med. **9**, 837232 (2022)
7. Dey, R., Salem, F.M.: Gate-variants of gated recurrent unit (GRU) neural networks. In: 2017 IEEE 60th International Midwest Symposium on Circuits and Systems (MWSCAS), pp. 1597–1600. IEEE (2017)

8. El-Sappagh, S., Ali, F., Abuhmed, T., Singh, J., Alonso, J.M.: Automatic detection of alzheimer's disease progression: an efficient information fusion approach with heterogeneous ensemble classifiers. Neurocomputing **512**, 203–224 (2022)
9. Graves, A., Graves, A.: Long short-term memory. In: Supervised Sequence Labelling with Recurrent Neural Networks, pp. 37–45 (2012)
10. Ioffe, S., Szegedy, C.: Batch normalization: accelerating deep network training by reducing internal covariate shift. In: International Conference on Machine Learning, pp. 448–456. PMLR (2015)
11. Islam, M.A., Majumder, M.Z.H., Hussein, M.A.: Chronic kidney disease prediction based on machine learning algorithms. J. Pathol. Inf. **14**, 100189 (2023)
12. Kinar, Y., et al.: Development and validation of a predictive model for detection of colorectal cancer in primary care by analysis of complete blood counts: a binational retrospective study. J. Am. Med. Inf. Assoc. **23**(5), 879–890 (2016)
13. Levey, A.S., et al.: A new equation to estimate glomerular filtration rate. Ann. Intern. Med. **150**(9), 604–612 (2009)
14. Li, D., et al.: Involution: inverting the inherence of convolution for visual recognition. In: Proceedings of the IEEE/CVF Conference on Computer Vision and Pattern Recognition, pp. 12321–12330 (2021)
15. Liang, G., Wang, H.: I-CNET: leveraging involution and convolution for image classification. IEEE Access **10**, 2077–2082 (2021)
16. López Ibáñez, B., Vinas, R., Torrent-Fontbona, F., Fernández-Real Lemos, J.M.: Handling missing phenotype data with random forests for diabetes risk prognosis. In: López, B., Herrero, P., Martin, C.(eds.) AID: Artificial Intelligence for Diabetes: 1st ECAI Workshop on Artificial intelligence for Diabetes at the 22nd European Conference on Artificial Intelligence (ECAI 2016), The Hague, Holland, 30 August 2016, Proceedings, pp. 39–42. European Conference on Artificial Intelligence (ECAI) (2016)
17. Martinsson, J., Schliep, A., Eliasson, B., Meijner, C., Persson, S., Mogren, O.: Automatic blood glucose prediction with confidence using recurrent neural networks. In: 3rd International Workshop on Knowledge Discovery in Healthcare Data, KDH@ IJCAI-ECAI 2018, 13 July 2018, pp. 64–68 (2018)
18. Martínez-Agüero, S., et al.: Interpretable clinical time-series modeling with intelligent feature selection for early prediction of antimicrobial multidrug resistance. Futur. Gener. Comput. Syst. **133**, 68–83 (2022). https://doi.org/10.1016/j.future.2022.02.021
19. Mayer, T., Cabrio, E., Villata, S.: Transformer-based argument mining for healthcare applications. In: ECAI 2020, pp. 2108–2115. IOS Press (2020)
20. McDonald, C.J., et al.: Loinc, a universal standard for identifying laboratory observations: a 5-year update. Clin. Chem. **49**(4), 624–633 (2003)
21. Mirabnahrazam, G., et al.: Predicting time-to-conversion for dementia of alzheimer's type using multi-modal deep survival analysis. Neurobiol. Aging **121**, 139–156 (2023)
22. Moncada-Torres, A., van Maaren, M.C., Hendriks, M.P., Siesling, S., Geleijnse, G.: Explainable machine learning can outperform cox regression predictions and provide insights in breast cancer survival. Sci. Rep. **11**(1), 6968 (2021)
23. Raju, N.G., Lakshmi, K.P., Praharshitha, K.G., Likhitha, C.: Prediction of chronic kidney disease (CKD) using data science. In: 2019 International Conference on Intelligent Computing and Control Systems (ICCS), pp. 642–647. IEEE (2019)
24. Razavian, N., Marcus, J., Sontag, D.: Multi-task prediction of disease onsets from longitudinal laboratory tests. In: Machine Learning for Healthcare Conference, pp. 73–100. PMLR (2016)

25. Rossing, P., et al.: Kdigo 2022 clinical practice guideline for diabetes management in chronic kidney disease. Kidney Int. **102**(5), S1–S127 (2022)
26. Tangri, N., et al.: A predictive model for progression of chronic kidney disease to kidney failure. JAMA **305**(15), 1553–1559 (2011)
27. Tymoshenko, K., Somasundaran, S., Prabhakaran, V., Shet, V.: Relation mining in the biomedical domain using entity-level semantics. In: ECAI 2012, pp. 780–785. IOS Press (2012)
28. Xie, Z., Nikolayeva, O., Luo, J., Li, D.: Building risk prediction models for type 2 diabetes using machine learning techniques. Prev. Chronic Dis. **16** (2019)
29. Yansari, R.T., Mirzarezaee, M., Sadeghi, M., Araabi, B.N.: A new survival analysis model in adjuvant tamoxifen-treated breast cancer patients using manifold-based semi-supervised learning. J. Comput. Sci. **61**, 101645 (2022)
30. Zhao, J.: An early prediction model for chronic kidney disease. Sci. Rep. **12**(1), 2765 (2022)

Segmentation of Cytology Images to Detect Cervical Cancer Using Deep Learning Techniques

Betelhem Zewdu Wubineh[(✉)] (ID), Andrzej Rusiecki (ID), and Krzysztof Halawa (ID)

Faculty of Information and Communication Technology, Wroclaw University of Science and Technology, Wroclaw, Poland
{betelhem.wubineh,andrzej.rusiecki,krzysztof.halawa}@pwr.edu.pl

Abstract. Cervical cancer is the fourth most common cancer among women. Every year, more than 200,000 women die due to cervical cancer; however, it is a preventable disease if detected early. This study aims to detect cervical cancer by identifying the cytoplasm and nuclei from the background using deep learning techniques to automate the separation of a single cell. To preprocess the image, resizing and enhancement are adopted by adjusting the brightness and contrast of the image to remove noise in the image. The data is divided into 80% for training and 20% for testing to create models using deep neural networks. The U-Net serves as baseline network for image segmentation, with VGG19, ResNet50, MobileNet, EfficientNetB2 and DenseNet121 used as backbone. In cytoplasmic segmentation, EfficientNetB2 achieves a precision of 99.02%, while DenseNet121 reaches an accuracy of 98.59% for a single smear cell. For nuclei segmentation, Efficient-NetB2 achieves an accuracy of 99.86%, surpassing ResNet50, which achieves 99.85%. As a result, deep learning-based image segmentation shows promising result in separating the cytoplasm and nuclei from the background to detect cervical cancer. This is helpful for cytotechnicians in diagnosis and decision-making.

Keywords: Cytoplasm · Deep learning · Nuclei · Segmentation · U-Net

1 Introduction

Cervical cancer describes the development of malignant tumors of normal cells that initially covered the upper part of the cervix [1], which is the fourth most common cancer among women [2–4]. Every year, more than 200,000 women die of cervical cancer; approximately three-quarters of these deaths occur in developing countries [5], mainly due to a lack of medical resources and experts. However, it is one of the preventable diseases if detected early through screening [6]. Cytological tests, such as Pap smear, are crucial screening procedures to identify abnormalities in the cervical region [7]. To find and identify nuclear and cytoplasmic atypia, the diagnostic process requires a cellular-level examination under a microscope by a cytologist or pathologist [8]. However, manual analysis is labor-intensive, error-prone, and time-consuming [9].

© The Author(s), under exclusive license to Springer Nature Switzerland AG 2024
L. Franco et al. (Eds.): ICCS 2024, LNCS 14835, pp. 270–278, 2024.
https://doi.org/10.1007/978-3-031-63772-8_25

Furthermore, due to factors such as the shortage of pathologists and regional economic differences, manual analysis has not been able to meet the urgent needs of patients. The presence of blood clots, mucus, overlapping cells, and other types of tissue and debris are factors that determine the quality of an image [2]. These may not be clearly visible to humans to identify abnormal cells from normal ones. Improving the screening capacity is the most effective way to reduce the incidence of cancer and save lives. This helps experts in diagnosis, reduces errors and workload, and speeds up the screening process [9]. Segmentation is one of the crucial tasks in the screening process because it can help better understand the morphological properties of cells by analyzing their constituent parts, such as the nucleus and cytoplasm [10].

The physical properties of the cytoplasm and nucleus are essential to determine whether a cell is normal or abnormal. Accurate cell segmentation helps experts identify normal and malignant cells within a Pap smear. Therefore, the screening process helps to detect abnormalities early before they become malignant [11]. To automate cervical cell segmentation and improve cervical cancer detection accuracy, deep learning-based computer-aided diagnostic techniques have been used [3]. The most challenging aspect of automating cervical cell screening is the precise segmentation of the nuclei and cytoplasm. In this study, we divide the image between the background and the cytoplasm and nuclei. In image segmentation and recognition, the U-Shape Network Structure (U-Net) has been shown to be extremely superior [12].

The aim of this study is to develop and evaluate deep learning (DL) models for the image segmentation of cytoplasmic and nucleic cervical cells, aiming to simplify and automate the separation of single cells. The remainder of this paper is organized as follows: Sect. 2 introduces the methods used, Sect. 3 presents the segmentation results and discusses the findings, and Sect. 4 provides concluding remarks.

2 Materials and Methods

In this section, the data collection, preprocessing technique, implementation detail and model development are discussed.

2.1 Dataset Collection

The dataset used for this study was collected from the Pomeranian Medical University in Szczecin, Poland. It consists of 419 cytological images with a resolution of 1130 × 1130 pixels in BMP format. In most images, the nuclei of abnormal cells appear large with blurred borders and low contrast between the cytoplasm and the background, making them highly susceptible to false predictions. The images from different classes along with their corresponding masks are depicted in Fig. 1.

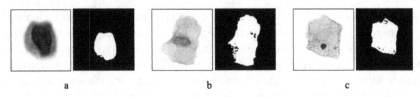

Fig. 1. Sample images with corresponding mask: a) HSIL, b) LSIL, and c) NSIL

In contrast, the nuclei of normal cervical cells are tiny and exhibit high contrast. The dataset includes images classified as high-grade squamous intraepithelial lesion (HSIL), low-grade squamous intraepithelial lesion (LSIL), and normal squamous intraepithelial lesion (NSIL), with 124, 61, and 234 images respectively.

2.2 Pre-processing of the Data

Medical images are often inconsistent, noisy, and incomplete; therefore, preprocessing becomes crucial to improve the performance of the model [13]. In the preprocessing phase, we begin by creating masks for the samples collected from the hospital. Masks are created using OpenCV. The initial step involves converting the image to grayscale. Subsequently, thresholding is applied to generate a binary mask using the Otsu method. Morphological operations, such as erosion and dilation, are performed to eliminate noise and enhance the cell boundary in the mask image. Finally, mask images are generated, with the cytoplasm represented in white and the background in black. Afterwards, the images are separated into the image data and masks (labels). Following this, the dataset is divided into a training set and a testing set with a ratio of 80% and 20%, respectively. All images are resized to a size of 224 × 224 pixels. Additionally, image enhancement techniques such as adjusting brightness and contrast are applied to reduce noise. Data augmentation is then used to improve the performance and generalization of DL models and to reduce overfitting problems. In this study, geometric transformations, such as rotation, flipping, and scaling were applied. Sample images after enhancement and augmented images are shown in Fig. 2.

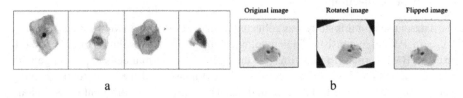

Fig. 2. Sample image a) after enhancement, b) data augmentation

2.3 Implementation Details

In this U-Net structure, the objective is to automatically extract a region of interest (ROI) around the tumor in the cervix. To train the model, it is necessary to define the experimental setup and hyperparameters. The DL framework used to implement the model is Keras with TensorFlow in Python. The experiment was conducted on a Windows 11 operating system, utilizing a 12th Gen Intel(R) Core (TM) i5-12500H processor, 32GB of RAM, and an NVIDIA GeForce RTX 4060 GPU model. The training hyperparameters used for the study are listed as follows: a batch size of 18 and training for 100 epochs without early stopping. The Adam optimizer [14] was employed with a learning rate set to 0.003. The Jaccard distance is utilized as the loss function, measuring the dissimilarity between predicted and actual sets. Minimizing the Jaccard distance aids in producing predictions that closely match the actual mask. Performance evaluation metrics for the model include accuracy, sensitivity, precision, Intersection over Union (IoU), and Dice coefficient.

2.4 Training the Model

In this section, we discuss model training using a U-shaped neural network that employs data consisting of five down-sampling modules and five up-sampling modules. The down-sampling (encoder) reduces the spatial dimension of the feature maps while increasing their depth, and the up-sampling (decoder) is to decode the encoded data, utilizing information from the concatenation, and increase the spatial dimension back to the original input size. The concatenation path helps to retain spatial information during the up-sampling process. As the image size is reduced on the encoder path, the decoder path increases the image size. The down sampling includes two convolutional layers, max pooling, batch normalization, and activation function. The filter size is 3×3 for each convolution layer and 2×2 for each max pool layer. The input image is $224 \times 224 \times 3$, and the number of filters increases in each block: 64, 128, 256, 512 and 1024. The central block has 1024 filters for 2 convolution layers. In the decoder path, transposed convolution (up sampling), concatenation, convolution layers, batch normalization and activation functions are employed. As the image size increases, the number of filters is decreased: 1024, 512, 256, 112 and 64. The final output size is $224 \times 224 \times 1$ which is the segmented image.

We then train U-Net as the base, with VGG19, ResNet50, MobileNet, Efficient-NetB2, and DenseNet121 as the backbone. These pre-trained models are employed for feature extraction and are most widely used in literature. To connect the pre-trained model to the U-Net architecture, we follow the following procedures: In the encoder, the initial layer of the pre-trained model backbones captures features and reduces the spatial dimension of the input image. The layers with reduced spatial dimension connect to the deepest layer of pre-trained backbone, used for capturing abstract features and retaining high-level semantic information while reducing the spatial dimensions. Subsequently, the decoders up-sample the feature maps and reconstruct the spatial information. The transposed convolutions increase the spatial resolution, and the features of the corresponding layer in the encoder are concatenated during the up-sampling process. The cervical cell segmentation model is shown in Fig. 3.

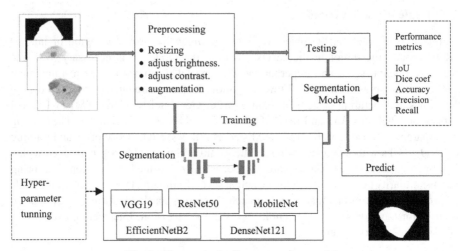

Fig. 3. The cervical cell segmentation model

In this U-Net architecture, there are two convolutional layers in each block: the first one includes a 3 × 3 convolutional layer, batch normalization, and a ReLU activation function. The second consists of a 3 × 3 convolutional layer, batch normalization, a ReLU activation function, and spatial dropout. For down-sampling and up-sampling, a filter size of 2 × 2 is used. In addition, a sigmoid activation function is applied in the last layer. These pretrained models serve as the backbone of the U-Net. The VGG 19 architecture comprises 16 convolutional layers and 3 fully connected layers, totaling 19 layers to learn weights [15]. ResNet50 can achieve a very deep network of up to 152 layers by inserting a skip connection, to pass the input from the previous layer to the next layer without altering it [16]. MobileNet is an architecture designed for mobile devices that combines efficient computation and separable convolution in depth [17] with a depth layer of 28 [18]. EfficientNet is a scaling method that uniformly scales all depth, resolution, and breadth parameters using a compound coefficient. EfficientNet substantially outperforms other convolutional networks in various tasks [19]. The baseline network, EfficientNetB0, has a subsequent network until EfficientNetB7 and EfficientNetB2 is one of these networks. DenseNet is a CNN architecture of 121 layers and is used to improve the information flow by connecting each layer to the other layer behind it. As a result, the decision is based on all layers rather than just the final layer [20]. Finally, we tested the performance of the model, which can be used to predict entire image masks and obtain relevant information about the images. This can be helpful for medical professionals who diagnose diseases and make decisions.

3   Results and Discussion

In this study, we segmented the pixels of the images into cytoplasm, nuclei, and background. Data were divided into training, validation, and testing sets. The U-Net-shaped structure served as the base model, and VGG19, ResNet50, MobileNet, EfficientNetB2, and DenseNet121 were used as backbones to segment the images. The backbones are typically used to extract the features. Accuracy, sensitivity, precision, IoU, and dice coefficient were used to evaluate performance and compare different models. Table 1 presents a comparison of the performance of cytoplasm segmentation using various approaches.

Table 1. Results of cytoplasmic segmentation

Methods	Accuracy	Sensitivity	Precision	IoU	Dice coef
U-Net	97.03	78.6	93.01	97.03	84.77
U-Net + Vgg19	98.64	93.91	92.83	97.9	80.53
U-Net + Resnet50	97.43	82.66	97.28	97.89	87.81
U-Net + MobileNet	97.67	79.75	96.61	97.18	78.52
U-Net + EfficientNetB2	98.10	83.36	99.02	97.98	88.93
U-Net + Densenet121	98.59	91.68	94.56	98.48	87.5

In Table 1, U-Net itself is used as the original network and serves as a backbone for the other pretrained architectures to segment the image. U-Net, as a base model, achieves better performance in the cervical cell segmentation task using EfficientNetB2 as the backbone, with a precision value of 99.02% and an accuracy of 98.59% in DenseNet121. The result of the segmented nuclei is shown in Table 2. In nuclei segmentation, both EficientNetB2 and ResNet50 produced impressive results with accuracies of 99.86% and 99.85%, respectively. The U-Net architecture, with pretrained models, was employed to extract crucial features related to cervical cancer. U-Net was used to segment the Pap smear images, with the aim of identifying specific areas within cervical cells such as the cytoplasm and nuclei.

Table 2. Nuclei segmentation result

Methods	Accuracy	Sensitivity	Precision	IoU	Dice coef
U-Net	99.25	80.01	95.10	99.50	84.95
U-Net + Vgg19	99.69	86.23	99.76	99.60	84.25
U-Net + Resnet50	99.85	92.07	95.72	99.77	87.20
U-Net + MobileNet	99.78	91.63	95.17	99.71	90.34
U-Net + EfficientNetB2	99.86	97.75	98.13	99.80	89.33
U-Net + Densenet121	99.82	87.51	98.66	99.76	88.09

The ground truth (actual mask) comprises three components of pixels: the background, the cytoplasm, and the nucleus. Figure 4 shows the results obtained from various classes of segmentation of the test set.

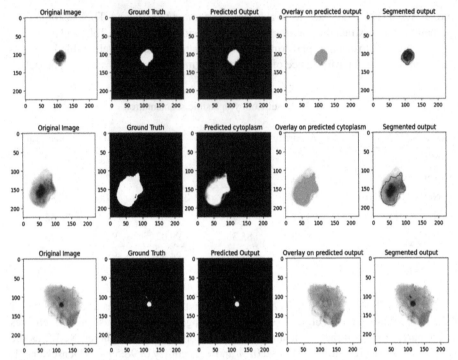

Fig. 4. Results of the test set in order of original image, actual mask, predicted mask, overlay on predicted cytoplasm/nuclei, and segmented cytoplasm/nuclei

The first two rows are the resulting cytoplasm, whereas the third is for the nuclei segmentation. The first column displays images of cervical cells. The second column presents the actual masks for the cytoplasm and nuclei, while the third column displays the model's prediction for the cytoplasm and nuclei. In the fourth column, overlays are applied to the predicted outputs, and the fifth column shows the segmented output of the cytoplasm and nuclei, respectively. The predicted column closely resembles the actual images, indicating that the model correctly separated the cytoplasm and nuclei from the background images and removed unwanted information. These results aid medical experts in identifying abnormal cells from normal ones. To fine-tune the hyperparameters, we explored different epoch values, and 100 epochs consistently produced superior outcomes.

4   Conclusions

In this study, we used the U-Net shaped structure with transfer learning algorithms as a backbone for accurate segmentation of cells in Pap smear images to detect cervical cancer in the early stages. Deep learning-based methods require large amounts of data for effective training; however, in medical imaging there is a lack of data that may impact the model's performance. When the number of images is small, transfer learning algorithms are effective in improving model performance. This involves transferring weights from a pretrained model to a specific task. We evaluated U-Net as a baseline with EfficientNetB2, which yields an accuracy of 99.02% and 99.86% for the cytoplasm and nuclei, respectively. Furthermore, DenseNet121 is the basis for this new dataset, achieving a precision value of 99.02% and an accuracy value of 98.85% in Resnet50. These results are promising for segmenting cervical cells to identify the cytoplasm and nuclei of the background. The result of this study is helpful for radiologists in making decisions in the cervical screening system.

References

1. Xia, M., Zhang, G., Mu, C., Guan, B., Wang, M.: Cervical cancer cell detection based on deep convolutional neural network. In: Chinese Control Conference, CCC, July 2020, pp. 6527–6532

2. Del Moral-Argumedo, M.J., Ochoa-Zezzati, C.A., Posada-Gómez, R., Aguilar-Lasserre, A.A.: A deep learning approach for automated cytoplasm and nuclei cervical segmentation. Biomed. Signal Process. Control **2023**(81), 104483 (2022)

3. Youneszade, N., Marjani, M., Pei, C.P.: Deep learning in cervical cancer diagnosis: architecture, opportunities, and open research challenges. IEEE Access **2023**(11), 6133–6149 (2022)

4. Alisha, S., Vinitha Panicker, J.: Cervical cell nuclei segmentation on pap smear images using deep learning technique. In: 2022 IEEE 3rd Global Conference for Advancement in Technology, GCAT 2022, pp. 1–5

5. Nguendo, Y.H.B., Tchinda, F.C.: Descriptive epidemiology of uterine cervix cancer at the medical oncology unit of the Yaoundé general hospital-Cameroon. GSC Biol. Pharmaceut. Sci. **9**(1), 083–091 (2019)

6. Zeleke, S., Anley, M., Kefale, D., Wassihun, B.: Factors associated with delayed diagnosis of cervical cancer in Tikur Anbesa specialized hospital, Ethiopia, 2019: Cross-sectional study. Cancer Manag. Res. **13**, 579–585 (2021)

7. Harangi, B., Toth, J., Bogacsovics, G., Kupas, D., Kovacs, L., Hajdu, A.: Cell detection on digitized pap smear images using ensemble of conventional image processing and deep learning techniques. In: International Symposium on Image and Signal Processing and Analysis, ISPA, Sept 2019, pp. 38–42

8. Wan, T., Xu, S., Sang, C., Jin, Y., Qin, Z.: Accurate segmentation of overlapping cells in cervical cytology with deep convolutional neural networks. Neurocomputing **365**, 157–170 (2019)

9. Li, G., Sun, C., Xu, C., Zheng, Y., Wang, K.: Cervical cell segmentation method based on global dependency and local attention. Appl. Sci. **12**(15) (2022)

10. Conceição, T., Braga, C., Rosado, L., Vasconcelos, M.J.M.: A review of computational methods for cervical cells segmentation and abnormality classification. Int. J. Mole. Sci. **20**(20) (2019)

11. Shanthi, P.B., Hareesha, K.S., Kudva, R.: Automated detection and classification of cervical cancer using pap smear microscopic images: a comprehensive review and future perspectives. Eng. Sci. **19**, 20–41 (2022)

12. Zhou, W., Chen, F., Zong, Y., Zhao, D., Jie, B., Wang, Z., et al.: Automatic detection approach for bioresorbable vascular scaffolds using a U-shaped convolutional neural network. IEEE Access **7**, 94424–94430 (2019)

13. Bnouni, N., Amor, H.B., Rekik, I., Rhim, M.S., Solaiman, B., Essoukri, N., et al.: Boosting CNN Learning by Ensemble Image Preprocessing Methods for Cervical Cancer, 2021, pp. 264–269

14. Park, J., Yang, H., Roh, H.J., Jung, W., Jang, G.J.: Encoder-weighted W-net for unsupervised segmentation of cervix region in colposcopy images. Cancers **14**(14) (2022)
15. Sharma, N., Gupta, S., Koundal, D., Alyami, S., Alshahrani, H., Asiri, Y., et al.: U-Net model with transfer learning model as a backbone for segmentation of gastrointestinal tract. Bioengineering **10**(1) (2023)
16. Ji, Q., Huang, J., He, W., Sun, Y.: Optimized deep convolutional neural networks for identification of macular diseases from optical coherence tomography images. Algorithms **12**(3), 1–12 (2019)
17. Widiansyah, M., Rasyid, S., Wisnu, P., Wibowo, A.: Image segmentation of skin cancer using MobileNet as an encoder and linknet as a decoder. J. Phys. Conf. Ser. **1943**(1) (2021)
18. Howard, A.G., Zhu, M., Chen, B., Kalenichenko, D., Wang, W., Weyand, T., et al.: MobileNets: Efficient Convolutional Neural Networks for Mobile Vision Applications, 2017 [online]. Available http://arxiv.org/abs/1704.04861
19. Waziry, S., Wardak, A.B., Rasheed, J., Shubair, R.M., Yahyaoui, A.: Intelligent facemask coverage detector in a world of chaos. Processes **10**(9), 1–12 (2022)
20. Gottapu, R.D., Dagli, C.H.: DenseNet for anatomical brain segmentation. Procedia Comput. Sci. **140**, 179–185 (2018)

Federated Learning on Transcriptomic Data: Model Quality and Performance Trade-Offs

Anika Hannemann[1,2(✉)], Jan Ewald[2], Leo Seeger[2], and Erik Buchmann[1,2]

[1] Department of Computer Science, Leipzig University, Leipzig, Germany
[2] Center for Scalable Data Analytics and Artificial Intelligence (ScaDS.AI)
Dresden/Leipzig, Leipzig University, Leipzig, Germany
{hannemann,ewald,seeger,buchmann}@cs.uni-leipzig.de

Abstract. Machine learning on large-scale genomic or transcriptomic data is important for many novel health applications. For example, precision medicine tailors medical treatments to patients on the basis of individual biomarkers, cellular and molecular states, etc. However, the data required is sensitive, voluminous, heterogeneous, and typically distributed across locations where dedicated machine learning hardware is not available. Due to privacy and regulatory reasons, it is also problematic to aggregate all data at a trusted third party. Federated learning is a promising solution to this dilemma, because it enables decentralized, collaborative machine learning without exchanging raw data.

In this paper, we perform comparative experiments with the federated learning frameworks TensorFlow Federated and Flower. Our test case is the training of disease prognosis and cell type classification models. We train the models with distributed transcriptomic data, considering both data heterogeneity and architectural heterogeneity. We measure model quality, robustness against privacy-enhancing noise, computational performance and resource overhead. Each of the federated learning frameworks has different strengths. However, our experiments confirm that both frameworks can readily build models on transcriptomic data, without transferring personal raw data to a third party with abundant computational resources.

Keywords: Federated Learning · Cell Type Classification · Disease Prognosis

1 Introduction

Machine learning has the potential for a paradigm shift in healthcare, towards medical treatments based on individual patient characteristics [9,32]. For example, precision medicine uses biomarkers, genome, cellular and molecular data, and considers the environment and the lifestyle of patients [17,19]. Machine learning on large scale genomic and transcriptomic data enables the identification of disease subtypes, prediction of disease progression and selection of targeted therapies. Therefore, models need to be trained on large, diverse patient cohorts (sample size) with high-resolution genetic characterization (number of features).

© The Author(s), under exclusive license to Springer Nature Switzerland AG 2024
L. Franco et al. (Eds.): ICCS 2024, LNCS 14835, pp. 279–293, 2024.
https://doi.org/10.1007/978-3-031-63772-8_26

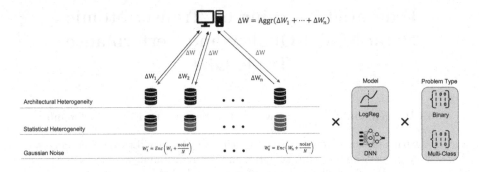

Fig. 1. Key challenges of Federated Learning.

This is challenging: The data is commonly distributed across multiple health-care institutions that may not possess the high-performance computing resources needed to build large deep-learning models. The sensitive nature of genomic and transcriptomic data presents privacy challenges. Genomic mutations and markers could even allow to re-identify individuals and their relatives [27]. This disallows to freely share such data with centralized aggregators.

Federated learning (FL), as shown in Fig. 1, allows for decentralized model training across multiple data sets without requiring to transfer sensitive raw data [24,29]. Only model parameters are exchanged between the aggregator and participating clients, and the overall computational burden is effectively shared. Adding noise [3,5,8] to shared parameters could further increase privacy.

In this paper, we investigate the technical and conceptual challenges that arise when implementing FL on transcriptomic data. Figure 1 illustrates our four key challenges: *Architectural Heterogeneity* refers to different numbers of clients with varying computational capabilities. *Statistical Heterogeneity* relates to data distributions and sizes. *Gaussian Noise* addresses the impact of applying noise to the data, e.g., to achieve Differential Privacy, with different models (Logistic Regression and Sequential Deep Learning) and problem types (Binary and Multi Label). Finally, *Resource Consumption* addresses storage, communication overhead and training times. To explore these challenges, we have conducted comparative experiments with two state-of-the-art FL frameworks – TensorFlow Federated (TFF) and Flower (FLWR) – and transcriptomic data. In particular, we make the following contributions:

- We train disease prognosis and cell type classification models with TFF and FLWR using hyperparameter tuning, and we measure the model quality.
- We analyze the effects of the number of clients, the amount of training data and data distribution on the global model quality.
- We measure the impact of Gaussian noise, locally applied to the weights, on the global model quality.
- We compare memory consumption, run-times and network traffic of TFF and FLWR from both the client's and the aggregator's perspective.

We have demonstrated that FL frameworks can be readily applied to precision medicine applications. Even more, we obtained an excellent global model with an AUC of up to 0.98 for disease prognosis and cell type classification with transcriptomic data. This performance is robust in the presence of diminishing data quality, increasing clients and diverse data distributions, and it reduces the necessary computational resources for the individual medical institution.

Paper Structure: Section 2 introduces related work. Section 3 describes our methodology. Section 4 presents our experimental results. Section 5 concludes.

2   Related Work

This section reviews related work in the fields of FL and its applications in precision medicine, and discusses technical challenges and FL frameworks.

Federated Learning and Applications. FL [24] allows distributed clients $N_1, ..., N_n$ to collaboratively train a model M_{glob} without sharing raw data. Instead of centralizing the data sets like in traditional machine learning, each client N_i trains a local model M_i on its own data set. The model parameters are then shared with a central aggregator, which aggregates them to create a global model M_{glob}. This process is repeated as more data is collected, with clients continuously updating their local models and forwarding the updated parameters to the aggregator. Raw data remains with the respective clients and is never transmitted.

FL is used for collaboration across medical institutions, hospitals, health care insurers or other entities [4,10,20,34]. Some problems relevant to medicine were addressed by [9,32] who trained a federated model to model Alzheimer's and Parkinson's disease. Beguier *et al.* [5] present a differentially private and federated cancer occurrence prediction based on genomic data. For practical implications and benchmarking of frameworks for FL, however, there is less literature available. The multi-class data set we use in experiments [16], to our knowledge, has not been modeled by any FL system. The authors of the binary data set propose a collaborative learning method named swarm learning without an aggregator [30]. This method achieves very good model quality, but does not take into account many challenges that arise when bringing FL in to production. We identified four major challenges with heterogeneity in data distribution, participating clients, consumption of computational resources and privacy:

Technical and Conceptual Challenges. The issue of **statistical heterogeneity** of training data in FL encompasses both the distribution and size of the data. This challenge involves dealing with the non-IID (non-independent and identically distributed) nature of distributed data, which can lead to skewed or biased model training [18,26]. Additionally, the size of data sets of each client can vary significantly, where smaller data sets may not adequately represent the population, impacting model quality and generalizability. Fu *et al.* [12] showed that not only data heterogeneity can influence the model's quality, but also the varying number of clients which we call **architectural heterogeneity**. Navigating these aspects of statistical heterogeneity is crucial for ensuring the robustness and efficacy of FL models. In the context of transcriptomic data and health-care institutions both issues are very common since hospitals vary greatly in their sizes and specialization or different laboratories introduce bias due to small differences in sequencing protocols and machinery [29].

Furthermore, while FL is designed to enhance **privacy** by training models locally and sharing only model updates, ensuring the privacy and security of these updates against potential inference attacks remains a critical concern. Multiple works showed, that there is no formal privacy guarantee for FL without additional privacy-enhancing techniques [18]. Recent publications [7,33] showed that baseline FL is vulnerable to reconstruction attacks, while others [13,25] successfully performed Membership Inference Attacks (MIA). Multiple works [3,5,8] explored Differential Privacy in a FL scenario for medical data. Differential Privacy protects the privacy of individual data points in a training data set while allowing ML models to benefit from the overall information. It adds controlled noise to data or model parameters, making it difficult to infer specifics about individual entries. However, the application of noise usually comes with information loss. For transcriptomic data in combination with clinical patient data, this is a dilemma since biomarker signals are often weak for multi-factorial diseases. Hence, privacy levels need to be carefully chosen to find a trade-off between model quality, which is highly critical in medical applications, and privacy.

Finally, **resource consumption** presents another challenge for FL. The diverse and potentially resource-limited nature of participating clients in a FL network can lead to inefficiency and delay [12,18]. Furthermore, the communication required for model updates and synchronization in FL adds to network bandwidth demands, which can be a bottleneck in resource-limited environments.

FL Frameworks. The field of federated learning is rapdily evolving, and there are many existing open-source FL frameworks, such as TensorFlow Federated (TFF) [2,14], Flower (FLWR) [6], PySyft [35], FATE [22] and FedML [15]. These frameworks vary in terms of their features, ease of use, and specific use cases. The choice of a federated learning framework typically depends on the specific requirements and constraints of the application.

Both PySyft and TFF are well established and benefit from a large community support. While TFF is based on the TensorFlow ecosystem, PySyft works primarily with PyTorch. Both PySyft and FATE provide multiple optional privacy-enhancing methods such as Differential Privacy and Secure Multi Party Computation. FLWR is designed to be framework-agnostic and can work with various machine learning frameworks, including TensorFlow, PyTorch, and others. In terms of abstraction level, Flower's API is more high-level and is, therefore, supposed to be more user friendly than TFF and PySyft. While FLWR and FATE only allow simulation and cluster deployment, FedML provides a flexible and generic API and allows on-device training. Also, FedML can be used for various network topologies such as Split Learning, Meta FL and Transfer-Learning.

3   Methodology

This section explains our experimental concept, the data sets we used, and the architectures of the machine learning models for our experiments.

3.1   Experimental Concept

To quantify the impact of our four key challenges on FL with transcriptomic data, we measure the quality of a global model obtained by FL on centralized data first. With this baseline, we conduct comparative experiments to explore the effects of *Architectural* and *Statistical Heterogeneity* on the model quality. We explore the impact of *Gaussian noise* to enhance privacy and measure its effect on the model. Furthermore, we assess the *Resource Consumption* at the aggregator and the clients.

To draw robust conclusions, we vary problem type and model architecture. In particular, we conduct experiments not only on a binary-labeled data set but also on a multi-class data set. This aligns with clinical research, which typically covers a variety of diseases and research questions rather than a single condition. In multi-class problems, the complexity increases as the model must differentiate between multiple, often overlapping conditions.

We decided to use the FL frameworks **TensorFlow Federated** (TFF) [2,14] and **Flower** (FLWR) [6]. Our choice was driven by multiple factors: We prioritize documentation and usability. Furthermore, we are interested in exploring horizontal FL with frameworks that have programming interfaces at different levels of abstraction. Finally, frameworks with the same communication protocol allow to compare the network performance.

3.2   Data Sets

We use two data sets. The Acute Myeloid Leukemia data set [31] was previously obtained from 105 studies, resulting in 12,029 samples with binary labels. We call it the **binary-labeled data set**. It consists of gene expressions by microarray and RNA-Seq technologies from peripheral blood mononuclear cells (PBMC) of patients with either a healthy condition or acute myeloid leukemia (AML).

The **multi-class data set** includes expression profiles generated by single-cell RNA-Seq for cell types of the human brain, in particular the middle temporal gyrus (MTG). The data set was published by [16], who isolated sample nuclei from eight donors and generated gene expression profiles by single-cell RNA-Seq for a total of 15928 cells (samples) describing 75 distinct cell subtypes. We reduced the number of classes (cell types) to make the data set more suitable to experiment with class imbalance. For that we selected only the five most abundant cell types (classes) leading to 6931 cells (samples) for training. We preprocess both data sets as in previous analyses and benchmarks [16,17,30,31].

3.3   Model Architectures

We experiment with a **logistic regression model** and a **deep-learning model**. Following [30], the deep-learning model uses a sequential neural network architecture. It consists of a series of dense layers, each with 256, 512, 128, 64, and 32 units, all activated using the 'relu' activation function. Dropout layers with dropout rates of 0.4 and 0.15 prevent overfitting. The configuration of the output layer is based on the number of classes in the data set.

Table 1. Optimized Hyperparameters

Data Set	Model	Hyperparameters
Binary	Deep Learning	Adam, L2: 0.005, Epochs: 70
	Logistic Regression	SGD, L2: 0.001, Epochs: 8
Multi Class	Deep Learning	Adam, L2: 0.005, Epochs: 30
	Logistic Regression	SGD, L2: 1.0, Epochs: 10

For our baseline and for hyperparameter tuning, we train both models on centralized data. In particular, the hyperparameter space was randomly searched to find Cross Entropy as optimal loss function with a batch size of 512. Hyperparameters that differ in the respective combinations of model and data are summarized in Table 1. We denote the rounds of training based on the local epochs, so that the total number of epochs remains a constant. Assume one round of training and two clients using 100 local training epochs. With two rounds of training, this would be 50 local epochs for both clients.

4   Experiments

In this section, we describe our experimental setup and our analysis results regarding model quality, data quality and resource consumption.

4.1   Experimental Setup

We perform all experiments using one CPU core from an AMD(R) EPYC(R) 7551P@ 2.0 GHz - Turbo 3.0 GHz processor and 31 Gigabyte RAM for each client. The network is a 100 Gbit/s Infiniband. We measure the network traffic with tshark [1]. No GPU is used during the experiments. To ensure resource parity among different frameworks and the central model, each training process is bound exclusively to one CPU core. Our experiments are implemented in Python. For preprocessing and data loading, we used the libraries Pandas [23] and Scikit-learn [28]. Both the logistic regression model and the deep-learning model were implemented using Keras, with default settings and federated algorithms from FLWR and TFF [2,6,14]. The default of FLWR is a federated averaging strategy in a client-aggregator setup. Additionally, we compute the average score of each metric for every client. The default building function in TFF uses a robust aggregator without zeroing and clipping of values as model aggregator. The clients of TFF all use the eager executor of TFF and are loading the data with a custom implementation of the data interface of TFF [2]. In this configuration, FLWR and TFF implement the federated averaging algorithm with a learning rate set to 1.0 [24]. The code to our experiments can be found at [21].

For our experiments, we tested combinations of Logistic Regression (LogReg) and Sequential Deep Learning (DL) models together with binary problems (Binary) and multi-label problems (Multi). In the figures, we abbreviate the combinations of models and problem types as Binary LogReg, Multi LogReg, Binary DL and Multi DL. For each combination, we tested 3, 5, 10, 50 clients and 1, 2, 5, 10 rounds of training, and we measured model quality and computational resources used. To analyze the effect under investigation, we iterated over the respective other parameter and reported the averaged result, i.e., when varying the number of clients, we conducted tests for each training round configuration and presented the cumulative effect observed across all training rounds.

4.2 Model Quality

To explore the impact of heterogeneity on the global model, we use a 5-fold cross validation and compute the Area Under the Curve (AUC). A higher AUC value (closer to 1) indicates a better model, that distinguishes between the classes more effectively. We compare the AUC of the centralized baseline with the AUC obtained with FL and varying numbers of clients and training rounds. In the following, we explain our key findings.

Finding 1: Boosting Training Rounds Does Not Always Enhance Model Quality. We assumed from previous work (cf. Sect. 2) that an increasing number of training rounds improves the quality of the global model. To examine the impact of frequency of weight updates among clients during training, we varied the numbers of training rounds and kept the total number of epochs constant. This approach is consistent with our round configurations optimized with hyperparameter tuning, as described in Sect. 3.

Consider Fig. 2). The left two columns of diagrams show the results of our experiments with varying numbers of rounds, the right ones varying numbers of clients over the respective other variable. The top line of diagrams were obtained with deep learning models, the bottom line with logistic regression. We find that the AUC of the global model does not improve significantly with the number of rounds for logistic regression. Deep learning benefits slightly with more rounds of weight updates (see Fig. 2). For non-balanced data sets, matching most real-world FL scenarios, updating rounds prove to be more effective to mitigate class-imbalance across different clients (see Fig. 3). Thus, healthcare institutions should carefully chose the number of rounds in FL applications, based on the data distribution and machine learning algorithms used.

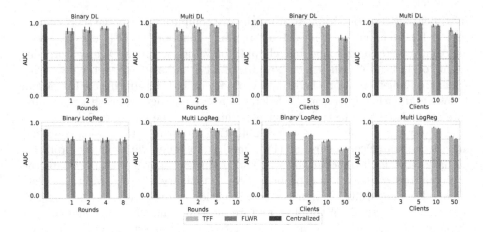

Fig. 2. AUC with respect to an increasing training rounds and clients

Finding 2: Fewer Clients and More Data Increases Model Quality.
To analyze the effect of the number of clients, we split the training data in
disjoint subsets and distributed them to clients. Each subset has the same size
and class distribution as the whole data set. The data on each client is then split
into training and test data with a ratio of 80:20. The test data of all clients is
combined and the aggregated FL model is evaluated on this test data. Then,
both FL frameworks were used to train a global model.

For a small number of clients, Fig. 2 reports that the AUC of FL is similar to
our baseline (AUC 0.98), proving that FL can reach centralized model quality
in real-world scenarios. A slight decrease for 10 clients can be discovered which
is followed by a drop in AUC in the extreme case of 50 clients. This results from
the reduced size of local training data: As the number of clients increases, the
data set is divided among them, resulting in a reduction in the training data
available for local training. The limited data size does indeed reflect a possible
scenario where small healthcare institutions want to attend to a FL scenario.
We conclude that clients should only allowed when they provide a substantial
number of samples on which local model training can be performed.

Finding 3: Model Quality Is Driven by Models, Not by Frameworks.
Figure 2 shows that the logistic regression model has an overall lower model
quality than the deep learning model for the multi-class problem, as its highest
AUC value is 0.90 for both frameworks. In contrast, the deep learning model has
an AUC of 0.99 for TFF and 0.98 for FLWR. This emphasizes the superiority
of deep learning for this data set, and underlines that the model (and their
hyperparameterisation) must fit to the problem.

Benchmarks for a direct comparison of model quality are rare, especially for
transcriptomic data. We wanted to learn if there is a difference between FLWR
and TFF across the tested scenarios. As Fig. 2 illustrates, if all parameters and
aggregation algorithms for FLRW and TFF are configured in the same way,

both frameworks deliver a similar model quality. This observation holds for various configuration. Healthcare institutions are therefore free to select FL frameworks based on functionality (e.g. privacy-support), usability and computational resource demand, instead of concerning model quality.

Finding 4: Class Imbalance Impairs Federated Learning. A challenge for FL is that data points are usually not independent and identically distributed (IID), leading to statistical heterogeneity. With transcriptomic data, a balanced class distribution among all clients seems unlikely. Therefore, we explored the effect of data imbalance on the global model's quality. We investigate IID- and non-IID-data by methodically increasing the imbalance to find a sweet spot of imbalance and model quality.

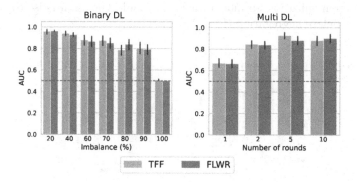

Fig. 3. AUC with respect to an increasing class imbalance and training rounds over all imbalance configurations

We start our experiments with a number of clients equal to the number of classes, and all classes are equally distributed among the clients. Subsequent experiments increase the number of samples from one class while reducing the number of samples from another class. This process is repeated independently for each client and class, until each client contains samples from a single class. Due to lack of space, we present a visualization of only a subset of the conducted experiments (Fig. 3). Detailed visualizations and results can be found at [21].

We compare the resulting AUC with an equally-distributed data set. As Fig. 3 shows, deep learning can indeed fight class imbalance. However, if the imbalance exceeds a certain threshold, a sudden drop in AUC can be expected. The threshold depends on the data set and machine learning algorithm. Our results indicate that a non-IID distribution not necessarily results in poor model quality. But, the model quality in the presence of non-IID is strongly dependent on the problem type and data set. This can be further increased with the appropriate model selection. As Fig. 3 shows, deep learning is robust with a drop in AUC at 90% imbalance. Whereas more training rounds does not improve the robustness for logistic regression, it does for deep learning. We also investigated the effect of multiple rounds to a non-IID setting. Again, we increased the training rounds over all configurations in data distributions and report the average. The increase in a number of training rounds leads to an improvement of model quality for deep learning with non-IID data (see Fig. 3), but does not affect the quality of logistic regression, regardless of the problem type [21]. Thus, healthcare institutions need to consider that the model quality depends on minimal number of samples per class. This number depends on aspects like the training algorithm.

4.3 Data Quality and Privacy

There are many anonymization approaches such as Differential Privacy, which apply noise to the data. Because TFF and FLWR have different levels of support for Differential Privacy, we have chosen a general approach to investigate the impact of anonymization on model quality: We add Gaussian noise to the local parameters before aggregation (Fig. 4).

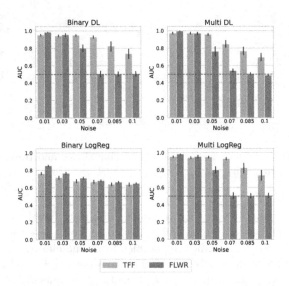

Fig. 4. AUC with respect to an increasing local Gaussian noise

Finding 5: Adding Noise to the Weights Has a Strong Impact on Model Quality. Our experiments use five clients and varying noise parameters 0.01, 0.03, 0.05, 0.07, 0.085, 0.1. We observe that all model and problem types deal with some noise. However, at some point AUC drops to approx. 0.07–0.085. TFF copes with noise better than FLWR, possibly because TFF uses regularization techniques that mitigate the impact of noisy updates. Increasing the training rounds does not improve the model quality, but slightly decreases AUC. Thus, if healthcare institutions want to apply differential privacy, sophisticated approaches are required to ensure model quality.

4.4   Computational Resources

When analyzing the computational resources of a federated system, both the local and the global perspective are relevant. We investigate the aggregated training time, memory consumption and network traffic for the clients and the aggregator. Again, we present our main findings and refer to [21] for detailed results.

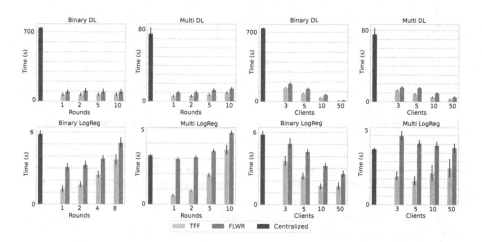

Fig. 5. Local training time with respect to increasing training rounds and clients

Finding 6: FL Does Not Increase Individual Training Times. We measure the training time for each client individually over multiple training rounds. Recall that we keep the total number of training epochs and the total number of samples constant, i.e., the more clients, the fewer local training rounds and the smaller the sample sizes per individual client. Thus, we assume that more clients result in smaller training times per client. Figure 5 confirms this. The figure is organized in the same way as Fig. 2, i.e., models in rows and problem types in columns.

TFF provides a much faster local training compared to FLWR, because of differences in their implementation. The difference between centralized training and federated training is less distinct for logistic regression than for training deep learning models.

Thus, healthcare institutions benefit more from federated training for complex machine learning approaches with long training times such as deep learning models. From a global perspective, the increased network traffic (see Fig. 7) might slightly increase the total training times. This depends on the number of round and clients, and the network capacity and latency of the coordinator.

Finding 7: Memory Consumption Is Effectively Shared. We measured the aggregated memory consumption for both the clients and the overall system. As the number of clients or rounds increases, the overall resource consumption

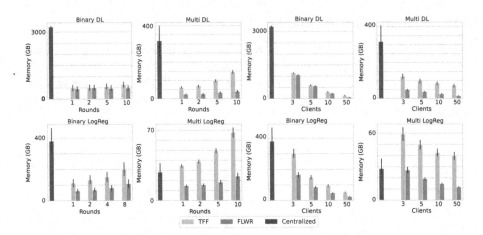

Fig. 6. Local memory usage with respect to increasing training rounds and clients

increases due to the increase in coordination effort. However, the client-wise resource consumption decreases, which is an advantage for healthcare institutions.

We observed that both the clients and the global FL system as a whole require more memory with deep learning than with logistic regression. This was expected, because deep learning uses much more parameters than logistic regression. Second, TFF uses much more memory than FLWR (Fig. 6, again with models in rows and problem types in columns). We conclude that FL saves healthcare institutions a significant amount of memory, at the cost of a slight increase in global training time and global memory requirements. Further, FL frameworks show distinct differences in their training times and memory consumption revealing potential for optimization of FL tools.

Finding 8: The Network Load Is Not a Bottleneck. To assess the network load, we assume that the amount of data transmitted and received is determined by the data serialization method of the framework, and is not influenced by hardware or interference from other clients. Therefore, we experiment with a fixed number of 10 clients. In accordance to the increased memory usage of TFF, TFF comes with higher network traffic as well. Additionally, since deep learning needs to share more parameters than logistic regression, the network traffic rises from 4 MB (peak for LogReg) up to 30 MB (peak for DL).

For comparison, an average household has a bandwidth of 209 Mbps and can easily handle the network demands [11]. We conclude that the network traffic is acceptable for healthcare institutions, and may pose a problem only for the training of very large neural networks such as foundation models. The frameworks have different demands on computing resources, but this is not a basis for selection in most medical scenarios, and it does not restrict the applicability of FL on transcriptomic data.

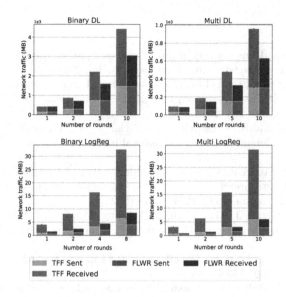

Fig. 7. Network traffic with respect to an increasing number of rounds

5   Conclusions

We have analyzed the challenges of applying Federated Learning to transcriptomic data regarding architectural heterogeneity, statistical heterogeneity, Gaussian noise and resource consumption. This is important for application areas such as precision medicine, where sensitive patient data is distributed among clients, which do not possess the computational resources for traditional machine learning approaches. In particular, we tested two real-world data sets and use-cases with varying numbers of training rounds and clients, and we compared a centralized baseline with two FL frameworks.

Our analysis shows, that for multi-factorial problems and high number of features, deep learning models outperform logistic regression models in terms of model quality. Increasing the number of training rounds does not greatly improve the global model quality, showing the high effectiveness of weigh aggregation in FL. However, hyperparameter tuning has a large impact. Transcriptomic data is robust to some class imbalance, especially using deep learning models. Problem type and data set are key factors for robustness, also with respect to the amount of training data. Privacy-preserving Gaussian noise can lead to a drastic loss in model quality. FL saves memory and training time for individual clients: The more clients, the lower the individual load. Flower consumes less computational resources than TensorFlow Federated, requires less knowledge about FL, but is also less customizable. The network traffic we measured seems acceptable for typical health institutions. Finally, our findings confirm that FL is applicable and beneficial for disease prognosis and cell type classification using transcriptomic data.

References

1. tshark: command line network protocol analyzer (2024). https://www.wireshark. org/docs/man-pages/tshark.html. Accessed 20 Jan 2024
2. Abadi, M., et al.: {TensorFlow}: a system for {Large-Scale} machine learning. In: 12th USENIX Symposium on Operating Systems Design and Implementation (OSDI 16), pp. 265–283 (2016)
3. Adnan, M., Kalra, S., Cresswell, J.C., Taylor, G.W., Tizhoosh, H.R.: Federated learning and differential privacy for medical image analysis. Sci. Rep. **12**(1), 1953 (2022)
4. Antunes, R.S., André da Costa, C., Küderle, A., Yari, I.A., Eskofier, B.: Federated learning for healthcare: Systematic review and architecture proposal. ACM Trans. Intell. Syst. Technol. (TIST) **13**(4), 1–23 (2022)
5. Beguier, C., Terrail, J.O.D., Meah, I., Andreux, M., Tramel, E.W.: Differentially private federated learning for cancer prediction. arXiv preprint arXiv:2101.02997 (2021)
6. Beutel, D.J., et al.: Flower: a friendly federated learning research framework. arXiv preprint arXiv:2007.14390 (2020)
7. Boenisch, F., Dziedzic, A., Schuster, R., Shamsabadi, A.S., Shumailov, I., Papernot, N.: When the curious abandon honesty: federated learning is not private. In: 2023 IEEE 8th European Symposium on Security and Privacy (EuroS&P), pp. 175–199. IEEE (2023)
8. Choudhury, O., et al.: Differential privacy-enabled federated learning for sensitive health data. arXiv preprint arXiv:1910.02578 (2019)
9. Danek, B.P., et al.: Federated learning for multi-omics: a performance evaluation in Parkinson's disease. bioRxiv (2023)
10. Dayan, I., et al.: Federated learning for predicting clinical outcomes in patients with covid-19. Nat. Med. **27**(10), 1735–1743 (2021)
11. Fair Internet Report: Internet in the USA - stats and figures. https:// fairinternetreport.com/United-States. Accessed 19 Dec 2023
12. Fu, L., Zhang, H., Gao, G., Zhang, M., Liu, X.: Client selection in federated learning: principles, challenges, and opportunities. IEEE Internet Things J. (2023)
13. Ganju, K., Wang, Q., Yang, W., Gunter, C.A., Borisov, N.: Property inference attacks on fully connected neural networks using permutation invariant representations. In: Proceedings of the 2018 ACM SIGSAC Conference on Computer and Communications Security, pp. 619–633 (2018)
14. Google: TensorFlow federated: machine learning on decentralized data, 25 November 2023. https://www.tensorflow.org/federated
15. He, C., et al.: FedML: a research library and benchmark for federated machine learning. arXiv preprint arXiv:2007.13518 (2020)
16. Hodge, R.D., et al.: Conserved cell types with divergent features in human versus mouse cortex. Nature **573**(7772), 61–68 (2019)
17. Hodson, R.: Precision medicine. Nature **537**(7619), S49 (2016)
18. Kairouz, P., et al.: Advances and open problems in federated learning. Found. Trends® Mach. Learn. **14**(1–2), 1–210 (2021)
19. Kosorok, M.R., Laber, E.B.: Precision medicine. Ann. Rev. Stat. Its Appl. **6**, 263–286 (2019)
20. Lee, G.H., Shin, S.Y.: Federated learning on clinical benchmark data: performance assessment. J. Med. Internet Res. **22**(10), e20891 (2020)

21. Seeger, L.: Benchmarking federated (2024). https://github.com/leoseg/BenchmarkingFederated. Accessed 20 Jan 2024

22. Liu, Y., Fan, T., Chen, T., Xu, Q., Yang, Q.: FATE: an industrial grade platform for collaborative learning with data protection. J. Mach. Learn. Res. **22**(1), 10320–10325 (2021)

23. McKinney, W., et al.: Data structures for statistical computing in python. In: Proceedings of the 9th Python in Science Conference, vol. 445, pp. 51–56, Austin, TX (2010)

24. McMahan, B., Moore, E., Ramage, D., Hampson, S., Arcas, B.A.: Communication-efficient learning of deep networks from decentralized data. In: Artificial Intelligence and Statistics, pp. 1273–1282. PMLR (2017)

25. Melis, L., Song, C., De Cristofaro, E., Shmatikov, V.: Exploiting unintended feature leakage in collaborative learning. In: 2019 IEEE Symposium on Security and Privacy (SP), pp. 691–706. IEEE (2019)

26. Mendieta, M., Yang, T., Wang, P., Lee, M., Ding, Z., Chen, C.: Local learning matters: rethinking data heterogeneity in federated learning. In: Proceedings of the Conference on Computer Vision and Pattern Recognition, pp. 8397–8406 (2022)

27. Oestreich, M., Chen, D., Schultze, J.L., Fritz, M., Becker, M.: Privacy considerations for sharing genomics data. EXCLI J. **20**, 1243 (2021)

28. Pedregosa, F., et al.: Scikit-learn: machine learning in Python. J. Mach. Learn. Res. **12**, 2825–2830 (2011)

29. Pfitzner, B., Steckhan, N., Arnrich, B.: Federated learning in a medical context: a systematic literature review. ACM Trans. Internet Technol. (TOIT) **21**(2), 1–31 (2021)

30. Warnat-Herresthal, S., et al.: Swarm learning for decentralized and confidential clinical machine learning. Nature **594**(7862), 265–270 (2021)

31. Warnat-Herresthal, S., et al.: Scalable prediction of acute myeloid leukemia using high-dimensional machine learning and blood transcriptomics. iScience **23**(1) (2020)

32. Wu, J., et al.: Integrating transcriptomics, genomics, and imaging in Alzheimer's disease: a federated model. Front. Radiol. **1**, 777030 (2022)

33. Zhao, J.C., Sharma, A., Elkordy, A.R., Ezzeldin, Y.H., Avestimehr, S., Bagchi, S.: Secure aggregation in federated learning is not private: leaking user data at large scale through model modification. arXiv preprint arXiv:2303.12233 (2023)

34. Zhou, J., et al.: PPML-Omics: a privacy-preserving federated machine learning method protects patients-privacy in omic data. bioRxiv (2022)

35. Ziller, A., et al.: PySyft: a library for easy federated learning. In: Rehman, M.H., Gaber, M.M. (eds.) Federated Learning Systems. SCI, vol. 965, pp. 111–139. Springer, Cham (2021). https://doi.org/10.1007/978-3-030-70604-3_5

Visual Explanations and Perturbation-Based Fidelity Metrics for Feature-Based Models

Maciej Mozolewski$^{(\boxtimes)}$ [ID], Szymon Bobek[ID], and Grzegorz J. Nalepa[ID]

Jagiellonian Human-Centered AI Lab, Mark Kac Center for Complex Systems Research, Institute of Applied Computer Science, Faculty of Physics, Astronomy and Applied Computer Science, Jagiellonian University, profession Stanisława Łojasiewicza 11 Street., 30-348 Krakow, Poland
m.mozolewski@doctoral.uj.edu.pl,
{szymon.bobek,grzegorz.j.nalepa}@uj.edu.pl

Abstract. This work introduces an enhanced methodology in the domain of *eXplainable Artificial Intelligence* (*XAI*) for visualizing local explanations of black-box, feature-based models, such as LIME and SHAP, enabling both domain experts and non-specialists to identify the segments of *Time Series* (*TS*) data that are significant for machine learning model interpretations across classes. By applying this methodology to electrocardiogram (*ECG*) data for anomaly detection, distinguishing between healthy and abnormal segments, we demonstrate its applicability not only in healthcare diagnostics but also in predictive maintenance scenarios. Central to our contribution is the development of the *AUC Perturbational Accuracy Loss metric* (*AUC-PALM*), which facilitates the comparison of explainer fidelity across different models. We advance the field by evaluating various perturbation methods, demonstrating that perturbations centered on time series prototypes and those proportional to feature importance outperform others by offering a more distinct comparison of explainer fidelity with the underlying black-box model. This work lays the groundwork for broader application and understanding of *XAI* in critical decision-making processes.

Keywords: XAI · Visualizations · Anomaly Detection · Time Series · AUC-PALM · ECG · Dynamic Time Warping Barycenter Averaging · Time Series classification · Deep Learning · RNN-autoencoder · reconstruction loss · SHAP · LIME · Healthcare Analytics · Feature Importance · Model Interpretability

1 Introduction

In many applications, domain experts possess expectations regarding the features that should generally influence the classification of cases into specific classes. Our focus is on time series data, particularly with two classes: normal and anomalous. Medical data, such as ECG, which constitutes time series data, serves as an example. However, it is crucial to note that medical data is not limited to time series but also includes images, textual, and tabular data. Analyzing these data types and explaining models working on them is often more straightforward due to their inherent characteristics, which sometimes make visualization and interpretation more accessible. This is true particularly in medical imaging, where visualization and interpretation are more straightforward due to the data's visual nature [11, 14]. This ease of analysis contrasts with the complexities of time series data, where explaining model decisions poses significant challenges.

Interpretability in feature-based black-box explainers, such as Local Interpretable *Model-agnostic Explanations* (*LIME*) and *SHapley Additive exPlanations* (*SHAP*), presents a significant challenge, especially when aiming to discern the average influence of features on predictions for a given class. The study by [15] underscores the applications of XAI in healthcare, highlighting the critical need for transparency, fairness, and accuracy in AI-driven decision-making processes.

Visualizing explanations for time series data is particularly challenging due to the complexity of capturing and representing temporal relationships and dynamics effectively. The study by [21] developed a variety of metrics, such as fidelity, monotonicity, stability, and interpretability, to validate and evaluate the effectiveness of explanation techniques. Our visualizations enable experts to ascertain which segments of the series, in this case, ECG, the model focuses on concerning the normal and abnormal classes. This insight is crucial as it allows for a deeper understanding of the model's decision-making process, potentially leading to improved diagnostic and predictive outcomes.

However, visualization alone, i.e., understanding the model mediated by explainers, is insufficient. It's imperative to ensure that our explainers faithfully represent the black-box model they aim to elucidate. We employ a perturbation method, widely recognized in literature, to achieve this. Yet, there are numerous proposals on how to calculate the *AUC Perturbational Accuracy Loss metric* (*AUC-PALM*), especially for time series data where introducing perturbations thoughtfully is paramount. Our method introduces perturbations around the prototype of a given class and examines whether perturbations should be proportional or inversely proportional to feature importance.

In conclusion, our work contributes to the field by providing a methodology that not only enhances the interpretability of black-box models through visual explanations but also ensures the fidelity of these explanations through a novel application of the *AUC-PALM* metric. By focusing on the specific challenges presented by time series data, we offer insights that are broadly applicable, underscoring the importance of continuous research and diverse applications in XAI, as indicated by comprehensive reviews [10, 20].

Our main contributions are:

- We highlight the importance of visualizing *Time Series* (*TS*) data to improve understanding with local explainers such as SHAP and LIME, making model interpretations clearer.
- Utilizing a metric described in literature [3], the *AUC Perturbational Accuracy Loss metric* (*AUC-PALM*), we adapted it for time series analysis, allowing for a finer distinction in model fidelity, crucial for better model evaluation. Our developed perturbation methods improve the *AUC-PALM* measure for any XAI algorithm that attributes importance to features. These methods are fitted for the analysis of *TS* data.
- Our method is universally applicable to *Time Series* data, with a special benefit for healthcare due to its extensive use in this field. This is showcased through its application to *ECG* data, providing precise and understandable model explanations.

The paper is structured as follows: Sect. 2 reviews current TS classifiers research. Section 3 describes our method and a use-case study with the ECG dataset and Deep InceptionTime Model. Section 4 presents visualizations and results using *AUC-PALM*. Section 5, and Sect. 6 concludes the work and future development perspectives.

2   Related Works

This section focuses on *XAI evaluation metrics* and *Time Series* (TS) classification. Authors [21], presents a comprehensive overview of evaluation metrics for ML explanations. Metrics for model-based and example-based explanations primarily evaluate interpretability and simplicity, while attribution-based explanations primarily evaluate fidelity and soundness. In [8], the authors discuss methods for evaluating various aspects of explainable AI, such as user satisfaction, trust, reliance, curiosity, and system performance. In [16] authors proposes a suite of multifaceted metrics to objectively compare different explainers based on correctness, consistency, and confidence. The paper shows that the proposed metrics are computationally inexpensive and can be used across different data modalities. Work [1] introduces the concept of robustness for interpretability, arguing that it is a crucial feature. The authors show that current methods lack robustness and propose methods to enforce robustness in existing interpretability approaches. Finally, the aim of the paper [6] is to help researchers to map existing tools and apply evaluation metrics when developing an XAI system.f The work summarizes the state-of-the-art review in XAI evaluation metrics and highlights challenges and future developments.

In [19], the authors present the first extensive literature review on XAI for TS classification. They propose a taxonomy for explanation methods and highlight open research directions. In "A Model-Agnostic Approach to Quantifying the Informativeness of Explanation Methods for Time Series Classification" [13], the authors propose a model-agnostic approach for quantifying and comparing different saliency-based explanations for TS classification. The authors list a number of explainers for classifiers based on deep neural networks. They also use perturbations as an evaluation measure. With perturbation-based analysis they show that the discrimination of TS parts plays a critical role in classification accuracy. They distinguish 2 approaches to perturbation: applied only to discriminative region (*Type 1*) and applied only to non-discriminative region (*Type 2*).

2.1   Evaluation Metrics for eXplainable AI - Challenges and Prospects

There exists a multitude of techniques in the field of *XAI* designed to interpret and understand the decision-making processes of AI models. Some prominent *XAI* techniques include *Local Interpretable Model-agnostic Explanations (LIME)*, *SHapley Additive exPlanations (SHAP)*, and *Counterfactual Explanations* [4]. These approaches aim to provide human-understandable explanations to clarify AI system behavior.

Evaluation metrics in *XAI* are categorized into subjective, based on human preferences, and objective, achieved through formal definitions [6], which is depicted in

Fig. 1. XAI methods come with metrics specific to the method or focused on behavior of model being explained and its other aspects like cognitive complexity or computational cost affecting algorithm execution for large datasets. For evaluating explanation effectiveness in task-specific methods, metrics like non-representativeness or diversity are critical [18]. Counterfactual explanations demand metrics for diversity of changes and feasibility [12]. Recommendation systems require tailored, task-specific metrics [17], while model-related metrics usually focus on feature importance e.g., sensitivity, monotony [18]. Some methods use oversimplified *post-models*, evaluated based on size, complexity, or accuracy [5]. Data modality (e.g. images, text, tabular data), type and structure of explanations influence selection of metrics [5, 13, 16]. The evaluation of explanations consistency and stability is vital for model trustworthiness [3], with particular challenges and requirements highlighted in space exploration contexts [2]. Comprehensive reviews indicate a need for continuous research and diverse applications in XAI [10, 20].

The *Intelligible eXplainable AI* (InXAI) framework provides tools for computing metrics such as *Perturbational Accuracy Loss*, *Stability*, and *Consistency* for explanation models like SHAP and LIME[1]. This framework is significant as the code within the InXAI package is actively developed with the goal of merging it into the master branch.

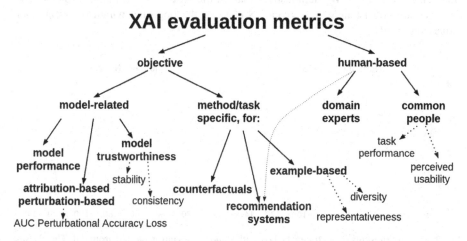

Fig. 1. Explanatory evaluation metrics are a relatively young field. There are 2 groups of explanatory evaluations - those based on objective metrics, calculated numerically, and human-centric, i.e. reflecting the final utility for the user. Sometimes the division between these 2 groups is blurred, such as in the case of XAI evaluation metrics for recommendation algorithms. The metrics highlighted in blue are part of the INXAI package. (Color figure online)

2.2   InceptionTime Deep Network for Time Series Classification

InceptionTime is a highly successive model [9] for classification of TS. It is an ensemble of deep Convolutional Neural Network (CNN) models inspired by the Inception-v4

---

[1] Github: https://github.com/sbobek/inxai.

architecture. InceptionTime shows high accuracy and scalability in TS classification tasks. The network has its implementation in *PyTorch*[2].

Considering the gaps identified in the state-of-the-art analysis of explanations for *Time Series* data, we propose a methodology with two key aspects: visualization and the adaptation of the *AUC Perturbational Accuracy Loss metric*, aiming to address these shortcomings effectively.

3 Methodology

We present proposal of a methodology for evaluating eXplainable Artificial Intelligence (XAI) algorithms for anomaly detection machine-learning models in Time Series (TS). Our focus is on local explanations in the form of feature importance, as they are the most general, popular, and versatile approaches in the domain of XAI. Visuals include among others *AUC-PALM*. This metric assesses fidelity of the explainer (its consistency with the explained model) via perturbing TS data proportionally or inversely to the importance of the given explainer. Formula for perturbing TS takes into account both the feature importances at the observation level for the given the explainer and the class prototype. Class prototypes were obtained with Dynamic Time Warping (DTW) Barycenter Averaging. We demonstrate that, in the case of TS, it matters whether the damages are consistent with a given TS class or the opposite, which offers another indication of fidelity.

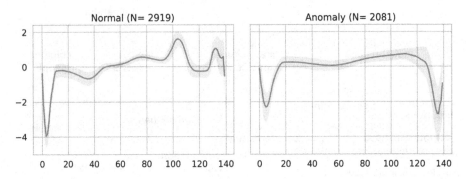

Fig. 2. The plot presents mean for each class with standard deviation plotted around it. The classes associated with pathological patterns (*R-on-T PVC*, *PVC*, *SP or EB* and *UB*) were recoded as *Anomaly* class.

3.1 Dataset

The dataset used in the experiments is derived from [7]. It was prepared by Y. Chen, E. Keogh[3]. The preparation process included two steps: extraction of each heartbeat,

---

[2] InceptionTime (in Pytorch) - GitHub: https://github.com/TheMrGhostman/InceptionTime-Pytorch.

[3] See: http://timeseriesclassification.com/description.php?Dataset=ECG5000.

and making each heartbeat equal length using interpolation. The *"ECG5000"* dataset consists of *5,000* univariate TS, divided into *4,500* for training and *500* for testing. These TS, acquired through ECG, contain *140* timesteps each. The dataset comprises *5* different classes (with *2* major ones), each TS has *1* dimension ECG.

Each sequence corresponds to a single heartbeat from a single patient with congestive heart failure. The dataset includes five classes of heartbeats: Normal (*N*), R-on-T Premature Ventricular Contraction (*R-on-T PVC*), Premature Ventricular Contraction (*PVC*), Supra-ventricular Premature or Ectopic Beat (*SP or EB*), and Unclassified Beat (*UB*). Since we utilize a binary classifier in the article, we applied the re-encoding of all ECG classes associated with pathological patterns into a single *Anomaly* class. This is illustrated in Fig. 2.

Table 1. Confusion Matrix for the classifier on the test set across all classes.

Predicted	True class	#
Normal	Normal	306
Anomaly	R on T	164
	SP	14
	PVC	8
Normal	SP	4
	PVC	3
	UB	2
Anomaly	UB	1

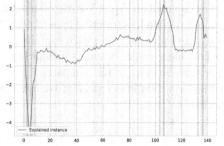

Fig. 3. Example LIME importances for single observation as plotted by *LimeTimeSeriesExplainer package*. One can observe usage of number of features equal to the length of the Time Series (140).

3.2 RNN-Autoencoder Extension

We developed a deep learning model using the Transfer Learning technique. We utilized a pre-trained RNN-autoencoder model implemented in *PyTorch*. The model was available online and trained on the data discussed in Sect. 3.1, specifically on the *Normal* class from the test subset. This allowed the author of the original code, Valkov V.[4], in *Collab notebook*[5] to determine a cut-off threshold for the cumulative reconstruction loss, beyond which there was a high probability of belonging to the *Anomaly* class.

In contrast, we extended the autoencoder with a classification head, consisting of Linear, Dropout, and Sigmoid layers. The input for the classification head was the difference between the input and the reconstruction, i.e., the reconstruction loss vector with the length of the TS (140 samples in each series). Incorporating the reconstruction loss vector allowed us to later use this vector as one of the explainers and compare it

---

[4] See Github: https://github.com/curiousily/Getting-Things-Done-with-Pytorch.

[5] Time Series Anomaly Detection using LSTM Autoencoders with PyTorch in Python: https://curiousily.com/posts/time-series-anomaly-detection-using-lstm-autoencoder-with-pytorch-in-python.

with SHAP and LIME. The model was fine-tuned with frozen autoencoder weights, on the test subset, this time for both classes: *Normal* and *Anomaly*. We achieved an accuracy of 0.98 on the test set. One can observe with *Confusion Matrix* in Table 1 that the classifier mistakes for 3 classes recoded to *Anomaly* (SP, PVC, UB) for just 9 observations, classifying them as *Normal*.

3.3   Visualisation of Importances

SHAP was calculated using the DeepExplainer package[6]. For LIME, we employed the LimeTimeSeriesExplainer package[7], with the number of features (*num_features*) set equal to the length of the series, allowing us to obtain an explainer comparable to SHAP, which is depicted on Fig. 3. We used the output from the RNN-autoencoder for obtaining reconstruction loss (referred to as "*LOSSES*""). We developed visualizations to display feature importances for each explainer, categorized by class. These visualizations show both the average TS and its standard deviation. A novel aspect of our method is an area around the average TS, indicating the importance of each TS segment. This area's upper and lower outlines represent average values of positive and negative importances, respectively.

$$\Phi^e_{\text{down}} = \overline{\min(\Phi^e, 0)}$$
$$\Phi^e_{\text{up}} = \overline{\max(\Phi^e, 0)}$$
$$\Phi^e_{\text{under-line}} = \bar{TS} + \Phi^e_{\text{down}} \cdot const$$
$$\Phi^e_{\text{over-line}} = \bar{TS} + \Phi^e_{\text{up}} \cdot const$$

(1)

$\bar{TS}$ denotes the mean value of TS, while $\Phi^e_{\text{under-line}}$ and $\Phi^e_{\text{over-line}}$ are outlines of importances. The local explainer, Φ^e, matches the TS length. Φ^e_{down} and Φ^e_{up} represent the averages of negative and positive explainer values, respectively. We draw a range around $\bar{TS}$ on the plot, with lower and upper edges corresponding to Φ^e_{down} and Φ^e_{up}, respectively. A constant multiplier enhances visibility. Additionally, we calculated class prototypes using *tslearn*'s *barycenters.dtw_barycenter_averaging* function. These prototypes were visualized and used in our TS permutation method.

The resulting plot offers easy interpretation, particularly in binary classification. For example, with SHAP, the final part of the plot distinctly contributes to different classes. SHAP identifies cohesive areas, emphasizing their centers more than edges, while LIME finds almost the entire series area crucial. Comparing with reconstruction loss, both SHAP and LIME highlight series parts with significant reconstruction loss. These observations are shown in Fig. 4.

3.4   AUC Perturbational Accuracy Loss Metric

The next step involved comparing explainers using the *Area Under Curve Perturbational Accuracy Loss Metric* (*AUC-PALM*). This metric serves as a direct measure of

---

[6] Documentation: https://shap-lrjball.readthedocs.io/en/latest/generated/shap.DeepExplainer.html.

[7] GitHub: https://github.com/emanuel-metzenthin/Lime-For-Time.

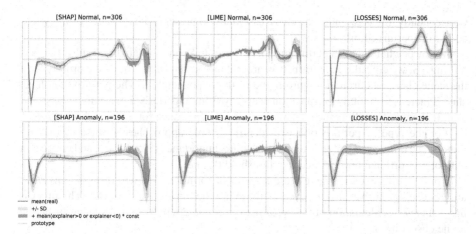

Fig. 4. Comparison of SHAP, LIME and reconstruction loss (denoted "LOSSSES") as explainers. One can observe, that they differ considerably. In the case of SHAP, the final part of the graph negatively contributes to the *Normal* class, and positively to the *Anomaly* class. The same is true for LIME, but in this case almost the entire area of the series is important for explanation. Both SHAP and LIME tend to indicate those areas of the series for which reconstruction loss is greatest. The red line furthermore indicates the prototype for the class. Examples are for test set.

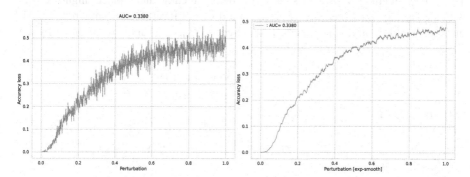

Fig. 5. Exponential smoothing of AUC Perturbational Accuracy Loss curves improves clarity, however does not change the metric value.

the decline in model performance, providing an intuitive and straightforward method to assess the impact of perturbations on the explanatory power of the models.

To address visualization, we developed plots where the noisy *AUC-PALM* is smoothed exponentially, still ensuring accurate calculation of the metric, as shown in Fig. 5. The graphical representation of *AUC-PALM* showcases the accuracy loss on the Y-axis against the degree of perturbation on the X-axis. From this graph, the area under the curve is calculated, representing the *AUC-PALM*. This area is a crucial metric for evaluating the robustness of the explanation models in relation to the changes introduced in the input data. The calculation of *AUC-PALM* involves normalizing both the X and Y axes to 100%, making the maximum possible area under the curve equal to 1.0.

Consequently, the area under the curve essentially represents the average accuracy loss across the perturbed instances.

We faced challenges in perturbing Time Series (TS) and visualizing results. For perturbing TS, we considered perturbation around the original TS, around the prototype of the same class, and around the prototype of the opposite class. The perturbation level should increase smoothly to transition accuracy from unperturbed to random classification. It should also reflect the explainer's feature-importance for a given observation.

For the type of perturbation, we identified three approaches: inverse to feature-importance ("*inverted*"), consistent with feature-importance ("*straight*"), and consistent but not proportional beyond a certain threshold ("*zoned*").

For TS perturbations, we tested a number of formulas. We will discuss "*inverted*" and "*zoned*" perturbations in detail, as those are the most complex cases. The process has 2 stages: 1. calculation of the perturbation vector, 2. application of the perturbation to the observation.

Stage 1. For feature importance visualization in TS, we normalize the absolute value of the explainer's output to create a unit length vector, F_{norm}. In the "*inverted*" case, perturbations involve subtracting this vector from the unit vector ("*[V1]*") or dividing the unit vector by F_{norm}, with a small constant to avoid division by zero ("*[V3]*"). The "*zoned*" method zeros values below the *80th* percentile and replaces higher values with a constant, while the "*straight*"r method does not invert feature importances. All methods lead to a normalized perturbation vector, P_{norm}. This is represented by the Equation 2.

$$F_{norm} = \frac{F}{\|F\|}$$

$$[\text{around prototype}] : R = Pr$$
$$[\text{around observation}] : R = O$$

$$[V1] : P = 1 - |F_{norm}|$$
$$[V3] : P = \frac{1}{|F_{norm}+c|}$$

$$(2)$$

$$P_{final} = R \circ P_{norm}$$

$$P_{norm} = \frac{P}{\|P\|}$$

$$O_P = O + [i \cdot \text{rand}(\{1, -1\}) \text{ for } i \text{ in } P_{final}] \cdot \alpha$$

$$(3)$$

Stage 2. Perturbation application varies based on whether it's around a class prototype or relative to the observation itself. We define the "*reference*" as this chosen TS. P_{final} - the perturbation vector - is obtained by elementwise multiplication of the *importance vector* and *reference vector*. The final perturbed TS, O_P, is the sum of the input observation and perturbation vector, each component multiplied by a random sign and a scalar for perturbation strength. See the Formula 3.

As being stated, the reference level can be either the prototype of the class to which the TS instance belongs: *[around prototype]*, or the observation itself: *[around observation]*. The prototype can be either from the same class or from the opposite class. The prototype is calculated using the Dynamic Time Warping (DTW) Barycenter Averaging method, although this is not shown in the formula. The final perturbation P_{final} is calculated as an element-wise product of this reference R and the P_{norm} from the previous formula. The perturbation is applied to the O to obtain the perturbed observation O_P in such a way that each component of the observation has added perturbation components with a random sign. Furthermore, it is multiplied by the α, which determines the strength of the perturbation and is changed from a small value to a large one as the *AUC-PALM* is drawn (calculated).

4   Results

4.1   Joint Evaluation of *Normal* and *Anomaly* classes

We analyzed the *AUC-PALM* graphs and values for the test subset, including both *Normal* and *Anomaly* classes, to understand the impact of perturbation strategies on explainer performance. Figures 6 and 7 show that both *"inverted"* and *"straight"* perturbation methods affect SHAP, LIME, and reconstruction losses differently. The *"straight"* method results in larger *AUC-PALM* differences and quicker convergence to *0.5*, indicating a random classifier. The *"zoned"* case, performing similarly to *"straight"*, is not shown. SHAP outperformed in *AUC-PALM* results, followed by LIME, with reconstruction losses being slightly less effective but still a viable explainer.

4.2   Insights from Individual Class Analysis

Since SHAP proved to be a more accurate explainer according to the *AUC-PALM* on the combined *Normal* and *Anomaly* classes, providing better feature importances, we investigated whether such a relationship also occurs when analyzing the *Normal* and *Anomaly* classes separately. The Table 2 provides a detailed overview of the results from various experiments performed on two classes of data: *Anomaly* and *Normal*. A range of tests were conducted for each data class, taking into account different types of perturbations. The outcomes are represented in relation to three distinct methods: LIME, reconstruction losses ("REC. LOSSES"), and SHAP.

We adopt the ratio of variance to mean as a measure of the differentiation strength between different explainers. Notably, the highest ratio of variance to mean is observed for *"straight"* condition, both for *Anomaly* and *Normal* class. For *Anomaly*, perturbation around same class gave the highest result, while for *Normal* around opposite class. Nevertheless, perturbing *Anomaly* class around the *"opposite class"* is also a viable and almost equally potent alternative, only slightly lagging behind the leading solutions. It is apparent from the results that perturbation centered around the anomalous class yields the most significant effect. These findings underscore the utility of tailored perturbation strategies in maximizing the differentiation between various explainers. This indicates that applying perturbations in this particular way provides the most valuable insights into performance of explainers. Hence, it can be inferred which approach towards perturbations is the most effective.

The Fig. 8 presents the top three results from the Table 2 summarizing the experiments. As can be observed, under the *"straight"* condition and *"opposite class"* for the *Normal* class, all three explainers are well separated. In the case of *Anomaly*, the

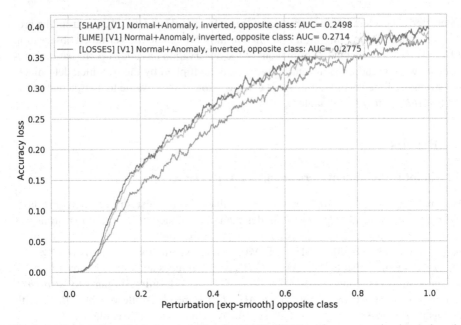

Fig. 6. An "inverted" method of perturbation for LIME, SHAP, and reconstruction losses-based explainer. The lower the area AUC, the better the explainer. SHAP performs the best. Results obtained on test set for Normal and Anomaly classes altogether. The line is smoothed exponentially.

var/mean metric indicated that the best perturbation is around the same class, although the chart shows that the "*opposite class*" provides more information. For completeness, a chart for one of the worst var/mean values is also presented. In this case, the chart does not provide significant insight. At the same time, it is noticeable that explanations for the *Normal* class are better than those for the *Anomaly* class, reflecting the fact that the autoencoder was trained only on the *Normal* class.

It is important to note that directly comparing the *AUC-PALM* metric values between classes carries some risk. The perturbation coefficient α is uncalibrated, so if the model's accuracy does not decline to the same level at the end of the plot, comparing the area under the curve will not be reliable. Moreover, the presented graph is derived from a different computation process than the one collected in the Table 2. It can be observed that the calculated *AUC-PALM* values exhibit high stability and are consistent up to the third decimal place for SHAP and reconstruction losses (here referred to as autoencoder losses), and up to the second decimal place for LIME. The notebook accompanying this paper can be found on our Github[8].

---

[8] Jupyter .ipynb notebook for paper: https://github.com/mozo64/tsxai/blob/main/06_time_seri es_anomaly_detection_ecg_clear.ipynb.

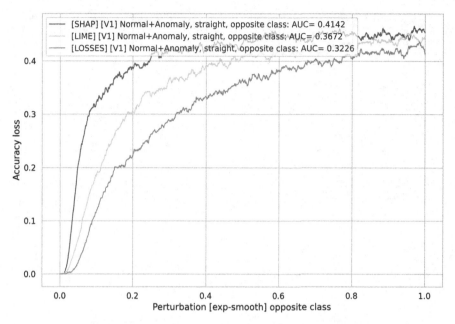

Fig. 7. A "straight" method of perturbation for LIME, SHAP, and reconstruction losses-based explainer. The higher the area AUC, the better the explainer. SHAP is the best once again. Results obtained on test set for Normal and Anomaly classes altogether. The line is smoothed exponentially.

Table 2. Results of experiments with various types of perturbations

Class	Proportionality	Prototype	*LIME*	*REC. LOSSES*	*SHAP*	mean	var	var/mean
Anomaly	inverted	around obs.	0,1853	0,2045	0,1697	0,1865	0.0003032	0.001626
		opposite class	0,1771	0,1784	0,1721	0,1759	0.00001103	0.00006273
		same class	0,1769	0,1988	0,1584	0,1780	0.0004090	0.002298
	straight	around obs.	0,4014	0,3745	0,4093	0,3950	0.0003336	0.0008445
		opposite class	0,3138	0,2767	0,3380	0,3095	0.0009525	0.003077
		same class	0,4193	0,3558	0,4306	0,4019	0.001626	0.004045
	zoned	around obs.	0,3481	0,3563	0,3515	0,3520	0.00001717	0.00004877
		opposite class	0,2680	0,2691	0,2934	0,2768	0.0002068	0.0007469
		same class	0,3444	0,3537	0,3480	0,3487	0.00002201	0.00006313
Normal	inverted	around obs.	0,3269	0,3378	0,3332	0,3326	0.00003020	0.00009081
		opposite class	0,3294	0,3376	0,2957	0,3209	0.0004944	0.001541
		same class	0,3354	0,3403	0,3388	0,3382	0.000006347	0.00001877
	straight	around obs.	0,4134	0,3566	0,3975	0,3892	0.00085962	0.002209
		opposite class	0,3993	0,3484	0,4606	0,4028	0.003161	0.007847
		same class	0,4027	0,3555	0,4029	0,3870	0.0007461	0.001928
	zoned	around obs.	0,3944	0,3592	0,3733	0,3756	0.0003127	0.0008326
		opposite class	0,4054	0,3568	0,3956	0,3860	0.0006613	0.001713
		same class	0,3903	0,3564	0,3755	0,3741	0.0002888	0.0007720

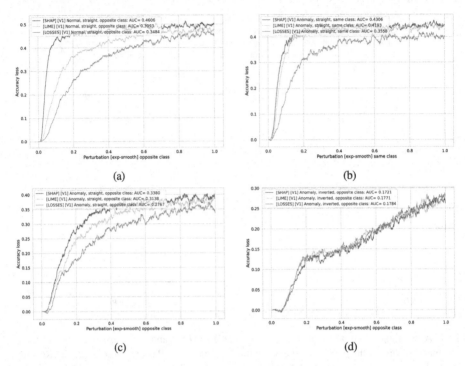

Fig. 8. SHAP, LIME and reconstruction losses-based explainer compared with best perturbation method according to *var/mean* metric (charts *(a)–(c)*). The higher the area under curve, the better the explainer. For both *Normal* and *Anomaly* class, the best is SHAP. The chart *(d)* represents the worst perturbation method, for reference. The line is smoothed exponentially.

5   Discussion

Our study underscores the role of visualization in elucidating feature importances, as detailed in Sect. 3.3. These visualizations provide immediate insights into the segments of time series critical for distinguishing between normal and anomalous classes. Often, explanations overlook the fact that different segments of the series may contribute to anomalies, while others may indicate normalcy. In our visualizations, we have also considered the character of the data, creating visualizations that are significantly more useful for medical experts than the default ones provided by libraries such as *SHAP* or *LIME*. Moreover, these visualizations are compatible with other local, feature-based explanations. This innovation is adaptable to any black-box, feature-based explanation method, enhancing its utility across various applications.

We have presented an approach to the evaluation of explanation quality in feature-importance attribution algorithms, such as *SHAP* and *LIME*, with a special emphasis on per-class analysis in *Time Series Anomaly Detection*. Our work is important for Time Series analysis, and demonstrated through ECG data analysis described in Sect. 3.1, showcasing the applicability of our methods also in healthcare scenarios.

Central to our analysis is the exploration of various perturbation scenarios to understand their impact on model performance in *Time Series* (TS) data. By employing *Dynamic Time Warping* (DTW) *Barycenter Averaging* for prototyping, we navigate through perturbations around class prototypes and the observation itself, enhancing our understanding of feature-importance-based explanations (see Sect. 3.4). Our methodology introduces novel perturbation methods, such as *"straight"* (proportional to feature importance) and *"zoned"* (above selected threshold), alongside *"inverted"* (inversely proportional) approaches, to discern the efficacy of explainers, with the *"straight"* method proving superior in explainer differentiation. This approach not only facilitates rapid identification of class-dependent areas in binary classifiers but also extends to multi-class scenarios by treating them as a stack of binary classifiers, thereby underscoring its universality across different classifier architectures, including non-deep learning models (see Sect. 3.2). Furthermore, we explore the use of *reconstruction loss* from autoencoder-based models as a baseline for comparison with feature importance-based explainers.

6 Conclusion

This study has explored the important role of visualizations in conveying to domain experts which parts of a time series are globally significant for predicting a given class, whether normal or anomalous. These visual aids provide a rapid insight into the model's decision-making process, based on local explanations for feature-based models in both machine learning (ML) and deep learning (DL) anomaly detection contexts.

While literature describes various methods for perturbation, our approach to calculating *AUC Perturbational Accuracy Loss metric* (*AUC-PALM*) allows for a more explicit differentiation between different explainers, facilitating informed decisions on which is more effective in specific scenarios.

We systematically investigated perturbations: around the prototype of the given class, the prototype of the opposite class, and directly around the observation itself (without a prototype), presenting formulas for calculating *AUC-PALM* in each case. In our experiments using SHAP, LIME, and reconstruction loss as a baseline, the most significant distinction was observed with perturbations around the prototype of the opposite class (or, for simplicity, anomalous class would suit for most cases), proportional to feature importance-a condition we termed *"straight."* This highlighted SHAP's superior performance and pointed out the reconstruction loss's limitations. Our findings underscore the necessity for nuanced and accessible explanatory tools in the field of anomaly detection. By providing clear visualizations and a robust metric for explainer evaluation, we aim to bridge the gap between complex data patterns and actionable insights, enabling more effective decision-making in critical applications.

Future directions include refining our approach by training class-specific autoencoders and conducting experimental studies with participants to evaluate the effectiveness of our techniques in real-world machine-learning problems, i.e. predictive maintenance scenarios, paving the way for more interpretable and reliable anomaly detection in *Time Series* data.

Acknowledgment. This paper is funded from the XPM (Explainable Predictive Maintenance) project funded by the National Science Center, Poland under CHIST-ERA programme Grant Agreement No. 857925 (NCN UMO-2020/02/Y/ST6/00070). The research has been supported by a grant from the Priority Research Area (DigiWorld) under the Strategic Programme Excellence Initiative at Jagiellonian University. We acknowledge the use of OpenAI's ChatGPT-4 for reviewing and improving the language and style of this manuscript.

References

1. Alvarez-Melis, D., Jaakkola, T.S.: On the robustness of interpretability methods (2018)
2. Barkouki, T., Deng, Z., Karasinski, J., Kong, Z., Robinson, S.: XAI design goals and evaluation metrics for space exploration: a survey of human spaceflight domain experts (2023). https://doi.org/10.2514/6.2023-1828
3. Bobek, S., Bałaga, P., Nalepa, G.J.: Towards model-agnostic ensemble explanations. In: Paszynski, M., Kranzlmüller, D., Krzhizhanovskaya, V.V., Dongarra, J.J., Sloot, P.M.A. (eds.) ICCS 2021. LNCS, vol. 12745, pp. 39–51. Springer, Cham (2021). https://doi.org/10.1007/978-3-030-77970-2_4
4. Bobek, S., Mozolewski, M., Nalepa, G.J.: Explanation-driven model stacking. In: Paszynski, M., Kranzlmüller, D., Krzhizhanovskaya, V.V., Dongarra, J.J., Sloot, P.M.A. (eds.) ICCS 2021. LNCS, vol. 12747, pp. 361–371. Springer, Cham (2021). https://doi.org/10.1007/978-3-030-77980-1_28
5. Bobek, S., Nalepa, G.J.: Local universal rule-based explanations (2023)
6. Coroamă, L., Groza, A.: Evaluation metrics in explainable artificial intelligence (XAI). In: Guarda, T., Portela, F., Augusto, M.F. (eds.) ARTIIS 2022. CCIS, vol. 1675, pp. 401–413. Springer, Cham (2022). https://doi.org/10.1007/978-3-031-20319-0_30
7. Goldberger, A.L., et al.: PhysioBank, PhysioToolkit, and PhysioNet: components of a new research resource for complex physiologic signals. Circulation **101**(23), E215-20 (2000)
8. Hoffman, R.R., Mueller, S.T., Klein, G., Litman, J.: Metrics for explainable AI: challenges and prospects (2019)
9. Ismail Fawaz, H., et al.: Inceptiontime: finding alexnet for time series classification. Data Mining Knowl. Discov. **34**, 1–27 (2020). https://doi.org/10.1007/s10618-020-00710-y
10. Kadir, M.A., et al.: Evaluation metrics for XAI: a review, taxonomy, and practical applications. ResearchGate (2023). https://www.researchgate.net/publication/366917148Evaluation_Metrics_for_XAI_A_Review_Taxonomy_and_Practical_Applications
11. Li, M., Jiang, Y., Zhang, Y., Zhu, H.: Medical image analysis using deep learning algorithms. Front. Public Health **11**, 1273253 (2023). https://doi.org/10.3389/fpubh.2023.1273253. https://www.frontiersin.org/articles/10.3389/fpubh.2023.1273253/full
12. Nauta, M., et al.: From anecdotal evidence to quantitative evaluation methods: a systematic review on evaluating explainable AI. ACM Comput. Surv. **55**(13s), 1–42 (2023). https://doi.org/10.1145/3583558
13. Nguyen, T.T., Le Nguyen, T., Ifrim, G.: A model-agnostic approach to quantifying the informativeness of explanation methods for time series classification. In: Lemaire, V., Malinowski, S., Bagnall, A., Guyet, T., Tavenard, R., Ifrim, G. (eds.) AALTD 2020. LNCS (LNAI), vol. 12588, pp. 77–94. Springer, Cham (2020). https://doi.org/10.1007/978-3-030-65742-0_6
14. Parmar, C., Barry, J.D., Hosny, A., Quackenbush, J., Aerts, H.J.: Data analysis strategies in medical imaging. Clin. Cancer Res. **24**(15), 3492–3499 (2018). https://doi.org/10.1158/1078-0432.CCR-18-0385

15. Band, S.S., et al.: Application of explainable artificial intelligence in medical health: a systematic review of interpretability methods. Inform. Med. Unlocked **40**, 101286 (2023). https://doi.org/10.1016/j.imu.2023.101286. https://www.sciencedirect.com/science/article/pii/S2352914823001302

16. Santhanam, G.K., Alami-Idrissi, A., Mota, N., Schumann, A., Giurgiu, I.: On evaluating explainability algorithms (2020). https://openreview.net/forum?id=B1xBAA4FwH

17. Sisk, M., Majlis, M., Page, C., Yazdinejad, A.: Analyzing XAI metrics: summary of the literature review (2022). https://doi.org/10.36227/techrxiv.21262041

18. Sun, J., Shi, W., Giuste, F.O., Vaghani, Y.S., Tang, L., Wang, M.D.: Improving explainable AI with patch perturbation-based evaluation pipeline: a covid-19 X-ray image analysis case study. Sci. Rep. **13**(1), 19488 (2023). https://doi.org/10.1038/s41598-023-46493-2

19. Theissler, A., Spinnato, F., Schlegel, U., Guidotti, R.: Explainable AI for time series classification: a review, taxonomy and research directions. IEEE Access **10**, 100700–100724 (2022). https://doi.org/10.1109/ACCESS.2022.3207765

20. Vilone, G., Longo, L.: Notions of explainability and evaluation approaches for explainable artificial intelligence. Inf. Fusion **76**, 89–106 (2021). https://doi.org/10.1016/j.inffus.2021.05.009

21. Zhou, J., Gandomi, A., Chen, F., Holzinger, A.: Evaluating the quality of machine learning explanations: a survey on methods and metrics. Electronics **10**, 593 (2021). https://doi.org/10.3390/electronics10050593

Understanding Survival Models Through Counterfactual Explanations

Abdallah Alabdallah[1]([✉]), Jakub Jakubowski[2], Sepideh Pashami[1], Szymon Bobek[3], Mattias Ohlsson[1], Thorsteinn Rögnvaldsson[1], and Grzegorz J. Nalepa[3]

[1] Center for Applied Intelligent Systems Research (CAISR),Halmstad University, Halmstad, Sweden
abdallah.alabdallah@hh.se
[2] Department of Applied Computer Science, AGH University of Science and Technology, Krakow, Poland
[3] Faculty of Physics, Astronomy and Applied Computer Science, Institute of Applied Computer Science, and Jagiellonian Human-Centered AI Lab (JAHCAI), and Mark Kac Center for Complex Systems Research, Jagiellonian University, Kraków, Poland

Abstract. The development of black-box survival models has created a need for methods that explain their outputs, just as in the case of traditional machine learning methods. Survival models usually predict functions rather than point estimates. This special nature of their output makes it more difficult to explain their operation. We propose a method to generate plausible counterfactual explanations for survival models. The method supports two options that handle the special nature of survival models' output. One option relies on the Survival Scores, which are based on the area under the survival function, which is more suitable for proportional hazard models. The other one relies on Survival Patterns in the predictions of the survival model, which represent groups that are significantly different from the survival perspective. This guarantees an intuitive well-defined change from one risk group (Survival Pattern) to another and can handle more realistic cases where the proportional hazard assumption does not hold. The method uses a Particle Swarm Optimization algorithm to optimize a loss function to achieve four objectives: the desired change in the target, proximity to the explained example, likelihood, and the actionability of the counterfactual example. Two predictive maintenance datasets and one medical dataset are used to illustrate the results in different settings. The results show that our method produces plausible counterfactuals, which increase the understanding of black-box survival models.

Keywords: Survival Analysis · Explainable Artificial Intelligence · Survival Patterns · Counterfactual Explanations

---

A. Alabdallah and J. Jakubowski—Contributed equally to this work.

1   Introduction

Survival models are a special type of machine learning models, which predict the probability of survival over time. This group of models is widely used in the healthcare domain but has also been applied to predictive maintenance tasks [3, 5, 27, 32]. Complex machine learning tasks, involving high dimensionality of data, often require advanced machine learning models, which are not interpretable by humans, to achieve satisfactory performance. Safety critical domains, like the two mentioned above, usually require the explanations of model's decisions [23].

Explainability in AI refers to the ability to understand and interpret how a machine learning model makes decisions or predictions. Depending on the type of task and data, different explanation methods can be applied. Our work focuses on counterfactual explanations, which is a local explainability method that answers the question of what should change in the input to observe a different output. In this paper, we present how counterfactual explanations can be used to explain survival models. In contrast to classification and regression, the output of a survival model typically includes a survival function or a hazard function, representing the probability of survival over time, which is more difficult to explain.

In this work, we propose a method for generating counterfactual explanations for survival models that supports two options in terms of the definition of the change in output. The first option depends on the Survival Score, that is, the area under the survival function, to search for counterfactual examples that would change this score by a predefined value. In the second option, we use Survival Patterns [1] that represent the survival behaviors of groups in the population significantly different from each other.

Based on such Survival Patterns, we search for counterfactual examples that would change the predicted survival functions of subjects to predefined patterns. Other important aspects, that we consider in our research, are the plausibility and actionability of counterfactual explanations. In this work, we utilize an outlier detection model to drive counterfactual explanations close to the data distribution. We also added a special term to the loss function to handle categorical variables. Lastly, the actionability of the example is controlled by masking features that cannot be changed in practice. Restricting some features from changing can cause, in some cases, the target Survival Pattern to be not reachable. However, our method generates counterfactual examples that are closest to the target pattern. To the best of our knowledge, this is the first work that utilizes both survival scores and survival patterns to generate plausible and actionable counterfactual examples for survival models. A full implementation of our method is available on our GitHub repository[1].

---

[1] https://github.com/abdoush/SurvCounterfactual.

2   Related Works

2.1   Survival Analysis Background

Survival models are a type of statistical and machine learning models that aim at modeling time to an event e.g., the machine failure or the patient's death. One of the problems in survival analysis is the presence of censoring. Some subjects will experience an event during the study, but others will survive beyond the study, which are called censored cases. Survival models usually predict functions, that is, survival or hazard functions. The survival function is the probability of surviving beyond a certain time t; i.e., the failure time T is greater than t: $S(t) = P(T > t)$. The Cox proportional hazards model (CPH) [6] is the first model to predict individualized hazard functions that depend on the individual's features $\mathbf{x}$. The CPH model assumes that hazards between subjects are proportional and independent of time. As a result, the survival curves of subjects do not intersect leading to unique area-under-curve for different survival curves, which is rarely the case in real life. To address these issues, machine learning models for survival analysis have been developed, such as Randon Survival Forests (RSF) [12], Survival Support Vector Machine [26] or deep learning approaches [2,14,20]. In contrast to CPH, these models are able to learn nonlinear relations between features and support nonproportional hazards. They usually offer improved performance in terms of accuracy but lack explanatory insight into their predictions.

2.2   Explainable AI and Counterfactual Explanations

The CPH model, due to its linearity, can be considered inherently explainable, which is a strong advantage in safety-critical domains. More complex machine learning models for survival analysis, require external explanations methods. Many such methods were proposed for classification or regression and extended to survival analysis. SurvLIME [16] extended the LIME [28] method, where it approximates the survival model locally with a CPH model. SurvSHAP(t) [18] extended SHAP [21] to explain survival functions that can capture time-dependent variable effects. SurvSHAP [1] used a proxy-based approach to explain the survival model with SHAP, using the patterns found in the output of the survival model to build the proxy model.

A promising XAI method is Counterfactual Explanations (CE), which belong to the family of example-based explanations aiming to find a 'similar' observation to the one we are explaining, but with a different model prediction. CE was originally proposed in [31], where the authors suggested creating explanations by minimizing the objective function consisting of two terms: the distance to the target output and the distance between the original observation and the counterfactual explanation. A major issue with this approach is that unrealistic explanations might be created. To deal with this problem, the distance between the explanation and the observed data can be minimized [7], a model which estimates the likelihood of point belonging to a data distribution, e.g. Autoencoder, can be used [8], or a generative model can be employed to generate candidates for explanations [24].

In [17], the authors propose a method for generating counterfactual explanations for survival models using the mean survival time as the target. Our method, while sharing similarities, enhances this approach by incorporating additional terms in the optimization to improve the likelihood and actionability of generated counterfactual examples. Moreover, our alternative option utilizes Survival Patterns which identify distinct risk groups based on the entire survival function rather than mean-time alone. This adaptation accommodates both proportional and non-proportional hazard models and facilitates specifying meaningful target changes between survival groups.

3   Research Methods

Our method is model-agnostic in the sense that it depends only on the output of the survival model. Furthermore, the method does not require access to the training data after the model is trained. We use the Particle Swarm Optimization (PSO) algorithm to optimize our objective function to meet four criteria: 1) achieving the desired change in the target output, 2) minimizing the change to the input, 3) the plausibility of the counterfactual example, and 4) the actionability of the counterfactual example. Our method provides two options with respect to the first criterion, i.e. achieving the desired change in the target output. The first option relies on the Survival Score, representing the area under the survival function (mean survival time). This treats the survival problem as a regression task, aiming to find a counterfactual that achieves a specified change to the survival score, detailed in Sect. 3.1. While effective for proportional hazards with non-intersecting survival curves, it has limitations in cases of nonproportional hazards, where distinct survival behaviors can yield the same area under the curve.

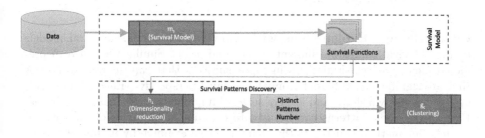

Fig. 1. Patterns discovery Workflow

To address nonproportional hazards, our second method option employs Survival Patterns. This involves grouping survival curves into distinct and significantly different Survival Patterns, each representing curves with similar survivability. Determining the optimal number of patterns and clustering curves in a lower-dimensional space transforms the problem into a classification task. The objective is to find a counterfactual example that changes the predicted survival curve to a predefined Survival Pattern. Further details on survival patterns are provided in Sect. 3.1.

3.1   Generation of Counterfactual Explanations

The objective function consists of three weighted terms. The purpose of each term is to fulfill one of the criteria described earlier.

$$\mathcal{L} = \alpha\mathcal{L}_y + \beta\mathcal{L}_x + \gamma\mathcal{L}_{LL} \tag{1}$$

The first term, $\mathcal{L}_y$, induces the desired target change, the second term, $\mathcal{L}x$, promotes proximity between the counterfactual and the original input, and the third term, $\mathcal{L}_{LL}$, ensures the counterfactual's likelihood. The last criterion, which is the actionability of the generated example, is realized by constraining the selected input feature from changing based on the domain expert knowledge. The following subsections provide details about each part of the objective function.

Counterfactual Explanations with Survival Scores. The survival score y, calculated as the area under the survival curve, represents the mean survival time of a subject. We reframe our explanation task as determining the necessary changes in observed features X to increase the predicted survival score by Y. These counterfactual explanations reveal the feature adjustments that can positively impact the expected survival time. In practical applications, these explanations guide adjustments to extend the machine's or patient's life. The term $\mathcal{L}_y$ in the objective function associated with the target value of the survival score $\mathcal{L}_y = |\mathbf{y} + \delta - \hat{\mathbf{y}}|$, where $\mathbf{y}$ is the original survival score, δ is the required change in the survival score, $\hat{\mathbf{y}}$ is the survival score of the counterfactual candidate, and $|.|$ indicates the absolute value.

Counterfactual Explanations with Survival Patterns. Survival models predict survival curves of different shapes and levels. Similarities between predicted curves reflect similarities in input subjects that can define risk groups that are significantly different from the survival perspective. Each of these risk groups has its own shape and level of survival function, which is called a Survival Pattern [1]. Such patterns can be used to explain the behavior of the survival model. Our method constructs the counterfactual example by finding the minimum change to the input features, which changes the prediction of the survival model from a source Survival Pattern to a specific target one.

Following what has been done in [1], Survival Patterns are discovered as follows. Let $X \in \mathbb{R}^{n \times p}$ be the input of the survival model m_s, and $S \in \mathbb{R}^{n \times m}$ where $S = m_s(X)$ be the output survival functions, where n is the number of examples, p is the number of features, and m is the number of timesteps in the predicted survival functions. The algorithm has three steps; see Fig. 1:

- **Lower Dimensional Representation**: Survival functions, which are discrete one-dimensional signals of probabilities over m time steps, $S \in \mathbb{R}^{n \times m}$, are transformed into a lower dimensional space using the function h_z. This results in $Z \in \mathbb{R}^{n \times r}$ where r is the number of dimensions of the new space (Z-space). In this work, we used Principal Components Analysis (PCA) and chose the number of components r, which maintains an explained variance over 99%.
- **Finding the Number of Survival Patterns**: Using the lower dimensional representation Z, the algorithm iteratively clusters the curves into $k \in \{2, 3, ..., K_{max}\}$ clusters using the k-means clustering algorithm. At each iteration, pair-wise comparisons between the resulting clusters are performed based on the log-rank test [25], which is a statistical test to assess the statistical difference between two groups from a survival point of view. The k^* that is selected is the largest k that achieves the maximum percentage of significantly different groups.
- **Survival Patterns Prediction Model**: In this step, a k-means clustering model g_c is fitted on the Z features, using the optimal number of Survival Patterns k^*.

The final Survival Patterns prediction model f, which will be used in the search for counterfactual examples, is composed of three models: $f(\mathbf{x}) = (g_c \circ h_z \circ m_s)(\mathbf{x})$. Based on the function f and the clusters' centers in the Z-space $c_i : i \in \{0, ..., k^* - 1\}$, it is possible to compute the distance between the survival curve of the proposed counterfactual $\mathbf{x}_{cf}$ and the survival curve of the center of the target pattern c_t in the Z-space, which will be used as the target part $\mathcal{L}_y$ of the loss function:

$$\mathcal{L}_y = \mathbb{1}((f(\mathbf{x}_{cf}) \neq t)\|(h_z \circ m_s)(\mathbf{x}_{cf}) - c_t\|_2 \tag{2}$$

where t is the desired target Survival Pattern, $\|.\|_2$ is the L_2 norm, and $\mathbb{1}(.)$ is the indicator function. The use of the indicator function will block the effect of this part of the loss function once the counterfactual crosses the boundaries of the target pattern.

Minimal Change to the Features. The second term of the objective function $\mathcal{L}_x(\mathbf{x}_{cf}) = \|\mathbf{x} - \mathbf{x}_{cf}\|_p$ aims at minimizing the distance between the original example $\mathbf{x}$ and the generated counterfactual $\mathbf{x}_{cf}$, where p is the order of the L^p-norm. In this work, we used L^1-norm to encourage sparsity in the difference between the explained example and the counterfactual one.

Likelihood of Counterfactual Explanations. To ensure the plausibility of counterfactual explanations, we utilize an Autoencoder model (AE) fitted to the training data. The model is used to determine the reconstruction error between the counterfactual candidate and its reconstructed version by the AE, which we call the anomaly score $\mathcal{L}_{AE} = \text{ReLU}(\|\mathbf{x}_{cf} - \mathbf{x}'_{cf}\|_p - A_t)$, where p is the order of the L^p-norm, x'_{cf} is the output of the AE model, and A_t is the anomaly threshold.

The threshold A_t is estimated based on the residuals of the test data set using the formula $Q3 + 1.5 * IQR$ where IQR is the interquartile range and $Q3$ is the upper quartile. A higher anomaly score indicates a candidate's deviation from the original data distribution. While autoencoders have been used for unlikely counterfactuals, our contribution involves introducing an anomaly threshold A_t with the ReLU function. If the reconstruction error is below this threshold, the loss term $\mathcal{L}_{AE}$ becomes zero, aiming to halt its impact when the counterfactual is sufficiently likely. This approach can result in counterfactual examples closer to the original subject. We also included an additional term $\mathcal{L}_{ohe}$ that ensures that the generated one-hot-encoded features have valid codes. These two terms, $\mathcal{L}_{AE}$ and $\mathcal{L}_{ohe}$, constitute our Likelihood Loss (LL), $\mathcal{L}_{LL} = \mathcal{L}_{AE} + \mathcal{L}_{ohe}$, that is responsible for determining the plausibility of the counterfactual.

Actionable Counterfactual Explanations. Counterfactual explanations achieve actionability by employing a boolean vector, provided by a domain expert, to mask uncontrollable features. Features designated by this vector remain unchanged, focusing the optimization algorithm on modifying other features to meet the objective. While constraints on input features may limit achieving the exact desired objective, our method aims to find the nearest attainable target in such cases.

3.2   Particle Swarm Optimization

Particle Swarm Optimization [15] (PSO) is a nonlinear function optimization method inspired by bird flocking behavior. In this research, PSO is employed to minimize the objective function for generating counterfactual explanations. Using N randomly initialized particles in the search space, each particle evaluates the objective function at its position. Particle positions are adjusted based on both individual and neighboring experiences, facilitating effective exploration and convergence to a near-optimal solution. The algorithm's performance can be enhanced by adjusting hyperparameters, a process detailed in the results section.

3.3   Datasets Description

Turbofan Engine. Commercial Modular Aero-Propulsion System Simulation (C-MAPSS) is a NASA-developed software for turbofan engine simulation. Saxena et al. [29] used C-MAPSS to create a dataset modeling engine degradation and failure, comprising four subsets with varying complexity. This analysis focuses on a subset with one operating mode and two failure modes, involving 100 engines. To align with survival analysis, we truncated the dataset after the 300th observation, marking these as no-event observations and removing subsequent data. The original 21 sensor measurements were reduced to 13 by eliminating redundant features based on the correlation coefficient.

Predictive Maintenance Dataset (PM). This is a dataset publicly available from [11]. The dataset contains information about machines that are provided by four providers labeled `Provider1` to `Provider4` and operated by three teams labeled `TeamA`, `TeamB`, and `TeamC`. The dataset also contains information about operating conditions (pressure, moisture, and temperature) measured through sensors. The aim is to study the lifetime of these machines under the aforementioned operating conditions.

Flchain Dataset [9]. This dataset is a publicly available medical dataset aimed at studying whether the Free Light Chain (FLC) assay is a good predictor of the survival probability of patients. For the sake of visualization, we only considered the three most important features, `age`, ΣFLC (which is the summation of the `kappa` and `lambda` features in the original dataset), and `creatinine`.

4 Results and Discussion

In this section, we conduct experiments on three datasets described in Sect. 3.3 to demonstrate the effectiveness of the proposed methods in generating counterfactual explanations. We explore various approaches tailored to different types of data and tasks.

At first, we conduct a comparison and hyperparameter tuning of Particle Swarm Optimization (PSO) and Simulated Annealing (SA) algorithms. Our findings indicate that utilizing PSO improves explanation generation across all three datasets.

The first experiment utilizes a Turbofan engine dataset, showcasing the generation of counterfactuals based on Survival Scores. In the second experiment, we used our method with Survival Patterns applied to the PM dataset, which has categorical features. The third experiment highlights the model's behavior when the target Survival Pattern is unattainable due to the presence of unactionable features. We employ the Flchain dataset that has the `Age` feature, which is naturally unactionable.

It is worth noting that in the first two experiments, we compared the results generated with and without using the LL term in the loss function. In the Survival Score option, not using the LL term makes our method similar to the method proposed in [17], which makes the first experiment a direct comparison between the two works. However, the comparison is indirect in the second experiment as we employ the Survival Patterns option which is not supported by [17].

4.1 Particle Swarm Optimization Vs. Simulated Annealing

In this section, we compare the convergence of the PSO algorithm with the SA algorithm, optimizing both and assessing convergence based on the final loss. For the PSO algorithm, key hyperparameters include the number of particles, cognitive coefficient (c1), social coefficient (c2), and inertia weight (w). First, we set c1, c2, and w based on empirically validated values [4], that is, $c_1 = 1.49618$, $c_2 = 1.49618$, and $w = 0.7298$, and optimized the number of particles to minimize computation time. Subsequently, we used these values to optimize c1, c2, and w for each dataset via a random search for 1000 iterations. We performed similar optimization for SA algorithm hyperparameters (T start, T end, iterations, and step), with the final values listed in Table 1. The PSO algorithm showed better convergence which is evident in the final loss distributions in Fig. 2. This superior performance of the PSO algorithm can be attributed to the fast computation of the loss function (0.13s per step) that allowed us to use a large number of particles in the PSO algorithm. Based on the previous results, we continue with the PSO algorithm in the following sections.

Table 1. PSO and SA Optimized hyperparameters.

	PSO				SA			
Dataset	particles	c1	c2	w	T start	T end	iterations	step
CMAPSS	900	1.780533	1.911480	0.247112	0.203940	0.001783	750	0.039156
PM	5000	0.641856	1.125175	0.013541	0.264709	0.010108	680	0.285497
FLCHAIN	200	0.119708	1.473350	0.222749	0.415107	0.028302	870	0.180351

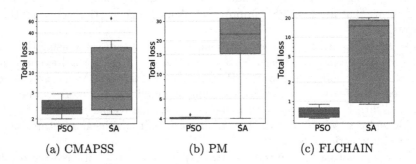

(a) CMAPSS   (b) PM   (c) FLCHAIN

Fig. 2. Loss comparison between PSO and SA on the three datasets

4.2   Survival-Scores-Based Counterfactual Explanations

This experiment illustrates the Survival Scores approach on the turbofan engine dataset. We employed the RSF model to predict the survival probability for each unit. Conducting an investigation, we randomly selected one unit, predicting its survival probability after 200 cycles. Our objective was to identify the required changes in feature values to increase its survival score by 30%. To gather statistics on the generated counterfactual examples, we conducted the experiment 20 times, both with and without utilizing the LL loss.

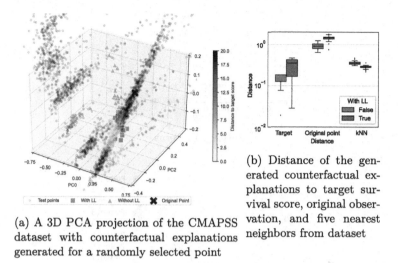

(a) A 3D PCA projection of the CMAPSS dataset with counterfactual explanations generated for a randomly selected point

(b) Distance of the generated counterfactual explanations to target survival score, original observation, and five nearest neighbors from dataset

Fig. 3. PM dataset results of counterfactual explanations with and without LL.

In Fig. 3a, a 3D PCA projection illustrates the dataset. The colors of the test observations denote proximity to the target survival score in counterfactual explanations. Desirable counterfactuals should be near the yellow region and relatively close to the original data. Counterfactuals without LL mostly lie outside the original distribution, making them less informative. Those with LL are closer to the desired region, suggesting higher explanatory validity.

In Fig. 3b, distances of counterfactual explanations to the target survival score, the original point, and the five nearest neighbors are presented. Including the LL Loss increased the distances to the original point and target survival score. However, the difference in the target score is negligible in terms of explanation validity, given its higher magnitude (10^2). Importantly, our goal was achieved as LL Loss inclusion resulted in explanations with improved proximity to the original data, measured by the proximity to the five nearest neighbors from the original points closest to the desired target score.

4.3   Survival-Patterns-Based Counterfactual Examples

This experiment aims to showcase the use of Survival Patterns in the generation of counterfactual examples. Particularly, we illustrate that our method is capable of handling one-hot-encoded categorical features.

We used the RSF model trained on the PM dataset. Our method recognized eight Survival Patterns in the prediction of the RSF model, as shown in Fig. 4a. It is worth noting that the numbers associated with the patterns do not reflect any kind of order; rather, they are assigned by the clustering algorithm.

We chose all the data points in Pattern 7 (The worst Survival Pattern) as the source pattern and generated counterfactual examples setting the target Pat-

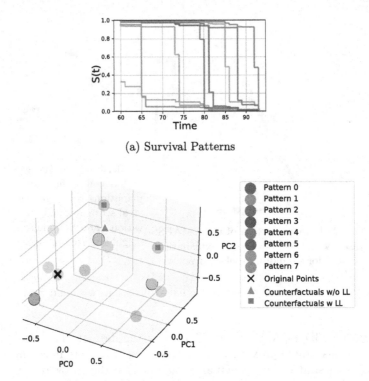

(a) Survival Patterns

(b) A 3D PCA projection of the dataset

Fig. 4. PM dataset results of counterfactual explanations with and without LL.

tern to Pattern 4 (The best Survival Pattern). For each point, we generated two counterfactual examples with and without using the Likelihood Loss. Figure 4b shows a three-dimensional PCA projection of the data colored with their respective Survival Patterns, with the counterfactual examples with and without LL. It is worth noting that each circle in the PCA plot represents many points very close to each other and belongs to a specific combination of categorical values. This means that any point far from these circles would have an invalid categorical value. Although the counterfactual examples without LL correctly changed the model's decision to the target pattern, they are unrealistic and far from the data distribution. While the counterfactual examples with LL are very close to the data distribution of the target pattern. In fact, the change from Pattern 7 to Pattern 4 requires only changing the categorical features from TeamC to TeamA or TeamB and from Provider3 to Provider2. This is what the algorithm did using the Likelihood Loss, which enabled it to generate examples with a valid one-hot encoding, shown as blue squares in Fig. 4b. Without Likelihood Loss, unrealistic examples with invalid one-hot encoding were generated, shown as a green triangle in Fig. 4b.

4.4 Actionability of Counterfactual Explanations

In this experiment, we show an example of actionable counterfactual explanations. This is done by restricting the changes in some features. This will also show a case where the predefined target Survival Pattern cannot be reached because of this restriction. RSF model is trained on the Flchain dataset, where our method identified ten Survival Patterns as shown in Fig. 5a. We chose three source examples from the worst Survival Pattern (pattern 2) and set the target pattern to the best Survival Pattern (pattern 9).

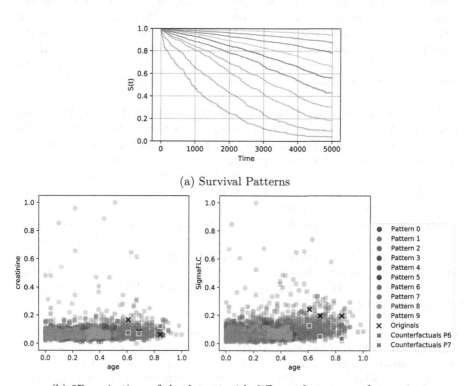

(a) Survival Patterns

(b) 2D projections of the dataset with different features on the y axis.

Fig. 5. Actionable counterfactual explanations masking the Age feature.

To generate actionable counterfactuals, we applied a mask to disallow the Age feature change. This condition made the target pattern unreachable. Our method relies on reaching the target pattern by minimizing the distance to the center of that pattern in the Z-space as shown in Eq. 2. This will get the counterfactual to the nearest-to-target pattern it can reach. Figure 5b shows two 2D projections of the data points colored with their respective patterns, the three selected source points (marked as ×s) from the source pattern (Pattern 2, colored in green) and the target pattern (Pattern 9, colored in cyan). However, because the target pattern is not reachable without moving along the Age feature, the method generated counterfactuals as close as possible to the target pattern. The respective counterfactuals are colored based on the patterns that they were able to reach (in this case, Patterns 6 and 7).

5   Conclusion

In this paper, we presented a method of generating plausible Counterfactual Explanations for black-box survival models. The proposed method finds the nearest plausible point to the explained observation that changes the output of the model. That is by changing the survival pattern or survival score of the studied example while maintaining the plausibility of the counterfactual example by minimizing the reconstruction loss of an Autoencoder model trained on the original data. The actionability is also guaranteed by restricting the changes in certain features.

We validated our method on three publicly available datasets. We generated counterfactual explanations for selected observations with and without the inclusion of Likelihood Loss. The results showed that not using the plausibility constraint can result in unlikely explanations. We also observed that restricting the change in some features can make the target pattern unattainable in some cases. However, in such a case, our method generates counterfactual explanations that are closest to the target pattern.

This work proposed a promising direction for explaining survival models using counterfactual explanations, as they can be easily interpreted by humans. A potential future work on this topic is to generate multiple diverse counterfactual explanations for a single subject. This is an important issue, which is a subject of research in Counterfactual Explanations [22] and can be used to strengthen the applicability of our method.

Acknowledgements. This research was funded by the CHIST-ERA XPM project, CHISTERA-19-XAI-012, and the CAISR+ project funded by the Swedish Knowledge Foundation. Project XPM is supported by the National Science Centre, Poland (2020/02/Y/ST6/00070), under CHIST-ERA IV program, which has received funding from the EU Horizon 2020 Research and Innovation Programme, under Grant Agreement no 857925.

References

1. Alabdallah, A., Pashami, S., Rögnvaldsson, T., Ohlsson, M. SurvSHAP: a proxy-based algorithm for explaining survival models with SHAP. In: 2022 IEEE 9th International Conference on Data Science and Advanced Analytics (DSAA), pp. 1–10 (2022). https://doi.org/10.1109/DSAA54385.2022.10032392
2. Alabdallah, A., Ohlsson, M., Pashami, S., Rögnvaldsson, T.: The concordance index decomposition: a measure for a deeper understanding of survival prediction models. Artif. Intell. Med. **148**, 102781 (2024). https://doi.org/10.1016/j.artmed.2024.102781

3. Alabdallah, A., Rognvaldsson, T., Fan, Y., Pashami, S., Ohlsson, M.: Discovering premature replacements in predictive maintenance time-to-event data. In: PHM Society Asia-Pacific Conference, vol. 4, no. 1 (2023). https://doi.org/10.36001/phmap.2023.v4i1.3609
4. Altarabichi, M.G., Nowaczyk, S., Pashami, S., Sheikholharam Mashhadi, P.: Fast genetic algorithm for feature selection - a qualitative approximation approach. Exp. Syst. Appl. **211** (2023). https://doi.org/10.1016/j.eswa.2022.118528
5. Chen, C., et al.: Predictive maintenance using cox proportional hazard deep learning. Adv. Eng. Inform. **44**, 101054 (2020)
6. Cox, D.: Regression models and life-tables. J. Roy. Statist. Soc. Ser. B (Methodol). **34**, 187–220 (1972)
7. Dandl, S., Molnar, C., Binder, M., Bischl, B.: Multi-objective counterfactual explanations. In: Parallel Problem Solving From Nature - PPSN XVI, pp. 448–469 (2020)
8. Dhurandhar, A., et al.: Explanations based on the missing: towards contrastive explanations with pertinent negatives. Adv. Neural Inf. Process. Syst. **31** (2018)
9. Dispenzieri, A., et al.: Use of nonclonal serum immunoglobulin free light chains to predict overall survival in the general population. Mayo Clin. Proc. **87**, 517–23 (2012)
10. Eberhart, R. C., Shi, Y.: Comparing inertia weights and constriction factors in particle swarm optimization. In: Proceedings of the 2000 Congress on Evolutionary Computation. CEC00 (Cat. No.00TH8512), La Jolla, vol. 1, pp. 84–88 (2000). https://doi.org/10.1109/CEC.2000.870279
11. Fotso, S., et al.: PySurvival: open source package for survival analysis modeling. https://www.pysurvival.io/
12. Ishwaran, H., Kogalur, U., Blackstone, E., Lauer, M.: Random survival forests. Ann. Appl. Stat. **2**, 841–860 (2008)
13. Kaplan, E., Meier, P.: Nonparametric estimation from incomplete observations. J. Am. Stat. Assoc. **53**, 457–481 (1958)
14. Katzman, J., Shaham, U., Cloninger, A., Bates, J., Jiang, T., Kluger, Y.: DeepSurv: personalized treatment recommender system using a Cox proportional hazards deep neural network. BMC Med. Res. Methodol. **18**, 24 (2018)
15. Kennedy, J., Eberhart, R.: Particle swarm optimization. In: Proceedings of ICNN 1995 - International Conference on Neural Networks, vol. 4, pp. 1942–1948 (1995)
16. Kovalev, M., Utkin, L., Kasimov, E.: SurvLIME: a method for explaining machine learning survival models. Knowl.-Based Syst. **203**, 106164 (2020)
17. Kovalev, M., Utkin, L., Coolen, F., Konstantinov, A.: Counterfactual explanation of machine learning survival models. Informatica **32**, 817–847 (2021)
18. Krzyziski, M., Spytek, M., Baniecki, H., Biecek, P.: SurvSHAP(t): time-dependent explanations of machine learning survival models. Knowl.-Based Syst. **262**, 110234 (2023)
19. Lang, J., Giese, M., Ilg, W., Otte, S.: Generating Sparse Counterfactual Explanations for Multivariate Time Series. arXiv (2022)
20. Lee, C., Zame, W., Yoon, J., Schaar, M.: DeepHit: a deep learning approach to survival analysis with competing risks. Proc. AAAI Conf. Artif. Intell. **32** (2018)
21. Lundberg, S., Lee, S.: A unified approach to interpreting model predictions. Adv. Neural. Inf. Process. Syst. **30**, 4765–4774 (2017)
22. Mothilal, R., Sharma, A., Tan, C.: Explaining machine learning classifiers through diverse counterfactual explanations. Proceedings of the 2020 Conference on Fairness, Accountability, and Transparency, pp. 607–617 (2020)
23. Pashami, S., et al.: Explainable Predictive Maintenance. arXiv:2306.05120 [cs.AI] (2023)

24. Pawelczyk, M., Broelemann, K., Kasneci, G.: Learning model-agnostic counterfactual explanations for tabular data. Proc. Web Conf. **2020**, 3126–3132 (2020)
25. Peto, R., Peto, J.: Asymptotically efficient rank invariant test procedures. J. Roy. Statist. Soc. Ser. A (Gen.) **135**, 185–207 (1972)
26. Pölsterl, S., Navab, N., Katouzian, A.: Fast training of support vector machines for survival analysis. In: Machine Learning and Knowledge Discovery in Databases, pp. 243–259 (2015)
27. Rahat, M., Kharazian, Z., Mashhadi, P.S., Rögnvaldsson, T., Choudhury, S. Bridging the gap: a comparative analysis of regressive remaining useful life prediction and survival analysis methods for pedictive maintenance. In: PHMAP Conference, vol. 4, no. 1 (2023). https://doi.org/10.36001/phmap.2023.v4i1.3646
28. Ribeiro, M., Singh, S., Guestrin, C.: Why should I trust you?: Explaining the predictions of any classifier. In: Proceedings of the 22nd ACM SIGKDD, pp. 1135–1144 (2016)
29. Saxena, A., Goebel, K., Simon, D., Eklund, N.: Damage propagation modeling for aircraft engine run-to-failure simulation. In: 2008 International Conference on Prognostics and Health Management, pp. 1–9 (2008)
30. Van Looveren, A., Klaise, J.: Interpretable counterfactual explanations guided by prototypes. In: ECML PKDD, vol. 2021, pp. 650–665 (2021)
31. Wachter, S., Mittelstadt, B., Russell, C.: Automated Decisions and the GDPR. Harvard Journal of Law and Technology, Counterfactual Explanations without Opening the Black Box (2017)
32. Yang, Z., Kanniainen, J., Krogerus, T., Emmert-Streib, F.: Prognostic modeling of predictive maintenance with survival analysis for mobile work equipment. Sci. Rep. **12** (2022)

Large Language Models for Binary Health-Related Question Answering: A Zero- and Few-Shot Evaluation

Marcos Fernández-Pichel(✉) [iD], David E. Losada [iD], and Juan C. Pichel [iD]

Centro de Investigación en Tecnoloxías Intelixentes (CiTIUS), Universidade de
Santiago de Compostela, Santiago de Compostela, Spain
{marcosfernandez.pichel,david.losada,juancarlos.pichel}@usc.es

Abstract. In this research, we investigate the effectiveness of Large
Language Models (LLMs) in answering health-related questions. The
rapid growth and adoption of LLMs, such as ChatGPT, have raised
concerns about their accuracy and robustness in critical domains such as
Health Care and Medicine. We conduct a comprehensive study comparing multiple LLMs, including recent models like GPT-4 or Llama2, on a
range of binary health-related questions. Our evaluation considers various context and prompt conditions, with the objective of determining
the impact of these factors on the quality of the responses. Additionally, we explore the effect of in-context examples in the performance of
top models. To further validate the obtained results, we also conduct
contamination experiments that estimate the possibility that the models have ingested the benchmarks during their massive training process.
Finally, we also analyse the main classes of errors made by these models
when prompted with health questions. Our findings contribute to understanding the capabilities and limitations of LLMs for health information
seeking.

Keywords: Binary Question Answering · Health · Large Language
Models

1 Introduction

The emergence of Large Language Models (LLMs) has induced significant
improvements in performance on various Natural Language Processing (NLP)
downstream tasks [28,29]. The appearance of BERT [7], GPT-2 [30], and GPT-3
[3], among others, has accelerated the development of LLMs. With the increasing reliance of users on online medical information [10], the reliability of these
models to provide correct responses to health-related information needs must
be put under scrutiny. The potential consequences of incorrect health-related
information can result in personal harm [26,36]. Hence, the evaluation of the
robustness of these models in this critical domain is of utmost importance. In
addition, it must be taken into account that the performance of LLMs is highly
dependent on the prompt and context provided by the questioner [3,13,20].

L. Franco et al. (Eds.): ICCS 2024, LNCS 14835, pp. 325–339, 2024.
https://doi.org/10.1007/978-3-031-63772-8_29

In this paper, we present a systematic evaluation of LLMs, exploring their potential to correctly answer health-related questions. To that end, we compare multiple LLMs and examine their performance on a range of binary health questions extracted from standardised Information Retrieval (IR) collections. Our evaluation considers a wide range of context and prompt conditions and we discuss the potential challenges and implications of using LLMs for health information needs. Our ultimate goal here is not to attain the highest possible performance but, rather, to gain insights into these AIs' responses given an assorted set of input conditions. Through the conducted experiments, we try to answer the following research questions:

- To what extent do LLMs provide correct answers to binary health-related questions? How different models perform for this task?
- To what extent does the provided context and demonstrations influence the models' answers?
- Are these models really responding to *unseen* questions? Is their effectiveness conditioned by some form of data contamination?
- What kinds of mistakes do these LLMs tend to make?

2   Related Work

Current LLMs have great potential for addressing health-related and medical information needs. However, their reliability for such critical task remains largely unknown, as most efforts have focused on general domain tasks. For instance, Jiang et al. [13] tried to optimise knowledge discovery in LLMs by generating high quality prompts (manual or automatic) and by exploiting ensemble methods. Liu et al. [19] focused their efforts on another critical aspect, the optimal configuration of in-context examples to enhance GPT-3's few shot capabilities. They found that this is specially crucial in Text Generation tasks. Other recent studies [17,20] performed systematic reviews of different models, prompts, metrics and tasks. The appearance of ChatGPT has also stimulated targeted studies to gauge the model's knowledge and utility for a number of tasks [1,2,34].

Several fine-tuned models have been specifically built for the medical domain [16,37]. However, existing evaluations of these models have been restricted to a single specialised topic, like genetics or radiology, and there is a lack of comparisons across multiple models [4,8,12,14,31,35]. Evaluating the accuracy of general-purpose LLMs for multiple types of medical queries has received little attention. A recent study [38] analysed the impact of prompts in health information seeking. However, the study was confined to a single LLM (ChatGPT) and the main goal was to evaluate prompts that incorporate supporting and contrary evidence obtained from a search engine.

Our contribution consists of a systematic evaluation of LLMs' capabilities to correctly answer health questions. We will only focus on these models' internal knowledge and we will assess their performance when prompted with different inputs and in-context examples. In contrast to previous studies, we will systematically compare several models and we will focus on general health questions, without restricting the analysis to specialised topics. Moreover, we include the recently released Llama2 model in our comparison[1]. We also report our endeavours to estimate if the models really generalise well or, by the contrary, they have seen these benchmarks during its pre-training process (i.e., we study the so-called data contamination [11,22,23,33]). Finally, we also provide an initial exploration of the most common mistakes (e.g., about a medical treatment).

3   Experimental Design

3.1   Models

We considered language models of different nature (close and open source) and architecture. We restrict the study to general-purpose models that are freely available to end-users. Thus, fine-tuned models like ChatDoctor [37] or BioBERT [16] are out of the scope of this research (as standard web users do not have the knowledge to install and invoke these tools). For a rigorous experimentation, we considered recent LLMs of different nature (including both proprietary and open source models):

- **GPT-3** is a series of models with a decoder-only structure with 175 billion parameters. Its training corpus is extensive, encompassing a variety of web sources and the entire Wikipedia, with information up to June 2021. These models were built on top of InstructGPT [25] and were fine-tuned with human feedback using reinforcement learning (RLHF). For these experiments we considered two different versions: text-davinci-002 (d002) and text-davinci-003 (d003).
- **ChatGPT** is similar to InstructGPT, but it meant a paradigm shift towards more conversational interaction [9]. Its training data goes up to September 2021. For these experiments we used gpt-3.5-turbo version (a snapshot from June 2023).
- **GPT-4**. It is a bot also designed for conversational purposes. It serves as a cutting-edge advancement in this field and surpasses ChatGPT's performance in various tasks that require human-like intelligence, such as passing an exam [24]. Its training data also goes up to September 2021. For these experiments we used gpt-4-8k version (a snapshot from June 2023).
- **Flan T5** is a sequence-to-sequence model developed by Google. It was fine-tuned on instruction-based datasets, which include a wide range of information collected up until 2022 [21]. For these experiments we used the flan-t5-xl version.
- **Llama2** is the most recent model developed by Meta AI. It was trained with over 1 million human annotations on conversational data. Its training data goes up to September 2022, but its fine-tuning also includes data up to July 2023. We used the llama-13b-chat version for these experiments.

---

[1] https://ai.meta.com/llama/.

```
<topic>
  <number>1234</number>
  <query>dexamethasone croup</query>
  <description>Is dexamethasone a good treatment for croup?</description>
  <narrative>Croup is an infection of the upper airway and causes swelling,
    which obstructs breathing and leads to a barking cough. As one kind of
    corticosteroids, dexamethasone can weaken the immune response and
    therefore mitigate symptoms such as swelling. A very useful document
    would discuss the effectiveness of dexamethasone for croup, i.e. a very
    useful document specifically addresses or answers the search topic's
    question. A useful document would provide information that would help
    a user make a decision about treating croup with dexamethasone, and
    may discuss either separately or jointly: croup, recommended treatments
    for croup, the pros and cons of dexamethasone, etc.</narrative>
  <disclaimer>We do not claim to be providing medical advice, and medical
    decisions should never be made based on the stance we have chosen.
    Consult a medical doctor for professional advice.</disclaimer>
  <stance>helpful</stance>
  <evidence>https://www.ncbi.nlm.nih.gov/pmc/articles/PMC5804741/</evidence>
</topic>
```

Fig. 1. A topic from the TREC 2021 Health Misinformation Track (Topic 101).

The first three models were tested through OpenAI's official Python API[2], while the two latter ones were tested through their Hugging Face implementation.

We are aware that there is a growing concern in the scientific community about evaluations performed on proprietary models. Sometimes, it is difficult to guarantee reproducibility of a model that suffers constant updates and where the technical intricacies (e.g. architectural design or training data) are unknown. However, we believe that the adoption of these conversational AI systems by the general population makes it necessary to put them under scrutiny. Regardless of the open or close nature of each platform, the reality is that systems such as ChatGPT are currently used by millions of users worldwide. Therefore, in this paper we made an effort for reproducibility by providing the code[3], the outputs of each round, and all the dates of the execution of the experiments[4].

3.2 Datasets

To conduct the evaluation, we used three different collections from the TREC Health Misinformation (HM) Track [5,6]. The collections consist of health-related topics, in the form of questions (e.g., "*Can wearing masks prevent COVID-19?*"), and web documents. For our experimentation, we only used the questions and their binary ground truth answers (yes/no), which represent the best understanding of current medical practice (gathered by the task organisers when creating the collection). Figure 1 shows an example of a topic. The 2020 questions are all related to COVID-19, while the 2021 and 2022 questions encompass general health information needs. The 2020 questions were released in mid

---

[2] https://openai.com/blog/openai-api.

[3] https://anonymous.4open.science/r/llm-binary-health-qa-8743.

[4] These experiments were run between September and December 2023.

2020 and, thus, we cannot discard that the LLMs have seen this benchmark within their training data. The 2021 questions were released in mid July and, thus, they might have been available for all models, except for GPT-3, whose training ended earlier. The 2022 questions, instead, could only have been seen by Flan T5 or Llama2. This therefore conforms an assorted set of health questions, with varying levels of difficulty for the models (depending on their exposure to this type of data and the level of specificity of the information needs). In any case, Sect. 6 further analyses the possibility of data contamination.

3.3   Contexts

As a core part of this research, we want to determine the effectiveness of these models for health information seeking. First of all, we try to see how well they would respond to non-expert end-users who give little or no context at all. This leads to the following input prompts:

- **no-context**: a prompt composed only of the medical question, i.e. "*Can Vitamin D cure COVID-19?*".
- **non-expert**: The text "*I am a non-expert user searching for medical advice online*" plus the corresponding question. This prompt might be representative of a regular user searching for medical advice.

In a second series of experiments, we also test more sophisticated prompts and, additionally, evaluate the effect of in-context examples. These artifacts are unlikely employed by normal users but, still, they can help to further understand and exploit the models' internal knowledge. We tested the following prompt:

- **expert**: The text "*We are a committee of leading scientific experts and medical doctors reviewing the latest and highest quality of research from PubMED. For each question, we have chosen an answer, either 'yes' or 'no', based on our best understanding of current medical practice and literature.*" plus the corresponding medical question. This prompt was designed by Waterloo's team in their participation in the TREC 2022 HM track [27]. The rationale is to bias the LLM towards reputed contents associated to high quality sources.

More elaborate prompt engineering techniques, like Chain-of-Thought (CoT) could further enhance performance, but this was left as future work. The models' temperature was set to 0, with the intention of minimising randomness in their responses. To perform an automatic evaluation, we restricted the response of the models to a single "yes" or "no" token via the model's APIs.

Table 1. Zero-shot experiments, proportion of correct answers of each model-prompt combination for the three TREC datasets.

prompt	TREC HM 2020						TREC HM 2021					
	d-002	d-003	Chat GPT	GPT 4	Llama2	FT5	d-002	d-003	Chat GPT	GPT 4	Llama2	FT5
no-context	0.84	0.91	0.84	0.79	**0.92**	0.24	**0.72**	0.76	0.68	**0.68**	**0.76**	0.44
non-expert	0.78	**0.92**	0.84	**0.90**	0.86	0.31	0.40	0.62	0.56	0.66	0.70	0.54
expert	**0.86**	0.9	**0.86**	0.86	**0.92**	**0.79**	0.36	**0.80**	**0.70**	0.66	0.72	**0.64**
avg.	0.83	0.91	0.85	0.85	0.90	0.39	0.47	0.72	0.64	0.67	0.72	0.54
std. dev.	0.04	0.01	0.01	0.06	0.03	0.30	0.20	0.09	0.08	0.01	0.03	0.10

prompt	TREC HM 2022					
	d-002	d-003	Chat GPT	GPT 4	Llama2	FT5
no-context	**0.76**	**0.76**	0.76	0.86	0.74	0.56
non-expert	0.48	0.72	0.80	0.86	0.68	0.54
expert	0.68	0.72	**0.90**	**0.88**	**0.84**	**0.74**
avg.	0.63	0.73	0.82	0.87	0.75	0.61
std. dev.	0.14	0.02	0.07	0.01	0.08	0.11

4   Zero-Shot Evaluation

As can be seen in Table 1, text-davinci-003 and Llama2 are the best performers for TREC HM 2020 and 2021 collections. With the TREC HM 2022 collection, GPT-4 and ChatGPT outstand. There are also some differences in performance among the selected prompts. As expected, the most robust context seems to be *expert* one. We hypothesise that this is due to the inclusion of keyphrases such as *"research from PubMed"* or *"medical practice and literature"*, which bias the model towards reputable sources of knowledge.

Although models are relatively stable, they still exhibit some variations depending on the input. This is concerning, as a model's effectiveness can range from 90% of correct answers to ≈ 75% of correct responses. The overall levels of effectiveness are remarkable but, still, these inconsistencies are a cause of discomfort. Even adopting the most consistent prompt (*expert*) we observe concerning outcomes. For example, GPT-4 suffers from poor performance (66%) in the 2021 dataset.

The three datasets vary in their level of difficulty. The 2020 health questions (related to COVID-19) appear to be easier for the LLMs. A plausible explanation for this phenomenon could be that the models might have already been exposed to these health questions during their massive training. We will further explore this possibility in Sect. 6. Another explanation could be that the highly relevant and significant nature of COVID-19 as a topic might have motivated a specialised curation process for the relevant data.

We also employed McNemar's test to assess the significance of the differences between the top-performing models [15]. Between ChatGPT and GPT-4, we found no significant difference in 7 out of 9 comparisons (3 collections × 3 prompts). The pair ChatGPT vs Llama2 revealed no difference in 7 out of 9 comparisons and GPT4 vs Llama2 revealed no significant difference at all. The pairwise comparisons d-003 vs ChatGPT, d-003 vs Llama2 and d-003 vs GPT-4 revealed more cases of statistical significance but, still, more than a half of the compared instances yielded a no significance result.

5   Few-Shot Evaluation

To perform an analysis of the effect of demonstrations, we focused on the test questions from TREC HM 2022. Each question was prompted to the models prepended by one-to-three demonstrations extracted from TREC HM 2021. We randomly chose three pairs of *(medical question, correct answer)* from the 2021 dataset as in-context examples and explored the effect of including them[5]. Past research [17] has shown that evaluating a narrow range of in-context examples is a solid choice.

As can be seen in Table 2, the effect of the demonstrations strongly depends on the model. For instance, both versions of GPT-3 (davinci-002 and davinci-003) and FlanT5 are the models that benefit the most from the inclusion of these in-context examples. For these models, some in-context variants led to statistical significant benefits. On the other hand, the best performing models under the zero shot-setting do not seem to benefit from the inclusion of demonstrations. Regarding types of prompts, the *expert* variant is the one that benefits most from the inclusion of the few-shot examples. In terms of the number of examples, the results suggest that prompting with more than one does not boost performance.

In Fig. 2, we plot the proportion of correct answers for Llama2 and text-davinci-002 models (expert prompting) with varying number of demonstrations. This evolution shows that the weakest model benefits the most, and more number of in-context examples does not always translate into better performance.

6   Data Contamination

LLMs have shown excellent performance in multiple NLPs but, in some cases, this might be attributed to the presence of golden truth data from the evaluated benchmarks within the LLM's training corpora. This is particularly concerning with proprietary LLMs that do not disclose information about their training data, as there is no direct way to consult the sources of the training data. A fair evaluation of these models needs to test their generalisation abilities beyond the training data. A system that directly *copies the answer* from an existing ground truth file should not be considered as intelligent. A really intelligent system is the one that learns about the world from the training corpora and, next, makes proper inferences to answer new questions. Making an analogy with education, a student who had access to the responses of the exam should fail while a student who studied all the relevant material and submitted correct answers should pass.

---

[5] We used these question-answer pairs (in this same order): (*Will wearing an ankle brace help heal achilles tendonitis?, No*), (*Does yoga improve the management of asthma?, Yes*), (*Is starving a fever effective?, No*).

Table 2. Few-shot experiments, proportion of correct answers of each model-prompt combination with three shot samples. For each row, if few shot surpasses the 0-shot is marked in bold and the symbol * marks those cases where McNemar's test ($\alpha = .05$) finds a significant difference between both variants.

prompt	d002				d003			
	0-shot	1-shot	2-shot	3-shot	0-shot	1-shot	2-shot	3-shot
no-context	0.76	0.7	**0.78**	**0.78**	0.76	**0.86**	**0.86**	**0.86**
non-expert	0.48	**0.64**	**0.74**	**0.76**	0.72	**0.82**	**0.82**	**0.82**
expert	0.68	**0.74***	**0.76***	**0.78***	0.72	**0.82***	**0.84***	**0.84***

prompt	FT5				ChatGPT			
	0-shot	1-shot	2-shot	3-shot	0-shot	1-shot	2-shot	3-shot
no-context	0.56	**0.66**	**0.64**	**0.7**	0.76	**0.82**	**0.88**	**0.84**
non-expert	0.54	**0.68***	**0.66***	**0.64**	0.8	0.8	**0.88**	**0.86**
expert	0.74	0.68*	0.72	0.72	0.9	0.84	0.88	0.88

prompt	Llama2				GPT-4			
	0-shot	1-shot	2-shot	3-shot	0-shot	1-shot	2-shot	3-shot
no-context	0.74	0.7*	**0.84**	**0.76**	0.86	0.84	0.86	0.86
non-expert	0.68	**0.72**	**0.74**	0.62*	0.86	0.86	**0.88**	**0.88**
expert	0.84	0.64*	**0.76**	0.6*	0.88	0.88	**0.92**	**0.9**

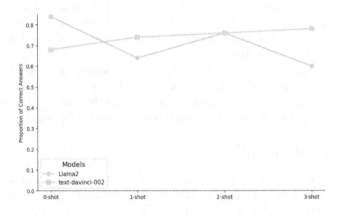

Fig. 2. Proportion of correct answers for Llama2 and text-davinci-002 models with expert prompting and different number of in-context examples.

Data contamination is an active area of research [11,22,23,33] that tries to estimate whether or not a NLP benchmark was ingested during the pre-training process. As part of our study, we have conducted data contamination estimation experiments to further validate the capacity of the LLMs to correctly answer medical and health questions.

Table 3. Results for the data contamination experiments across different models and datasets. For each guided vs general comparison, the symbol * marks those cases where the guided completion surpassed the general completion and Wilcoxon test ($\alpha = .05$) found a significant difference between both variants.

Model	Version	TREC HM 2020			TREC HM 2021		
		Levenshtein	BLEURT	ROUGE	Levenshtein	BLEURT	ROUGE
ChatGPT	General	0.45	0.41	0.27	0.46	0.48	0.29
	Guided	0.30	0.39	0.10	0.38	0.47	0.23
GPT-4	General	0.45	0.42	0.26	0.45	0.46	0.29
	Guided	0.44	0.50*	0.24	0.42	0.51*	0.25
Llama2	General	0.42	0.37	0.23	0.44	0.43	0.28
	Guided	0.43*	0.35	0.24	0.45	0.40	0.27

Model	Version	TREC HM 2022		
		Levenshtein	BLEURT	ROUGE
ChatGPT	General	0.48	0.41	0.26
	Guided	0.49	0.42	0.27
GPT-4	General	0.47	0.41	0.24
	Guided	0.47	0.43	0.25
Llama2	General	0.41	0.32	0.14
	Guided	0.40	0.25	0.15

We implemented Golchin et al. [11] heuristics, which were validated under controlled contamination experiments[6]. Essentially, the approach consists of i) prompting the model with a general instruction (with no information about the benchmark or specific identifiers of the task or split), ii) prompting the model with a guided instruction (which identifies the benchmark and task), and iii) comparing the responses against the golden truth text. If the model's output obtained from the guided prompt has higher similarity to the ground truth text compared to the model's output from the general prompt then this is evidence that the model might have seen the benchmark during its training. A statistical significance test that compares the two means of similarity can thus mark those cases where we can suspect that the model ingested the benchmark during its pre-training[7].

---

[6] The code of our data contamination experiments is available at: https://github.com/MarcosFP97/data-contamination.

[7] Note that this is a high precision but low recall method. Meaning that if the method estimates that the model ingested the dataset, we can state with some certainty that it did. However, if the heuristic does not provide evidence, we cannot state with certainty that the model did not ingest it.

In our case, the general prompt was: "*Complete the narrative field based on the query, question and answer fields such that the narrative provides an explanation for the answer to the given question. Query: {query}, Question: {question}, Answer: {answer}, Narrative:*". Meanwhile, the guided prompt was: "*You are provided with the query, question and answer fields of a topic from the TREC {year} Health Misinformation topic set. Complete the narrative field of the topic as exactly appeared in the dataset. Only rely in the original form of the topic in the dataset to complete the narrative field. Query: {query}, Question: {question}, Answer: {answer}, Narrative:*". As can be observed, the latter prompt specifies the name of the dataset and forces the model to produce the completion in its original form. These two completions produce a pair of texts that are compared against the narrative text from the actual benchmark. For computing the over-lappings (general completion vs golden truth and guided completion vs golden truth), we adopted the same metrics as in the original data contamination study [11]: BLEURT to estimate lexical similarity [32] and ROUGE-L [18] to estimate semantic relevance. Additionally, we also computed the Levenshtein distance, which measures the number of character permutations needed to transform the completion into the original data. This analysis was conducted for each available topic (i.e., each TREC question produced two completions and we report the average similarity across all topics).

We have performed this data contamination study for the most recent models, as can be seen in Table 3. Our results show some evidence that GPT-4 might have ingested TREC HM 2020 and TREC HM 2021 datasets, since we found statistically significant improvements for the guided completion with respect to the general one (in terms of semantic similarity). Levenshtein metric also shows some evidence that Llama2 might have been trained with the TREC HM 2020 collection. For ChatGPT we found no evidence that it has ingested any benchmark. Still, it performs similarly to GPT-4 and Llama2 in the TREC HM 2020 and 2021 collections (under the zero-shot setting) and there is no statistical difference between ChatGPT and these two models. This also seems to indicate that the good results by ChatGPT in the TREC HM 2020 collection are not due to contamination effects but, rather, to the peculiar characteristics of the COVID-19 topics. Furthermore, no model seems to have seen the TREC HM 2022 collection and many models performed effectively for this dataset (see Table 1). These results speak well of the capabilities of the LLMs models to correctly transfer the knowledge acquired during training and produce accurate answers to health and medical questions.

7   Error Analysis

To further understand the LLMs' behaviour for this task, we inspected the questions where none of the models provided a correct answer[8]. This analysis was

---

[8] We prompted again the models with these queries without restricting the form of token output. For the sake of simplicity, we used only the most recent models (as in the previous section).

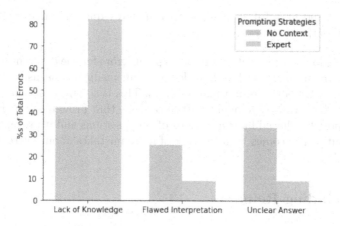

Fig. 3. Percentages of the different type of errors for the analysed prompting strategies.

done for the best performing prompt (*expert*) and for the *no-context* prompt, which arguably reflects the type of input submitted by a regular user.

We found that for the TREC HM 2020 collection, models answered incorrectly 8% and 6% of the questions with the *no-context* and *expert* prompts, respectively. For the TREC HM 2021, they failed in providing the correct answer for 12% of the cases for both prompting strategies. Finally, in the TREC HM 2022, they answered incorrectly 4% of the queries for both strategies. These results confirmed that TREC HM 2021 is the most difficult collection, with a larger percentage of errors. It also seems that providing no context derives in a greater or similar percentage of failed queries than using the expert prompt. After manually inspecting the models' outputs, we could organise the errors found into a taxonomy that represents the most common health advice mistakes:

- **Lack of knowledge about current medical consensus**: Sometimes, models provide answers that go against the medical consensus. For instance, to the question "Can Hydroxychloroquine worsen COVID-19?", ChatGPT answered "no, there is no evidence that hydroxychloroquine worsens covid-19..." while medical evidence says otherwise[9].
- **Flawed interpretation of the question**: Here, LLMs misinterpret the question. For example, "Can bleach prevent COVID-19? No, bleach should not be ingested...". But the ground truth has the most obvious interpretation of this question (the use of bleach for surface disinfection can actually prevent COVID-19). A human would hardly interpret the question in this way.
- **Unclear answer**: We include in this category the responses in which models did not provide a blunt answer. These cannot be counted as correct responses but the LLM's output is arguably useful, for example, "sit-ups can be both

---

[9] FDA cautions against use of hydroxychloroquine for COVID-19.

beneficial and harmful, depending on your individual circumstances and the way you perform the exercise...".

Figure 3 plots the percentage of each type of errors for the different prompting conditions. In summary, lack of knowledge about medical consensus is the most common error for both prompting strategies. This is a concerning outcome as it is the most dangerous type of error. It also seems that providing expert context mitigates possible flawed interpretations of the questions and it prevents unclear answers, but it also comes with the cost of more mistakes about current medical consensus.

8   Concluding Remarks

We have conducted an exhaustive evaluation on the ability of a set of LLMs in providing the correct answer to health and medical questions. We have evaluated the models with three different collections of medical question-answer pairs and prompted them with different contexts, ranging from intricate prompts to simpler prompts (close to those possibly submitted by non-expert users).

Under the zero-shot setting, the most sophisticated and modern models performed similarly. However, there are still some causes of discomfort, e.g. in some cases the models provide less than 70% of correct answers. This is a low figure for a critical task such as health information seeking. We also found out that intricate prompting strategies enhanced the performance compared with simpler contexts. From our point of view, this is an obstacle for the adoption of these models for health question answering. Note that end users are unlikely to produce very sophisticated prompts. We also discovered that including few-shot examples enhanced the performance even with the most complex prompt. However, the effect of the demonstrations is tied to the model, being the simpler ones the most benefited from the provided examples.

On the other hand, our data contamination experiments have shed light on the generalisation abilities of the models. We found no evidence that ChatGPT has ingested any of the collections of health questions and, additionally, our results indicate that no model has seen the TREC HM 2022 collection. This breaks a lance in favour of the models, as many of the health questions were new for them but, still, their performance was remarkable.

Finally, we conducted an error analysis in which we inspected the models' answers. We organised the errors into a taxonomy and identified that, in some cases, models provided advice that goes against the well-known medical consensus. This behaviour, which also happens with the most sophisticated prompts, is a barrier for the wide adoption of these models in their current form.

Limitations

We are aware that these conversational AI systems are highly sensible to the input prompt. Our study represents an initial exploration with some manually defined prompts and further prompt optimisation was left for future work. Our ultimate goal here was not to achieve the highest possible performance but, rather, to test these AIs with an assorted set of input conditions. Other strategies like chain-of-thought (CoT) prompting or prompt tuning were also left for future work. With this research we do not intend to pursue the replacement of human professionals that provide health advice. In fact, we firmly believe that human validation is crucial to learn more about these systems and to leverage AIs systems. For example, exploiting LLMs for automating certain documentation tasks (e.g., collecting and curating recommendations generated by AI systems). Finally, we are also aware that the evaluation of proprietary systems causes some concern in the scientific community (because of the lack of transparency about crucial aspects, such as their design and training data). However, as scientists, we cannot ignore the fact that these tools are used by millions of individuals worldwide and, thus, we need to evaluate the risks involved. In our study, we have made a special effort for transparency providing the code, outputs and dates of all experiments.

Acknowledgments. The authors thank: i) the financial support supplied by the Consellería de Cultura, Educación, Formación Profesional e Universidades (accreditation 2019–2022 ED431G-2019/04, ED431C 2022/19) and the European Regional Development Fund, which acknowledges the CiTIUS-Research Center in Intelligent Technologies of the University of Santiago de Compostela as a Research Center of the Galician University System, and ii) the financial support supplied by projects PLEC2021-007662 and PID2022-137061OB-C22 (Ministerio de Ciencia e Innovación, Agencia Estatal de Investigación, Proyectos de Generación de Conocimiento; supported by the European Regional Development Fund). Finally, David E. Losada thanks the financial support obtained from project SUBV23/00002 (Ministerio de Consumo, Subdirección General de Regulación del Juego).

Disclosure of Interests. The authors have no competing interests to declare that are relevant to the content of this article.

References

1. Ahn, C.: Exploring ChatGPT for information of cardiopulmonary resuscitation. Resuscitation **185**, 109729 (2023)
2. Biswas, S.S.: Potential use of chat GPT in global warming. Ann. Biomed. Eng. **51**, 1–2 (2023)
3. Brown, T., et al.: Language models are few-shot learners. Adv. Neural. Inf. Process. Syst. **33**, 1877–1901 (2020)
4. Chervenak, J., Lieman, H., Blanco-Breindel, M., Jindal, S.: The promise and peril of using a large language model to obtain clinical information: ChatGPT performs strongly as a fertility counseling tool with limitations. Fertil. Steril. **120**, 575–583 (2023)
5. Clarke, C., Maistro, M., Smucker, M.: Overview of the TREC 2021 health misinformation track. In: Proceedings of the Thirtieth Text REtrieval Conference, TREC (2021)
6. Clarke, C., Maistro, M., Smucker, M., Zuccon, G.: Overview of the TREC 2020 health misinformation track. In: Proceedings of the Twenty-Nine Text REtrieval Conference, TREC, pp. 16–19 (2020)

7. Devlin, J., Chang, M.W., Lee, K., Toutanova, K.: BERT: pre-training of deep bidirectional transformers for language understanding. arXiv preprint arXiv:1810.04805 (2018)
8. Duong, D., Solomon, B.D.: Analysis of large-language model versus human performance for genetics questions. Eur. J. Hum. Genet. **32**, 1–3 (2023)
9. Forbes: Introducing ChatGPT, November 2022. https://openai.com/blog/chatgpt. Acessed 4 Apr 2023
10. Fox, S.: Health topics: 80% of internet users look for health information online. Pew Internet & American Life Project (2011)
11. Golchin, S., Surdeanu, M.: Time travel in LLMs: tracing data contamination in large language models. arXiv preprint arXiv:2308.08493 (2023)
12. Holmes, J., et al.: Evaluating large language models on a highly-specialized topic, radiation oncology physics. arXiv preprint arXiv:2304.01938 (2023)
13. Jiang, Z., Xu, F.F., Araki, J., Neubig, G.: How can we know what language models know? Trans. Assoc. Comput. Linguist. **8**, 423–438 (2020)
14. Johnson, D., et al.: Assessing the accuracy and reliability of AI-generated medical responses: an evaluation of the Chat-GPT model (2023)
15. Lachenbruch, P.A.: Mcnemar test. Wiley StatsRef: Statistics Reference Online (2014)
16. Lee, J., et al.: BioBERT: a pre-trained biomedical language representation model for biomedical text mining. Bioinformatics **36**(4), 1234–1240 (2020)
17. Liang, P., et al.: Holistic evaluation of language models. arXiv preprint arXiv:2211.09110 (2022)
18. Lin, C.Y., Och, F.: Looking for a few good metrics: rouge and its evaluation. In: NTCIR Workshop (2004)
19. Liu, J., Shen, D., Zhang, Y., Dolan, B., Carin, L., Chen, W.: What makes good in-context examples for GPT-3? arXiv preprint arXiv:2101.06804 (2021)
20. Liu, P., Yuan, W., Fu, J., Jiang, Z., Hayashi, H., Neubig, G.: Pre-train, prompt, and predict: a systematic survey of prompting methods in natural language processing. ACM Comput. Surv. **55**(9), 1–35 (2023)
21. Longpre, S., et al.: The flan collection: designing data and methods for effective instruction tuning. arXiv preprint arXiv:2301.13688 (2023)
22. Magar, I., Schwartz, R.: Data contamination: from memorization to exploitation. arXiv preprint arXiv:2203.08242 (2022)
23. Nori, H., King, N., McKinney, S.M., Carignan, D., Horvitz, E.: Capabilities of GPT-4 on medical challenge problems. arXiv preprint arXiv:2303.13375 (2023)
24. OpenAI: GPT-4 technical report. arXiv:submit/4812508 (2023)
25. Ouyang, L., et al.: Training language models to follow instructions with human feedback. Adv. Neural. Inf. Process. Syst. **35**, 27730–27744 (2022)
26. Pogacar, F.A., Ghenai, A., Smucker, M.D., Clarke, C.L.: The positive and negative influence of search results on people's decisions about the efficacy of medical treatments. In: Proceedings of the ACM SIGIR International Conference on Theory of Information Retrieval, pp. 209–216 (2017)
27. Pradeep, R., Lin, J.: Towards automated end-to-end health misinformation free search with a large language model. In: Goharian, N., Tonellotto, N., He, Y., Lipani, A., McDonald, G., Macdonald, C., Ounis, I. (eds.) ECIR 2024. LNCS, vol. 14611, pp. 78–86. Springer, Cham (2024). https://doi.org/10.1007/978-3-031-56066-8_9
28. Radfar, M., Mouchtaris, A., Kunzmann, S.: End-to-end neural transformer based spoken language understanding. arXiv preprint arXiv:2008.10984 (2020)
29. Radford, A., Narasimhan, K., Salimans, T., Sutskever, I., et al.: Improving language understanding by generative pre-training (2018)

30. Radford, A., et al.: Language models are unsupervised multitask learners. OpenAI blog **1**(8), 9 (2019)
31. Samaan, J.S., et al.: Assessing the accuracy of responses by the language model ChatGPT to questions regarding bariatric surgery. Obes. Surg. **33**, 1–7 (2023)
32. Sellam, T., Das, D., Parikh, A.P.: BLEURT: learning robust metrics for text generation. arXiv preprint arXiv:2004.04696 (2020)
33. Sianz, O., Campos, J.A., García-Ferrero, I., Etxaniz, J., Agirre, E.: Did ChatGPT cheat on your test? (2023). https://hitz-zentroa.github.io/lm-contamination/blog/. Accessed 19 Jan 2024
34. Surameery, N.M.S., Shakor, M.Y.: Use chat GPT to solve programming bugs. Int. J. Inf. Technol. Comput. Eng. (IJITC) **3**(01), 17–22 (2023). ISSN 2455-5290
35. Thirunavukarasu, A.J., et al.: Trialling a large language model (ChatGPT) in general practice with the applied knowledge test: observational study demonstrating opportunities and limitations in primary care. JMIR Med. Educ. **9**(1), e46599 (2023)
36. Vigdor, N.: Man fatally poisons himself while self-medicating for coronavirus, doctor says, March 2020. https://www.nytimes.com/2020/03/24/us/chloroquine-poisoning-coronavirus.html. Accessed 9 June 2022
37. Yunxiang, L., Zihan, L., Kai, Z., Ruilong, D., You, Z.: ChatDoctor: a medical chat model fine-tuned on llama model using medical domain knowledge. arXiv preprint arXiv:2303.14070 (2023)
38. Zuccon, G., Koopman, B.: Dr ChatGPT, tell me what I want to hear: How prompt knowledge impacts health answer correctness. arXiv preprint arXiv:2302.13793 (2023)

Brain Tumor Segmentation Using Ensemble CNN-Transfer Learning Models: DeepLabV3plus and ResNet50 Approach

Shoffan Saifullah[1,2(✉)] [iD] and Rafał Dreżewski[1] [iD]

[1] Faculty of Computer Science, AGH University of Krakow, 30-059 Krakow, Poland
saifulla@agh.edu.pl,shoffans@upnyk.ac.id, drezew@agh.edu.pl
[2] Department of Informatics, Universitas Pembangunan Nasional Veteran
Yogyakarta, Yogyakarta 55281, Indonesia

Abstract. This study investigates the impact of advanced computational methodologies on brain tumor segmentation in medical imaging, addressing challenges like interobserver variability and biases. The DeepLabV3plus model with ResNet50 integration is rigorously examined and augmented by diverse image enhancement techniques. The hybrid CLAHE-HE approach achieves exceptional efficacy with an accuracy of 0.9993, a Dice coefficient of 0.9690, and a Jaccard index of 0.9404. Comparative analyses against established models, including SA-GA, Edge U-Net, LinkNet, MAG-Net, SegNet, and Multi-class CNN, consistently demonstrate the proposed method's robustness. The study underscores the critical need for continuous research and development to tackle inherent challenges in brain tumor segmentation, ensuring insights translate into practical applications for optimized patient care. These findings offer substantial value to the medical imaging community, emphasizing the indispensability of advancements in brain tumor segmentation methodologies. The study outlines a path for future exploration, endorsing ensemble models like U-Net, ResNet-U-Net, VGG-U-Net, and others to propel the field toward unprecedented frontiers in brain tumor segmentation research.

Keywords: Brain Tumor Segmentation · Deep learning · Image enhancement · Medical Imaging · Ensemble Methods

1 Introduction

Medical imaging, especially in brain tumor segmentation, has evolved significantly with the seamless integration of computational methods into neurology [1]. Despite the improved visualization of Magnetic Resonance Imaging (MRI), the complex nature of brain tumors necessitates advanced automated approaches [22]. Traditional manual segmentation introduces variability and biases, emphasizing the need for precise medical decisions in treatment planning, monitoring, and evaluating efficacy [9].

The rise of computational techniques, especially Convolutional Neural Networks (CNNs), has opened avenues to overcome manual brain tumor segmentation limitations [12]. Proficient in capturing intricate patterns, CNNs excel in delineating tumor regions [25]. However, challenges like data heterogeneity, class imbalance, and model generalization demand a shift to data-driven approaches [19]. Promising solutions like SegNet [30], LinkNet [39,40], and U-Net [3,38] with transfer learning [17,32] require rigorous validation and robust data augmentation. While ResNet50 and U-Net perform well, addressing brain tumor imaging complexity necessitates efficient methods [3]. Correlating segmentation with patient survival emphasizes the need for accurate methods. Researchers advocate for comprehensive CNNs, demonstrating effectiveness in automated detection and proposing robust CNN U-Net models for medical-grade applications.

The motivation for this research arises from the imperative to overcome limitations in traditional manual brain tumor segmentation. Integrating computational techniques, particularly deep-learning models, offers a transformative opportunity to expedite analyses and produce highly accurate and reproducible results. This research aims to explore the effectiveness of CNN-Transfer Learning models [35,37], proposing a novel approach that integrates DeepLabV3plus and ResNet50 for efficient and accurate brain tumor segmentation. Rigorous evaluation will compare the performance of this model against established benchmarks such as modified U-Net, SegNet, LinkNet, and others. Additionally, the research assesses the impact of image enhancement techniques, including a hybrid of Contrast-Limited Adaptive Histogram Equalization (CLAHE) and Histogram Equalization (HE), on segmentation accuracy [33,34]. This multifaceted approach aims to advance our understanding of computational methods in medical imaging and contribute insights that may enhance the precision and efficiency of brain tumor segmentation.

The structure of this paper includes the following sections: Sect. 2 offers an overview of related works, emphasizing strengths and limitations. Section 3 details proposed methods addressing identified challenges, including DeepLabV3plus and ResNet50. Section 4 presents results, compares performance, and discusses implications. Finally, Sect. 5 concludes with a summary of contributions and potential avenues for future research in brain tumor segmentation.

2   Related Works

Research on brain tumor segmentation and ensemble CNN models has proven them to be effective in various medical imaging applications [23]. Studies on glioblastoma segmentation and ensemble learning in hyperspectral image processing [20] demonstrate their versatility. The ViT-CNN ensemble model excels in classifying acute lymphoblastic leukemia [11], emphasizing its potential in medical diagnosis.

Significant strides have been achieved in brain tumor segmentation, deep learning, and medical image analysis. The BRATS benchmark [28] has been pivotal in unveiling computational challenges. Ensemble CNN models exhibit remarkable performance [5], emphasizing deep learning's prowess in categorizing brain cancers. The importance of data augmentation [15], advancements like the U-Net application [3,13,17], and innovations such as dockerized segmentation algorithms contribute to the expanding research landscape. Refinements in U-Net architecture [21] and a novel patch-based dictionary learning algorithm extend segmentation methodologies. Contributions span from automatic segmentation methods to developing deep learning CNN models for segmentation and classification [9].

Further advances in brain tumor detection, such as the TransConver network, showcased the potential of transformer and convolution parallel networks [18]. High accuracy, particularly on the BraTS 2018 dataset, underscores the importance of dataset-specific evaluations [26]. Employing CNN, deep learning, and AI algorithms achieves notable accuracies in categorizing and identifying brain tumors. Additional research focuses on classifying and segmenting tumors using pre-trained AlexNet, advancing transfer learning for brain tumor classification, and addressing challenges in automated tumor analysis [41]. Demonstrated effectiveness in transfer learning for brain tumor multi-classification highlights its applicability across diverse scenarios. Emphasizing the role of automated segmentation in enhancing research precision, speed, and reproducibility in medical imaging is also underscored [24].

This research stands out as state-of-the-art within this rich landscape, presenting a significant advancement in brain tumor segmentation using ensemble CNN-Transfer Learning Models [35], specifically the DeepLabV3plus and ResNet50 approach. The proposed method not only leverages the strengths observed in previous studies but also addresses limitations, contributing to the continual refinement and progress in medical image analysis.

3   Materials and Methods

3.1   Brain MRI Datasets and Preprocessing

In this study, a comprehensive dataset comprising 3064 TI-CE MRIs obtained from 233 patients forms the foundation of our research [6]. The dataset is meticulously categorized into three distinct classes: meningiomas (708 images), gliomas (1426 images), and pituitary tumors (930 images). Each image within the dataset possesses a corresponding ground truth, represented by masks, enabling the identification of abnormal regions. Figure 1 illustrates representative samples of brain MRI images accompanied by their respective ground truth masks, showcasing instances of Meningioma (a), Glioma (b), and Pituitary tumors (c).

To optimize the deep learning model's processing efficiency, a crucial preprocessing step involves resizing the images to a standardized resolution of 256×256 pixels [33,34]. This resizing facilitates computational efficiency and ensures consistency across diverse input images. Following the resizing process, image enhancement techniques are employed to elevate the overall quality of the images.

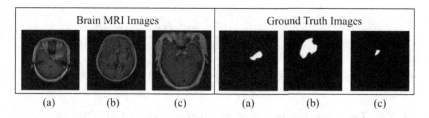

Fig. 1. Samples of brain MRI with the ground truth (masks) of (a) Meningioma, (b) Glioma, and (c) Pituitary.

The first scenario involves the application of Histogram Equalization (HE), a method widely recognized for enhancing image contrast. The second scenario employs Contrast Limited Adaptive Histogram Equalization (CLAHE) [34], which adapts the enhancement process to local regions, enhancing details in specific areas. The third and fourth scenarios involve hybrid approaches, combining HE and CLAHE in different sequences: HE-CLAHE and CLAHE-HE [34,36]. These variations exploit the synergistic effects of global and local contrast enhancements. HE operates by redistributing the intensity levels of the image histogram to cover the entire available range. The transformation function $T(rk)$ (Eq. (1)) for each pixel intensity rk in the original image is calculated based on the image histogram $H(i)$, probability pixels $P_x(i)$, and commutative distribution function $cdf_x(i)$.

$$T(rk) = (L-1) \cdot cdf_x(rk) \tag{1}$$

where L is the number of intensity levels. The commutative distribution function is computed by the sum of probabilities up to the intensity level rk (Eq. (2)).

$$cdf_x(rk) = \sum_{j=0}^{rk} P_x(j) \tag{2}$$

Finally, the probability pixels $P_x(j)$ are calculated as the ratio of pixels with intensity $H(i)$ to the total number of pixels (Eq. (3)).

$$P_x(j) = \frac{H(i)}{N} \tag{3}$$

where N is the total number of pixels. This process yields the distribution of the HE-transformed image $h(v)$, effectively enhancing contrast (Eq. (4)).

$$h(v) = \frac{cdf(v) - cdf_{min}}{n - cdf_{min}} \tag{4}$$

Unlike HE, CLAHE adapts its enhancement process with a clip limit (β), preventing over-enhancement and artifacts. The transformation function $T(rk)$ for each pixel intensity rk is given by Eq. (5). The clip limit β (Eq. (6)) controls the amount of contrast enhancement.

$$T(rk) = (L-1)\frac{cdf_x(rk)}{max\left(\beta - cdf_x(rk)\right)} \tag{5}$$

$$\beta = \frac{M}{n}\left(1 + \frac{\alpha}{100}(S_max - 1)\right) \tag{6}$$

3.2   CNN-Transfer Learning Approaches: DeepLabV3plus with ResNet50

Our proposed approach utilizes ensemble CNN-Transfer Learning for brain tumor segmentation, integrating the DeepLabV3plus model [31] with the ResNet50 backbone [27], as illustrated in Fig. 2. This ensemble design allows for enhanced predictive capabilities, utilizing the strengths of both models to improve segmentation accuracy and generalization. We adopted a standard data split of 80% for training and 20% for validation, training the model with a learning rate of 10^{-3} and the Adam optimizer over 50 epochs, with a mini-batch size of 32 iterations. The combined architecture optimally captures contextual information and deep feature extraction, resulting in superior segmentation accuracy.

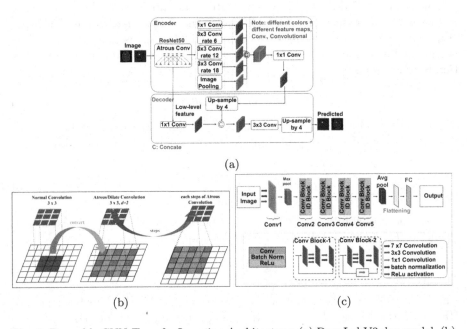

Fig. 2. Ensemble CNN-Transfer Learning Architecture: (a) DeepLabV3plus model, (b) the transformation from standard convolution to Atrous/Dilated convolution showcasing sequential processing of dilation pixels, (c) ResNet50 backbone utilized for accurate prediction in brain tumor segmentation.

DeepLabv3plus Layers are architectural innovations which play a central role in the proficiency of our ensemble CNN-transfer learning system for brain tumor prediction and segmentation (Fig. 2(a)). DeepLabv3plus brings advanced semantic segmentation capabilities to the model, explicitly emphasizing the precise delineation of object boundaries, a critical requirement in medical image analysis [2]. The architecture includes atrous convolution, also known as dilated convolution, which allows for an expanded receptive field without an increase in model parameters. This feature enables the accurate segmentation of brain tumors by capturing information from fine-grained to high-level features.

Atrous Spatial Pyramid Pooling (ASPP) is another essential element that utilizes parallel dilated convolutions to capture multiscale information effectively, ensuring recognition of both large and small tumor regions (Fig. 2(b)). Additionally, feature refinement and decoder modules contribute to model precision, allowing the distinction of tumor boundaries even in complex images. These layers significantly enhance the model's proficiency in brain tumor segmentation.

Further enhancing the understanding of Atrous Convolution, Fig. 2(b) illustrates the transformation from standard convolution to Atrous/Dilated convolution. The first two diagrams in Fig. 2(b) depict the fundamental shift in the convolutional operation, highlighting how Atrous convolution introduces gaps in the filter, enabling the model to capture information from a broader context. The third diagram in Fig. 2(b) illustrates the sequential processing of dilation pixels, emphasizing the stepwise integration of contextual information. This mechanism is pivotal in our approach, as it empowers the model to better discern fine details and boundaries in brain tumor images.

Figure 2(c) delves into the ResNet50 architecture, particularly emphasizing the Atrous Convolutional layer. This layer plays a crucial role in feature extraction, allowing the model to capture intricate spatial dependencies within brain tumor images. The utilization of Atrous Convolution enhances the network's receptive field, enabling the extraction of more contextual information for improved segmentation performance.

3.3 Performance Evaluation Metrics for Segmentation Approach

The evaluation of our segmentation approach relies on key performance metrics [35], each providing valuable insights into the model's effectiveness [16]. Accuracy (ACC) is the fundamental metric that quantifies the overall correctness of the model's predictions by considering true positives (TP), true negatives (TN), false positives (FP), and false negatives (FN). The accuracy computes the ratio of correctly classified pixels to the total number of pixels (Eq. (7)).

$$ACC = \frac{(TP + TN)}{(TP + TN + FP + FN)} \tag{7}$$

The loss function (L) serves as a measure of dissimilarity between predicted and ground truth masks. Often expressed as cross-entropy, the loss is calculated by Eq. (8), where y represents the ground truth, $\widehat{y}$ is the predicted mask, and N is the total number of pixels.

$$L(y, \widehat{y}) = -\frac{1}{N} \sum_{i=1}^{N} yi \cdot log(\widehat{y}i) + (1 - yi) \cdot log(1 - \widehat{y}i) \tag{8}$$

The Dice coefficient, denoted as Dice, assesses spatial overlap and is computed by Eq. (9). A higher Dice coefficient indicates better spatial alignment between the predicted and ground truth tumor regions.

$$Dice = \frac{(2xTP)}{(2xTP + FP + FN)} \tag{9}$$

The Jaccard index, or Intersection over Union (IoU), measures region similarity and is given by Eq. (10). This metric considers the common area between the predicted and actual tumor regions.

$$Jaccard = \frac{(TP)}{(TP + FP + FN)} \tag{10}$$

4 Results and Discussion

In this section, we present the results of our proposed approach, which integrates DeepLabV3plus and ResNet50 for brain tumor segmentation. Our model is evaluated using the BRATS dataset [6], and we compare its performance with established benchmarks from other researchers. Additionally, we explore the impact of image enhancement techniques, such as HE, CLAHE, and hybrid approaches, on segmentation accuracy.

4.1 Results of CNN-Transfer Learning for Brain Tumor Segmentation: DeepLabV3plus and ResNet50 Approach

This section presents an in-depth exploration of the DeepLabV3plus model with the ResNet50 backbone for brain tumor segmentation. It sheds light on the nuanced impact of various image enhancement strategies on its performance. Four pivotal performance metrics—accuracy, loss, Dice, and Jaccard indices— are meticulously employed to provide a holistic understanding of the model's proficiency across diverse preprocessing techniques. The evaluated model variants encompass image enhancement methods, including HE, CLAHE, and hybrid approaches (HE-CLAHE and CLAHE-HE), as summarized in Table 1.

Commencing with the baseline scenario without image enhancement ("–"), the model achieves an impressive accuracy of 0.9987. However, the discerning analysis of the Dice coefficient (0.9409) and Jaccard index (0.9048) unveils potential areas for improvement, specifically in capturing the intricate boundaries of brain tumors. Subsequent exploration of image enhancement strategies reveals HE as a pivotal technique, showcasing notable improvements. HE yields the accuracy of 0.9991, reduced loss (0.0020), and significant advancements in Dice (0.9618) and Jaccard (0.9270) indices, underlining its efficacy in contrast enhancement

Table 1. Comparison of brain tumor segmentation performance using different image enhancement techniques.

Image Enhancement	DeepLabV3plus with ResNet50			
	Accuracy	Loss	Dice	Jaccard
–	0.9987	0.0031	0.9409	0.9048
HE	0.9991	0.0020	0.9618	0.9270
CLAHE	0.9990	0.0024	0.9558	0.9160
HE-CLAHE	0.9991	0.0021	0.9599	0.9242
CLAHE-HE	0.9993	0.0019	0.9690	0.9404

for more precise tumor region segmentation. Building upon the success of HE, CLAHE contributes to further refinement, achieving the accuracy of 0.9990 and showcasing benefits in capturing local details, particularly in enhancing local features within brain tumor imaging data.

The combination of HE and CLAHE (HE-CLAHE) exhibits a synergistic effect, yielding an accuracy of 0.9991 and higher Dice (0.9599) and Jaccard (0.9242) indices. This combination underscores the importance of a judicious blend of global and local contrast enhancements for superior segmentation outcomes, with the sample of results in Fig. 3.

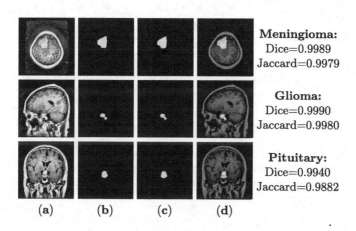

Meningioma:
Dice=0.9989
Jaccard=0.9979

Glioma:
Dice=0.9990
Jaccard=0.9980

Pituitary:
Dice=0.9940
Jaccard=0.9882

(a)   (b)   (c)   (d)

Fig. 3. Sample of segmentation results illustrating (a) brain MRI with CLAHE-HE, (b) ground truth (mask), (c) predictions from our proposed method, and (d) overlap of the original image.

Notably, reversing the order of enhancement techniques (CLAHE-HE) yields the highest overall performance, emphasizing the significance of the sequence in optimizing the model's ability to capture intricate details in brain tumor images. The results highlight that a careful consideration of preprocessing sequences

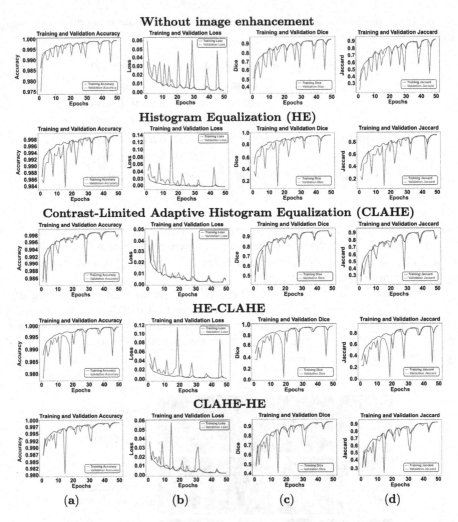

Fig. 4. Performance evaluation metrics of the proposed DeepLabV3plus with ResNet50 model based on (a) Accuracy, (b) Loss, (c) Dice, and (d) Jaccard indices, showcasing the impact of various image enhancement scenarios on brain tumor segmentation proficiency.

tailored to the characteristics of medical imaging data significantly influences the model's segmentation efficacy. The model achieves its peak performance with CLAHE-HE image enhancement, as confirmed by the graph evaluation in Fig. 4, demonstrating the robustness and stability of the proposed approach across diverse brain tumor segmentation scenarios.

4.2   Comparison of the Proposed Method with Other Approaches

This section provides a nuanced analysis of our proposed method's segmentation performance across specific brain tumor classes, including Meningioma, Glioma, and Pituitary, compared to previous studies, as shown in Table 2. The evaluation metrics considered are accuracy, Dice coefficient, and Jaccard index.

Table 2. Comparison of segmentation efficacy in individual classes of brain tumors for different methods.

Method	Dataset (Class)	Accuracy	Dice	Jaccard
Proposed Method	Meningioma	**0.9995**	**0.9809**	**0.9627**
SA-GA [14]		0.8777	–	–
Edge U-Net [21]		–	0.8880	0.7743
Multiclass CNN [7]		–	0.8940	–
Proposed Method	Glioma	**0.9995**	**0.9596**	**0.9233**
SA-GA [14]		0.9785	–	–
Edge U-Net [21]		–	0.9176	0.8747
Multiclass CNN [7]		–	0.7790	–
Proposed Method	Pituitary	**0.9995**	**0.9600**	**0.9240**
SA-GA [14]		0.9512	–	–
Edge U-Net [21]		–	0.8728	0.7985
Multiclass CNN [7]		–	0.8130	–

As detailed in Table 2, our proposed method demonstrates exceptional segmentation performance for Meningioma, achieving the remarkable accuracy of 0.9995. The Dice coefficient (0.9809) and Jaccard index (0.9627) further underscore the precision in delineating Meningioma boundaries. In contrast, the comparison with SA-GA, Edge U-Net, and Multi-class CNN reveals notable superiority, signifying the efficacy of our approach in capturing the intricate details of Meningioma structures. Our method outperforms existing models, showcasing its potential for accurate Meningioma segmentation in medical imaging.

In the case of Glioma, as illustrated in Table 2, our proposed method continues to excel with the accuracy of 0.9995, indicating robust segmentation capabilities. The Dice coefficient (0.9596) and Jaccard index (0.9233) further validate the accuracy and precision in capturing Glioma regions. Compared to SA-GA, Edge U-Net, and Multi-class CNN, our method consistently outperforms these models, emphasizing its effectiveness in Glioma segmentation. The superior performance underscores the potential clinical relevance of our method for accurate Glioma delineation in medical images.

As depicted in Table 2, our method achieves outstanding results in Pituitary segmentation, attaining the accuracy of 0.9995. The Dice coefficient (0.9600) and Jaccard index (0.9240) further highlight the efficacy of our approach in capturing Pituitary tumor structures. In contrast to SA-GA, Edge U-Net, and Multi-class CNN, our method demonstrates superior segmentation performance. The results affirm the applicability of our proposed method for precise Pituitary tumor segmentation, showcasing its potential impact on diagnostic accuracy.

The overall comparison across different brain tumor classes emphasizes the versatility and consistency of our proposed method. By outperforming existing models across multiple tumor types, our method showcases its robustness and potential applicability in diverse clinical scenarios. The superior segmentation outcomes across Meningioma, Glioma, and Pituitary tumors collectively contribute to the comprehensive effectiveness of our proposed method in brain tumor segmentation.

4.3 Comparative Analysis of Segmentation Performance Across All Data

This section thoroughly analyzes our proposed method's segmentation performance across diverse datasets, benchmarking it against various state-of-the-art models detailed in Table 3. Key evaluation metrics, including accuracy, Dice coefficient, and Jaccard index, offer a comprehensive understanding of the overall effectiveness of our approach. The results in Table 3 underscore the robust performance of our method, boasting an impressive accuracy of 0.9993, with a high Dice coefficient (0.9690) and Jaccard index (0.9404), validating its efficacy in precisely delineating brain tumor structures. This overall excellence highlights the versatility of our approach, positioning it as a compelling choice for diverse medical imaging applications.

Our proposed method consistently exhibits superior performance across all metrics compared to a range of existing models. Outperforming DeepLabV3plus ResNet18, SegNet CNN-Autoencoder, and others, our method showcases advancements in segmentation quality, achieving higher accuracy and excelling in both Dice and Jaccard metrics. These results underscore its effectiveness in capturing intricate details, emphasizing its potential for accurate and reliable brain tumor segmentation. Moreover, the method surpasses modified U-Net, SegNet, and LinkNet in accuracy, Dice, and Jaccard metrics, reinforcing its efficacy across a diverse range of existing models and highlighting its potential for widespread adoption in clinical and research settings.

Table 3. Comparison of segmentation efficacy across all data with different approaches.

Method	Accuracy	Dice	Jaccard
Proposed Method	**0.9993**	**0.9690**	**0.9404**
DeepLabV3plus ResNet18 [35]	0.9124	0.9340	0.9748
SegNet CNN-Autoencoder [4]	0.9917	0.7287	–
SA-GA [14]	0.9590	–	–
SegNet [30]	0.9340	0.9314	–
U-Net with ResNet [17]	0.9960	0.9011	–
U-Net [13]	0.9780	0.7800	–
Multiclass CNN [7]	–	0.8280	–
MAG-Net [8]	0.9952	0.7400	0.6000
Hybrid KFCM-CNN [29]	0.9760	0.8884	0.8204
U-Net based [3]	–	0.8900	0.8100
Cascaded Dual-Scale LinkNet [39]	–	0.8003	0.9074
SegNet-VGG-16 [30]	0.9340	0.9314	0.914
2D-UNet [38]	92.1600	0.8120	–
CNN with LinkNet [40]	–	0.7300	–
U-Net with adaptive thresholding [10]	0.9907	0.6239	–
O2U-Net [42]	0.9934	0.8083	–
CNN U-Net [32]	0.9854	–	0.8196

5   Conclusion

In conclusion, our study presents a pioneering approach to brain tumor segmentation, leveraging the DeepLabV3plus model with ResNet50 and incorporating diverse image enhancement techniques. We have identified the sequence-dependent impact of enhancement methods through meticulous analysis, highlighting the superior performance of the hybrid CLAHE-HE approach. Our model demonstrated exceptional proficiency in segmenting diverse brain tumor classes, achieving outstanding metrics with an accuracy of 0.9993, a Dice coefficient of 0.9690, and a Jaccard index of 0.9404, surpassing existing methods. The robustness of our proposed method across all data further underscores its effectiveness compared to state-of-the-art models.

While providing valuable insights for medical image analysis, this study acknowledges potential limitations and emphasizes the imperative need for future research to enhance generalizability and address variations in imaging protocols. As part of our future work, exploring ensemble models by integrating our proposed method with architectures such as modified U-Net, ResNet-U-Net, VGG-U-Net, and others could further elevate segmentation performance, contributing to the evolution of advanced medical imaging techniques.

Acknowledgement. This research was supported by the Polish Ministry of Science and Higher Education funds assigned to AGH University of Krakow and by PLGrid under grant no. PLG/2023/016757.

References

1. Ahamed, M.F., et al.: A review on brain tumor segmentation based on deep learning methods with federated learning techniques. Comput. Med. Imaging Graph. **110**, 102313 (2023). https://doi.org/10.1016/j.compmedimag.2023.102313
2. Akcay, O., Kinaci, A.C., Avsar, E.O., Aydar, U.: Semantic segmentation of high-resolution airborne images with dual-stream DeepLabV3+. ISPRS Int. J. Geo Inf. **11**(1), 23 (2021). https://doi.org/10.3390/ijgi11010023
3. Akter, A., et al.: Robust clinical applicable CNN and U-Net based algorithm for MRI classification and segmentation for brain tumor. Expert Syst. Appl. **238**, 122347 (2024). https://doi.org/10.1016/j.eswa.2023.122347
4. Badža, M.M., Barjaktarović, M.Č: Segmentation of brain tumors from MRI images using convolutional autoencoder. Appl. Sci. **11**(9), 4317 (2021). https://doi.org/10.3390/app11094317
5. Beliveau, V., Nørgaard, M., Birkl, C., Seppi, K., Scherfler, C.: Automated segmentation of deep brain nuclei using convolutional neural networks and susceptibility weighted imaging. Hum. Brain Mapp. **42**(15), 4809–4822 (2021). https://doi.org/10.1002/hbm.25604
6. Cheng, J., et al.: Enhanced performance of brain tumor classification via tumor region augmentation and partition. PLoS ONE **10**(10), e0140381 (2015). https://doi.org/10.1371/journal.pone.0140381
7. Díaz-Pernas, F.J., Martínez-Zarzuela, M., Antón-Rodríguez, M., González-Ortega, D.: A Deep learning approach for brain tumor classification and segmentation using a multiscale convolutional neural network. Healthcare **9**(2), 153 (2021). https://doi.org/10.3390/healthcare9020153
8. Gupta, S., Punn, N.S., Sonbhadra, S.K., Agarwal, S.: MAG-Net: multi-task attention guided network for brain tumor segmentation and classification. In: Srirama, S.N., Lin, J.C.-W., Bhatnagar, R., Agarwal, S., Reddy, P.K. (eds.) BDA 2021. LNCS, vol. 13147, pp. 3–15. Springer, Cham (2021). https://doi.org/10.1007/978-3-030-93620-4_1
9. Hoebel, K.V., et al.: Not without context-a multiple methods study on evaluation and correction of automated brain tumor segmentations by experts. Acad. Radiol. (2023). https://doi.org/10.1016/j.acra.2023.10.019
10. Isunuri, B.V., Kakarla, J.: Fast brain tumour segmentation using optimized U-Net and adaptive thresholding. Automatika **61**(3), 352–360 (2020). https://doi.org/10.1080/00051144.2020.1760590
11. Jiang, Z., Dong, Z., Wang, L., Jiang, W.: Method for diagnosis of acute lymphoblastic leukemia based on ViT-CNN ensemble model. Comput. Intell. Neurosci. **2021**, 1–12 (2021). https://doi.org/10.1155/2021/7529893
12. Jyothi, P., Singh, A.R.: Deep learning models and traditional automated techniques for brain tumor segmentation in MRI: a review. Artif. Intell. Rev. **56**(4), 2923–2969 (2023). https://doi.org/10.1007/s10462-022-10245-x
13. Kasar, P.E., Jadhav, S.M., Kansal, V.: MRI Modality-based brain tumor segmentation using deep neural networks. Research Square (2021). https://doi.org/10.21203/rs.3.rs-496162/v1

14. Kharrat, A., Neji, M.: Feature selection based on hybrid optimization for magnetic resonance imaging brain tumor classification and segmentation. Appl. Med. Inform. **41**(1), 9–23 (2019)
15. Kulkarni, S.M., Sundari, G.: Brain MRI classification using deep learning algorithm. Int. J. Eng. Adv. Technol. **9**(3), 1226–1231 (2020). https://doi.org/10.35940/ijeat.C5350.029320
16. Kumar, A.: Study and analysis of different segmentation methods for brain tumor MRI application. Multimedia Tools Appl. **82**(5), 7117–7139 (2023). https://doi.org/10.1007/s11042-022-13636-y
17. Kumar Sahoo, A., Parida, P., Muralibabu, K., Dash, S.: Efficient simultaneous segmentation and classification of brain tumors from MRI scans using deep learning. Biocybern. Biomed. Eng. **43**(3), 616–633 (2023). https://doi.org/10.1016/j.bbe.2023.08.003
18. Liang, J., Yang, C., Zeng, M., Wang, X.: TransConver: transformer and convolution parallel network for developing automatic brain tumor segmentation in MRI images. Quant. Imaging Med. Surg. **12**(4), 2397–2415 (2022). https://doi.org/10.21037/qims-21-919
19. Liu, X., Shih, H.A., Xing, F., Santarnecchi, E., El Fakhri, G., Woo, J.: Incremental learning for heterogeneous structure segmentation in brain tumor MRI. In: Greenspan, H., et al. (eds.) MICCAI 2023. LNCS, vol. 14221, pp. 46–56. Springer, Cham (2023). https://doi.org/10.1007/978-3-031-43895-0_5
20. Lv, Q., Feng, W., Quan, Y., Dauphin, G., Gao, L., Xing, M.: Enhanced-random-feature-subspace-based ensemble CNN for the imbalanced hyperspectral image classification. IEEE J. Sel. Topics Appl. Earth Observ. Remote Sens. **14**, 3988–3999 (2021). https://doi.org/10.1109/JSTARS.2021.3069013
21. Allah, A.M.G., Sarhan, A.M., Elshennawy, N.M.: Edge U-Net: brain tumor segmentation using MRI based on deep U-Net model with boundary information. Expert Syst. Appl. **213**, 118833 (2023). https://doi.org/10.1016/j.eswa.2022.118833
22. Metlek, S., Çetıner, H.: ResUNet+: a new convolutional and attention block-based approach for brain tumor segmentation. IEEE Access **11**, 69884–69902 (2023). https://doi.org/10.1109/ACCESS.2023.3294179
23. Mouhafid, M., Salah, M., Yue, C., Xia, K.: Deep ensemble learning-based models for diagnosis of COVID-19 from chest CT images. Healthcare **10**(1), 166 (2022). https://doi.org/10.3390/healthcare10010166
24. Najjar, R.: Redefining radiology: a review of artificial intelligence integration in medical imaging. Diagnostics **13**(17), 2760 (2023). https://doi.org/10.3390/diagnostics13172760
25. Neamah, K., et al.: Brain tumor classification and detection based DL models: a systematic review. IEEE Access, 1 (2023). https://doi.org/10.1109/ACCESS.2023.3347545
26. Nguyen, H.T.T., Pham, T.T.H., Le, H.T.: Application of deep learning in brain tumor segmentation. Sci. Technol. Dev. J. Eng. Technol. (2022). https://doi.org/10.32508/stdjet.v5i2.951
27. Polat, H.: Multi-task semantic segmentation of CT images for COVID-19 infections using DeepLabV3+ based on dilated residual network. Phys. Eng. Sci. Med. **45**(2), 443–455 (2022). https://doi.org/10.1007/s13246-022-01110-w
28. Ramasamy, G., Singh, T., Yuan, X.: Multi-modal semantic segmentation model using encoder based link-net architecture for BraTS 2020 challenge. Procedia Comput. Sci. **218**, 732–740 (2023). https://doi.org/10.1016/j.procs.2023.01.053

29. Rao, S.K.V., Lingappa, B.: Image analysis for MRI based brain tumour detection using hybrid segmentation and deep learning classification technique. Int. J. Intell. Eng. Syst. **12**(5), 53–62 (2019). https://doi.org/10.22266/ijies2019.1031.06

30. Rehman, A., Naz, S., Naseem, U., Razzak, I., Hameed, I.A.: Deep autoencoder-decoder framework for semantic segmentation of brain tumor. Aust. J. Intell. Inf. Process. Syst **15**(4), 53–60 (2019)

31. Roy Choudhury, A., Vanguri, R., Jambawalikar, S.R., Kumar, P.: Segmentation of brain tumors using DeepLabv3+. In: Crimi, A., Bakas, S., Kuijf, H., Keyvan, F., Reyes, M., van Walsum, T. (eds.) BrainLes 2018. LNCS, vol. 11384, pp. 154–167. Springer, Cham (2019). https://doi.org/10.1007/978-3-030-11726-9_14

32. Ruiz, C.B.: Classification and segmentation of brain tumor MRI images using convolutional neural networks. In: 2023 IEEE International Conference on Engineering Veracruz (ICEV), pp. 1–6. IEEE (2023). https://doi.org/10.1109/ICEV59168.2023.10329651

33. Saifullah, S., Dreżewski, R.: Enhanced medical image segmentation using CNN based on histogram equalization. In: 2023 2nd International Conference on Applied Artificial Intelligence and Computing (ICAAIC), pp. 121–126 (2023). https://doi.org/10.1109/ICAAIC56838.2023.10141065

34. Saifullah, S., Dreżewski, R.: Modified histogram equalization for improved CNN medical image segmentation. Procedia Comput. Sci. **225**(C), 3021–3030 (2023). https://doi.org/10.1016/j.procs.2023.10.295

35. Saifullah, S., Dreżewski, R.: Redefining brain tumor segmentation: a cutting-edge convolutional neural networks-transfer learning approach. Int. J. Electr. Comput. Eng. (IJECE) **14**(3), 2583 (2024). https://doi.org/10.11591/ijece.v14i3.pp2583-2591

36. Saifullah, S., Suryotomo, A.P.: Thresholding and hybrid CLAHE-HE for chicken egg embryo segmentation. In: 2021 International Conference on Communication & Information Technology (ICICT), pp. 268–273, June 2021. https://doi.org/10.1109/ICICT52195.2021.9568444

37. Saifullah, S., Suryotomo, A.P., Dreżewski, R., Tanone, R., Tundo: Optimizing brain tumor segmentation through CNN U-Net with CLAHE-HE image enhancement. In: Proceedings of the 2023 1st International Conference on Advanced Informatics and Intelligent Information Systems (ICAI3S 2023), pp. 90–101 (2024). https://doi.org/10.2991/978-94-6463-366-5_9

38. Sailunaz, K., Bestepe, D., Alhajj, S., Özyer, T., Rokne, J., Alhajj, R.: Brain tumor detection and segmentation: interactive framework with a visual interface and feedback facility for dynamically improved accuracy and trust. PLoS ONE **18**(4), e0284418 (2023). https://doi.org/10.1371/journal.pone.0284418

39. Sobhaninia, Z., Rezaei, S., Karimi, N., Emami, A., Samavi, S.: Brain tumor segmentation by cascaded deep neural networks using multiple image scales. In: 2020 28th Iranian Conference on Electrical Engineering (ICEE), pp. 1–4. IEEE (2020). https://doi.org/10.1109/ICEE50131.2020.9260876

40. Sobhaninia, Z., et al.: Brain tumor segmentation using deep learning by type specific sorting of images (2018). http://arxiv.org/abs/1809.07786

41. Wahlang, I., et al.: Brain magnetic resonance imaging classification using deep learning architectures with gender and age. Sensors **22**(5), 1766 (2022). https://doi.org/10.3390/s22051766

42. Zargari, S.A., Kia, Z.S., Nickfarjam, A.M., Hieber, D., Holl, F.: Brain tumor classification and segmentation using dual-outputs for U-Net architecture: O2U-Net. Stud. Health Technol. Inform. **305**, 93–96 (2023). https://doi.org/10.3233/SHTI230432

Focal-Based Deep Learning Model
for Automatic Arrhythmia Diagnosis

Abir Boulif[1,1](✉), Bouchra Ananou[1], Mustapha Ouladsine[1], and Stéphane Delliaux[2]

[1] Aix-Marseille University, University of Toulon, CNRS, LIS, 13397 Marseille, France
{abir.boulif,bouchra.ananou,mustapha.ouladsine}@univ-amu.fr
[2] Aix-Marseille University, INSERM, INRAE, C2VN, 13005 Marseille, France
stephane.delliaux@univ-amu.fr

Abstract. This paper approaches a new model for arrhythmia diagnosis based on short-duration electrocardiogram (ECG) heartbeats. To detect 8 arrhythmia classes efficiently, we design a Deep Learning model based on the Focal modulation layer. Moreover, we develop a distance variation of the SMOTE technique to address the problem of data imbalance. The classification algorithm includes a block of Residual Network for feature extraction and an LSTM network with a Focal block for the final class prediction. The approach is based on the analysis of variable-length heartbeats from leads MLII and V5, extracted from 48 records of the MIT-BIH Arrhythmia Database. The methodology's novelty consists of using the Focal layer for ECG classification and data augmentation with DTW distance (Dynamic Time Warping) using the SMOTE technique.

The approach offers real-time classification and is simple since it combines feature extraction, selection, and classification in one stage. Using data augmentation with SMOTE variant and Focal-based Deep learning architecture to identify 8 types of heartbeats, the method achieved an impressive overall accuracy, F1-score, precision, and recall of 98.61%, 94.08%, 94.53%, and 93.68% respectively. Additionally, the classification time per sample was only 0.002 s. Therefore, the suggested approach can serve as an additional tool to aid clinicians in ensuring rapid and real-time diagnosis for all patients with no exclusivities.

Keywords: Signal processing · Arrhythmia · Heartbeat · Electrocardiogram · Diagnosis · Classification · Deep learning · Healthcare sustainability

1 Introduction

According to the 2023 report of the World Heart Federation (WHF) [1], cardiovascular diseases (CVD) represent a global threat to the population. Deaths due to CVD have increased by 60% worldwide in the past 30 years. Therefore, it is essential to prioritize implementing tools to prevent premature heart attacks and strokes. To this end, arrhythmia detection proves important since it is the cause of most sudden cardiac arrests. Arrhythmia is a medical condition that results either in fast, slow, or irregular heartbeats [2]. Some of the most common types of heartbeats associated with arrhythmia are Premature Atrial and Ventricular contractions. To diagnose these heartbeats, we rely on the

L. Franco et al. (Eds.): ICCS 2024, LNCS 14835, pp. 355–370, 2024.
https://doi.org/10.1007/978-3-031-63772-8_31

analysis of the electrocardiogram (ECG), a non-invasive tool that records the electrical signals in the heart, to investigate symptoms of arrhythmia.

Various techniques in literature have been applied to categorize ECG signals automatically into heart rhythms or heartbeat classes. Heart rhythm describes the overall pattern of electrical activity in the heart, while heartbeat classes refer to specific types of individual heartbeats within a given rhythm. This work focuses on the classification of heartbeats. The classification is preceded by pre-processing, feature extraction, and feature selection. The pre-processing stage may comprise noise removal, data segmentation, data normalization, data reduction, and signal compression.

Noise includes power line interference, muscle noise, motion artifact, baseline wander, and high-frequency artifacts. Discrete Wavelet Transform (DWT) with its various wavelet distributions, is often used for noise removal [3–6]. More improved versions of wavelets were developed in [7, 8]. The Pan-Tompkins algorithm proposed by Pan and Tompkins in 1985 [9] was used for segmentation and QRS detection in [5, 10, 11] and for R-peak detection in [3, 6, 12]. Principal Component Analysis (PCA) is widely used for dimensionality reduction. It is also used for feature extraction [13]. Other methods can also be employed such as Discrete Wavelet Transform [14, 15], Higher Order Statistic (HOS) [5, 10, 16], Independent Component Analysis (ICA) [17], and Fast Fourier Transform (FFT) [14]. These are known to be hand-crafted methods for extraction. Deep learning with CNN layers, is also used for end-to-end extraction as in [18–20].

When dealing with imbalanced data, augmentation techniques can be used to address this issue. SMOTE [21, 22] and GANs [23] are effective methods in reducing overfitting during training, while other techniques only increase data volume by adding noise, without measurable improvement in dataset performance and variance [24].

To detect cardiac rhythms, several approaches have been employed, ranging from traditional machine learning algorithms to complex architectures. Support Vector Machines (SVM) achieved an accuracy of 98.91% [5], 94.30% [10], and 98.8% when combined with Genetic Algorithm [8]. Feed-Forward Neural Network reached respective accuracy of 98.90, 94.52, and 99.80% in [5, 10, 25]. Long-Short-Term-Memory (LSTM) finds numerous applications in time series, including the classification of ECG signals [4, 18]. Moreover, Convolutional Neural Networks (CNN) hold a significant position in deep learning and provide accurate results when detecting arrhythmias [3, 23, 24]. When combining CNN and LSTM, both temporal and spatial information are captured [21, 23]. More complex architectures were developed to address the problem of vanishing gradients [26], by combining a Residual Network and LSTMs to detect five heartbeats and achieve 99.4% accuracy. Wang [27] yielded an accuracy rate of 97.9% by leveraging the proficiency of three networks—CNN, LSTM, and bi-directional GRU.

The authors' prior work focused on using machine learning to diagnose Atrial Fibrillation (AF). In [28], a multi-dynamics analysis of the QRS complex using SVM and MKL models yielded sensitivity of 96.54% and 95.47%, respectively. Other previous research consisted of extracting features from R-wave derivatives to aid in medical decision-making, particularly for detecting AF [29–31].

This work focuses on classifying eight types of heartbeats. These include Normal beats (N), Atrial Premature beats (A), which can lead to sustained arrhythmias if frequent or persistent, and Premature Ventricular beats (V), which are common and may be associated with serious ventricular arrhythmias if left untreated. Left Bundle Branch Block beats (L), which may indicate an increased risk of cardiovascular events. Right Bundle Branch Block beats (R) are considered serious as they are often associated with structural heart diseases. Paced beats (p) are generated by a pacemaker, to help the heart muscle contract when the natural heart rate is too slow or when there is a heart block. Fusion of Ventricular and Normal beats (F), occur when the electrical signals of premature ventricular beat and normal beat coincide in time. Fusion of Paced and Normal beats (f), reflect the combination of the artificially paced beat and the natural beat initiated by the sinus node.

These heartbeats were chosen because they may cause major health problems if not addressed, and their morphology may be difficult to distinguish from other heartbeats.

A novel method based on Deep Learning classification has been developed to identify eight types of heartbeats. The short-duration signal is typically used when dealing with heartbeat diagnosis, and its use is advantageous as the algorithm focuses on dynamic sequences and feature extraction within a narrowed time frame. The use of a Residual network, which is a 12-layer CNN, for feature extraction and LSTM-Focal block for prediction can be generally employed to classify other time-series data. Therefore, this approach could be widely applied. The aspects of our study encompass:

- The processing of variable-length ECG heartbeats.
- The use of MLII (modified limb lead II) and V5 leads from the ECG.
- Data augmentation using new **DTW-SMOTE** variation.
- Feature extraction using a Residual Network.
- Heartbeat classification model using a **Focal Modulation layer**.

To our knowledge, no other papers in the literature used focal layers for ECG diagnosis.

2 Material and Methods

In this paper, a Deep Learning model based on Residual Network and LSTM-Focal architecture is designed to classify eight types of heartbeats. It is an end-to-end model, excluding any hand-crafted methods for feature extraction or selection. The Residual block, inspired by the Residual Network [32], which comprises 12 convolutional layers, is used for deep feature extraction. The LSTM and focal layers are mainly used for classification. The network inputs variable-length ECG heartbeats, ranging from 81 to 439 samples, and returns the heartbeat class. In terms of preprocessing, we applied noise removal, normalization, heartbeat segmentation, and data augmentation.

2.1 Assumptions

Our approach is based on the following assumptions:

- Heartbeat segmentation is realized based on R-peaks annotations.
- The ECG signal was de-noised, to ensure the exclusion of any significant interferences that could affect the classification accuracy.
- The majority of cardiac heartbeats, possess distinct patterns within ECG data.
- Each ECG sample contains one class heartbeat.
- The use of an End-to-end structure that combines feature extraction, selection, and heartbeat classification in one stage.

2.2 ECG Database

An open-access database, hosted on Physionet [33], is used in this paper since it is regularly utilized for arrhythmia research and contains annotated heartbeats. The MIT-BIH Arrhythmia database [34] is a collection of 48 half-hour excerpts of two-channel ambulatory ECG recordings that were acquired from 47 subjects. The recordings were digitized at a rate of 360 samples per second per channel, with an 11-bit resolution and covering a range of 10 millivolts. The database contains seventeen types of heartbeats.

Eight types of heartbeats are extracted from leads MLII and V5. The signal is first cleaned then normalized and segmented. Finally, it is passed through a data augmentation process to ensure the dataset is no longer imbalanced.

The test data is extracted before data augmentation, forming 30% of the original dataset, which contained 111,471 heartbeats in total, to assess efficiently the performance of the classification model with imbalanced unseen data.

We obtain a dataset of 475,248 heartbeats in total after augmentation. The dataset is then stratified equally between classes and split into the training set (80%) which contains 380,198 heartbeats and the validation set (20%) which contains 95,050 heartbeats.

2.3 Methods: Preprocessing

Noise Reduction

The main goal of noise reduction is to eliminate or minimize the random variations in the signal. This technique aims to enhance the clarity of the underlying information whilst improving the signal-to-noise ratio. The ECG signal contains different types of noise, each should be removed with a special filter.

Baseline wander is a low-frequency artifact that arises from charged electrodes or patient movement and breathing. A high-pass Butterworth five-order filter [35] with a cutoff frequency of 0.5 Hz is used to remove baseline wander. The Gain of the n-order filter is given by (1), where ω is the filter frequency and ω_c is the cutoff frequency.

$$G_n(\omega) = \frac{1}{\sqrt{1 + \left(\frac{\omega}{\omega_c}\right)^{2n}}} \tag{1}$$

Power-line interference is a high-frequency artifact caused by improper grounding of the ECG equipment. We smooth the signal with a moving average kernel, which is a Finite Impulse Response Filter [36] privileged when dealing with time series, with a width of one period of 50 Hz. The output of the FIR filter of n-order is given by (2):

$$y(k) = \sum_{i=0}^{n} b_i * x(k - i) \tag{2}$$

$x(k)$ is the input signal and b_i is the value of the impulse response at the ith instant.

Data Normalization
Applying Min-Max normalization results in smaller standard deviations, which helps to eliminate the impact of outliers. In addition, rescaling improves the backpropagation process during Deep Learning by speeding up the convergence rate. Figure 1 highlights the effect of data normalization and noise removal on the ECG signal.

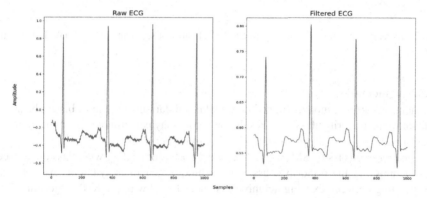

Fig. 1. Noisy versus de-noised and normalized ECG signal.

Heartbeat Segmentation
The Discrete Wavelet Transform (DWT) technique is used for dynamic heartbeat segmentation by detecting P-wave onsets and T-wave offsets based on the annotations of R-peaks provided by the MIT-BIH Arrhythmia database.

DWT enables signal decomposition by passing it through a series of low-pass and high-pass filters to extract the required information. It facilitates the capture of both spatial and temporal information in a signal. Figure 2 shows the delineation of cardiac heartbeat where the morphological and temporal characteristics can be depicted.

The segmented heartbeats vary in size, but to meet the requirements of the Deep Learning model, we apply zero padding to ensure uniform input size.

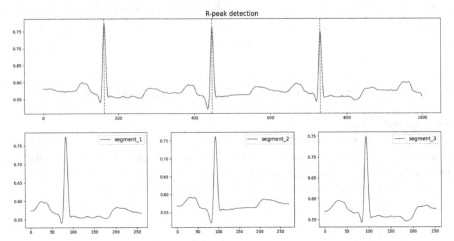

Fig. 2. R-peaks annotations from Physionet (Top), heartbeat segmentation with DWT (Bottom).

Data Augmentation

In machine learning, models trained on unbalanced datasets may have biased behavior in favor of the majority class. As a result, the model may perform poorly on the minority class, leading to limited generalization ability to unseen data.

To address this issue, SMOTE [37] is widely used for dealing with class imbalance, by oversampling the minority class. The technique generates synthetic examples by interpolating between existing neighbor instances found with the KNN algorithm. The classic version is based on the Euclidean distance to detect the KNN neighbors.

This paper presents a new variant of SMOTE that uses the DTW (Dynamic Time Warping) [38] similarity distance to detect the K nearest neighbors. The aim is to find the best mapping with minimal Euclidean distance matching. As the data consists of variable-size heartbeats, their distance cannot be measured with the classic Euclidean formula. However, DTW can handle series that have different lengths and may be warped in the time domain, to find the optimal alignment where the patterns like troughs and peaks can be correctly matched due to the one-to-many mapping.

The DTW algorithm computes the optimal alignment path dynamically by considering various alignments and selecting the one that minimizes the cumulative distance.

Let's consider two Times Series of the Normal heartbeat class, $X = (x_1, \ldots, x_N)$ and $Y = (y_1, \ldots, y_M)$ of lengths $N \in \mathbb{N}$ and $M \in \mathbb{N}$ respectively. The DTW distance is expressed by (3) where D is an N by M matrix defining the accumulated distance between x_i and y_j. It is computed dynamically as shown in (4).

$$DTW(X, Y) = \sqrt{D[x_N, y_M]} \tag{3}$$

$$D[x_i, y_j] = d(x_i, y_j) + \min(D[x_{i-1}, y_{j-1}], D[x_{i-1}, y_j], D[x_i, y_{j-1}]) \quad (4)$$

$d(x_i, y_j)$ is the Euclidean distance between elements x_i and y_j.

The optimal alignment path is then found by backtracking from $D[x_N, y_M]$ to $D[x_1, y_1]$, while following the minimum cumulative distance.

We apply DTW-SMOTE by resampling all the classes but the majority class. Table 1 shows the distribution of heartbeats after applying the DTW-SMOTE variant.

Table 1. Original data versus DTW-SMOTE data

Type of heartbeats	Original distribution	Smote distribution
Normal beat (N)	59,406	59,406
Right bundle branch block beat (R)	5032	59,406
Left bundle branch block beat (L)	4338	59,406
Paced beat (p)	4027	59,406
Premature ventricular beat (V)	2657	59,406
Atrial premature beat (A)	1700	59,406
Fusion of paced and normal beats (f)	451	59,406
Fusion of ventricular and normal beats (F)	418	59,406

To evaluate the performance of the DTW-SMOTE variant, we applied two Machine-Learning techniques for the classification of the augmented data using both classic SMOTE and DTW-SMOTE. The Decision Tree model applied to the SMOTE variant reached a better classification performance with an overall accuracy of 94.40% while the SVM model achieved almost similar results on both classic SMOTE data and DTW-SMOTE data. The evaluation metrics (in %) are detailed in Table 2.

Table 2. Evaluation metrics of decision tree and SVM on SMOTE and DTW-SMOTE data.

	Decision tree		SVM	
	SMOTE	DTW-SMOTE	SMOTE	DTW-SMOTE
F1-score	79.94	81.78	70.64	70.59
Precision	75.47	78.40	63.80	63.75
Recall	86.20	85.86	87.77	86.83
Accuracy	93.95	94.40	88.44	88.53

2.4   Methods: LSTM-Focal Classification Model

A deep neural network is designed to classify eight types of heartbeats. It consists of a Residual Network [32] for feature extraction and LSTM-Focal for classification.

Feature Extraction with Residual Network (ResNet)

The usage of residual blocks, which include skip connections, allows the network to directly learn residuals (the difference between input and output), making the model easier to train without encountering the vanishing gradient problem. The Residual Network used in this approach is a 12-layer convolutional neural network.

Convolution layers with different kernel sizes are performed on cardiac heartbeats to obtain feature maps, each one selecting various features from the ECG signal, i.e. peaks, troughs, waves, and local patterns. An adaptive average pooling layer is applied at the end of the model, to condense all of the feature maps into a single one, capturing all relevant information. We obtain as an output a 1-D vector of 512 features.

Heartbeat Classification with LSTM-Focal Based Model

The last block of the heartbeat classification model is the prediction block. The architecture of the global model is shown in Fig. 3.

The prediction block takes as an input the ResNet features and outputs the ECG heartbeat class. The architecture consists of two LSTM layers [39] that are very adept at capturing patterns and variations in ECG sequences with temporal dependencies.

The LSTM layers are then followed by a focal modulation layer which is a part of the FocalNets [40], designed by Microsoft in 2022, where the self-attention mechanism [41] is completely replaced by a focal modulation module for modeling token interactions in vision. This module allows better generalization ability through dynamic focal layers instead of static convolution kernels.

In the present approach, we implement only the focal modulation layer given that it represents the core of the FocalNet. This layer allows the model to selectively focus on specific parts of its input with more lightweight operations. The process is based on aggregation followed by interaction between the aggregated parts of the input. Unlike self-attention which gives priority to the interaction over the aggregation.

First, the output of the LSTM layers H is projected linearly and then split into query q, context Y^0, and gates G^l. These three components are passed through three major operations of the focal modulation layer:

- Hierarchical Contextualization: In this stage, the initial context $Y^0 \in \mathbb{R}^{H*W*C}$ is passed through a series of Depth-Wise Convolution (DWConv) and GeLU layers. These blocks are termed focal levels l and the output of each level is a context Y^l. The final feature map Y^{L+1} goes through a Global Average Pooling layer. The corresponding equations of the aforementioned steps are defined in (5)–(7).

$$q, Y^0, G = f(H) \tag{5}$$

$$Y^l = g^l\left(Y^{l-1}\right) \in \mathbb{R}^{H*W*C}, l \in \{1, \dots, L\} \tag{6}$$

$$Y^{L+1} = AvgPooling\left(Y^l\right) \in \mathbb{R}^C \tag{7}$$

f is the linear projection layer and g^l is the focal layer where l denotes the focal level. (H, W, C) is the output size corresponding to Height, Weight, and Channels. l denotes the focal level where $l \in \{1, \dots, L\}$.

- Gated Aggregation: Gates $G^l \in \mathbb{R}^{H*W*(L+1)}$ are employed to perform a weighted aggregation over the context Y^l. Next, the output Z is passed through a convolutional layer h to obtain the modulator M. The equations are shown in (8) and (9):

$$Z = \sum_{l=1}^{L+1} G^l \odot Y^l \in \mathbb{R}^{H*W*C} \qquad (8)$$

$$M = h(Z) \in \mathbb{R}^{H*W*C} \qquad (9)$$

- Interaction: where the initial query q interacts with the context aggregation. Another post-linear projection function is applied with Dropout to provide the final focal output according to (10).

$$F^{out} = linear(q \odot M) \qquad (10)$$

The output of the focal modulation is then normalized with the batch-normalization technique, which helps accelerating the learning process. Finally, a Fully-connected layer is added to the model to output the eight heartbeat classes.

Furthermore, this paper includes a comparison experiment using the LSTM-attention model, which employs an attention mechanism instead of the Focal module. This comparison serves to highlight the impact of the Focal layer on the classification result.

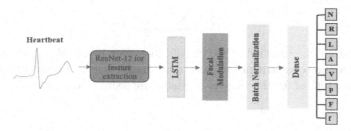

Fig. 3. The prediction block based on LSTM-focal layers.

Environment Setup

For the hardware characteristics, all the experiments were run on a cluster equipped with GPU node. The model of the GPU graphic card is NVIDIA GA102 [GeForce RTX 3090] with 192 G of RAM.

3 Results of Classification

For heartbeat recognition, a heartbeat fragment of 439 sample-long is fed to the input layer of the Residual Network in batches. Each time series accumulates 512 features. The output is then processed in the LSTM-Focal block to produce the heartbeat class.

Figure 4 shows the training and validation accuracy. The LSTM-Focal model is trained during 50 epochs, during which the learning is stabilized and reaches a maximum training accuracy of 99.94% and validation accuracy of 99.87%.

The model is then tested on unbalanced heartbeats. It succeeded in recognizing correctly 32,977 fragments out of 33,442, yielding an overall accuracy of 98.61%. The evaluation criteria for the test data are shown in Table 3. As can be observed, the model reached high F1-score performances for classes N, L, R, and p since they form the majority of the test data. The model yielded an overall precision, recall, and F1 score of 94.53%, 93.68%, and 94.08% respectively. The lowest recognition performance was observed for the F class which contained only 179 fragments during the test.

The LSTM-attention model is trained to assess the effectiveness of the Focal layer on the classification. Figure 4 shows that the model takes 15 epochs to stabilize, it reaches an overall training and validation accuracy of 99.31% and 99.5% respectively. According to Table 4, we conclude that the LSTM-Focal model reaches a better classification performance with an overall accuracy of 98.61% versus 97.65% for the LSTM-Attention model. The use of the Focal layer enhanced the ability of the model to correctly identify the positive heartbeats during the classification with an overall precision of 94.53% versus 89.94% for the model with the attention mechanism.

The time required for testing one heartbeat is measured by averaging the prediction time over all samples. It is shown in Eq. (11) where e represents the execution time at step t.

$$T = \sum_{t=0}^{n} \frac{e_t}{n} \tag{11}$$

We measure the variation of the classification time by calculating the mean and standard deviation over all iterations. As a result, in each time, we find a constant average classification time of 0.002 s with zero deviation for a single heartbeat.

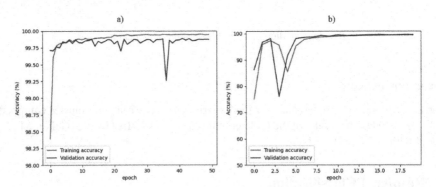

Fig. 4. Training and validation accuracy of LSTM-focal (a) and LSTM-attention (b) models.

Table 3. Evaluation metrics (in %) of the LSTM-focal model using test data.

Heartbeats	Precision	Recall	F1-score
Atrial premature beats (A)	91.01	91.63	91.32
Fusion of ventricular and normal beats (F)	85.09	76.54	80.59
Left bundle branch block beats (L)	96.85	97.47	97.16
Normal beats (N)	99.21	99.23	99.22
Right bundle branch block beats (R)	98.56	98.56	98.56
Premature ventricular beats (V)	95.47	94.29	94.88
Fusion of Paced and Normal beats (f)	90.86	92.27	91.56
Paced beats (p)	99.19	99.48	99.33
Macro average	94.53	93.68	94.08
Weighted average	98.60	98.61	98.60
Overall accuracy	98.61		

Table 4. Overall evaluation metrics (in %) of the LSTM-focal and the LSTM-attention models.

Model	Precision	Recall	F1-score	Accuracy
LSTM-focal	94.53	93.68	94.08	98.61
LSTM-attention	89.94	92.49	91.17	97.65

4 Discussion

In order to confirm the effectiveness of the proposed model, we compare it with some ECG diagnosis approaches in literature, that have used the MIT-BIH Arrhythmia database. The comparison criteria are listed in Table 5. They include the number of heartbeat classes, the feature set, the classification algorithm, and the overall accuracy.

Our model, indicates one of the best performances recorded in arrhythmia diagnosis. It outperformed certain models that used more complex architectures such as U-Net [44] and Google-Net [45], with an accuracy of 96.30% and 97.32% respectively versus 98.61% for our model. Other models that outperformed the current approach either used feature extraction before classification, as in [21], or used only one ECG lead.

The majority of state-of-the-art methods work on the classification of 5 types of heartbeats: N, V, L, R, and A. These are the most common in open-access databases. When compared to [3, 43, 46], the proposed method achieves a higher accuracy, particularly when using the same type of input (raw data). In comparison to techniques working with 17 classes [11, 43], the proposed method outperformed these models with an accuracy of 98.61% versus 89.95% and 91.33% respectively. This is due to the fact that training the model with a large number of imbalanced classes reduces performance and causes the model to fail to correctly distinguish minority classes.

Among the Machine Learning methods shown in Table 5, it can be said that the SVM are efficient for heartbeat classification and can outperform DL models when combined with appropriate feature extraction methods such as Wavelet Transform and PCA.

Table 5. Comparison of the proposed methods to literature methods.

Paper	No. of beats	Feature set	Classifier	Accuracy
Martis et al. [10]	5	HOS + PCA	Feed forward NN	94.52
Park et al. [11]	17	RR interval, R and P waves' positions and amplitudes	Decision tree	89.95
Li et al. [8]	5	DWT, PCA, LDA, KICA	SVM and genetic algorithm	98.80
Acharya et al. [3]	5	Raw data	CNN	94.03
Qin et al. [42]	6	Wavelet multi-resolution and PCA	One-versus-one SVM	99.70
Yang et al. [19]	5	PCA-Net	SVM	97.77
Yildirim et al. [43]	17	Raw data	CNN	91.33
Oh et al. [44]	5	Raw data	Modified U-network	97.32
Yildirim et al. [18]	5	Raw data	LSTM	99.23
Kim et al. [45]	5	Raw data	GoogleNet with 2 inceptions	96.30
Zubair et al. [46]	5	Raw data	CNN	96.36
Irfan et al. [21]	5	PCA	CNN + LSTM	99.35
The authors' approach	8	Raw data	LSTM-focal modulation	98.61

The present approach using the LSTM and Focal Modulation for classification seems to be efficient for most heartbeats due to the focus of the Focal layer on specific ECG features. Yet, we noticed that F heartbeat, which is the fusion of Normal and Ventricular beats, was misclassified into categories N or V. Also, p heartbeat was mainly misclassified into paced and Normal beats. Therefore, the classification of types f and F heartbeats led to the lowest accuracy. To encounter this problem, another feature should be extracted, to distinguish between the fused heartbeats.

5   Conclusion

In this paper, we propose a SMOTE variant based on DTW distance for data augmentation and a new Focal-based model for ECG heartbeat diagnosis. The DTW-SMOTE allows addressing the issue of data imbalance since the DTW similarity measure finds with more precision the neighbors to select for data generation. A Residual Network consisting of convolutions is employed for feature extraction. It is based on skip connections that avoid the loss of information derived from earlier layers, and take into account the spatial dimension of the ECG data. The model uses LSTM layers to capture any temporal dependencies and to keep in memory the long-term context information. Afterward, a Focal Modulation layer is introduced for more feature enhancement. Due to the use of dynamic kernels, the mechanism can effectively focus on features differently and help improve the heartbeat classification. The model reached an accuracy of 98.61% in detecting eight types of heartbeats, on the MIT-BIH Arrhythmia database.

To sum up, the essential elements of our work include:

- The use of variable-length ECG heartbeats, extracted from 2 leads (MLII and V5),
- Data augmentation using new **DTW-SMOTE** variation,
- End-to-end structure, including deep feature extraction and **Focal classification**.

As for the sustainable aspect of this study, the automatic diagnosis model can help with the early detection of heart disorders. This can alleviate the overall burden on healthcare systems.

Our model can be extended to recognize other types of heartbeats by using other databases. Additionally, the Focal layer can be applied for the classification of rhythm categories instead of heartbeat classes.

Disclosure of Interests. The authors have no competing interests to declare that are relevant to the content of this article.

References

1. World heart report 2023. World Heart Federation (2023). https://world-heart-federation.org/wp-content/uploads/World-Heart-Report-2023.pdf
2. What is an arrhythmia? National Heart Lung and Blood Institute (Last updated 2022). https://www.nhlbi.nih.gov/health/arrhythmias
3. Acharya, U.R., et al.: A deep convolutional neural network model to classify heartbeats. Comput. Biol. Med. **89**, 389–396 (2017). https://doi.org/10.1016/j.compbiomed.2017.08.022
4. Gao, J., Zhang, H., Lu, P., Wang, Z.: An effective LSTM recurrent network to detect arrhythmia on imbalanced ECG dataset. J. Healthcare Eng. **2019**, 1 (2019). https://doi.org/10.1155/2019/6320651
5. Elhaj, F.A., Salim, N., Harris, A.R., Swee, T.T., Ahmed, T.: Arrhythmia recognition and classification using combined linear and nonlinear features of ECG signals. Comput. Methods Progr. Biomed. **127**, 52–63 (2016). https://doi.org/10.1016/j.cmpb.2015.12.024
6. Rajagopal, R., Ranganathan, V.: Design of a hybrid model for cardiac arrhythmia classification based on Daubechies wavelet transform. Adv. Clin. Exper. Med. **27**(6), 727–734 (2018). https://doi.org/10.17219/acem/68982

7. Yildirim, Ö.: A novel wavelet sequence based on deep bidirectional LSTM network model for ECG signal classification. Comput. Biol. Med. **96**, 189–202 (2018). https://doi.org/10.1016/j.compbiomed.2018.03.016

8. Li, H., Yuan, D., Wang, Y., Cui, D., Cao, L.: Arrhythmia classification based on multi-domain feature extraction for an ECG recognition system. Sensors **16**(10), 1744 (2016). https://doi.org/10.3390/s16101744

9. Pan, J., Tompkins, W.J.: A real-time QRS detection algorithm. IEEE Trans. Biomed. Eng. BME **32**(3), 230–236 (1985). https://doi.org/10.1109/TBME.1985.325532

10. Martis, R.J., Acharya, U.R., Lim, C.M., Mandana, K.M., Ray, A.K., Chakraborty, C.: Application of higher order cumulant features for cardiac health diagnosis using ECG signals. Int. J. Neural Syst. **23**(04), 1350014 (2013). https://doi.org/10.1142/s0129065713500147

11. Park, J., Kang, K.: PcHD: personalized classification of heartbeat types using a decision tree. Comput. Biol. Med. **54**, 79–88 (2014). https://doi.org/10.1016/j.compbiomed.2014.08.013

12. Raj, S., Ray, K.C.: Automated recognition of cardiac arrhythmias using sparse decomposition over composite dictionary. Comput. Methods Programs Biomed. **165**, 175–186 (2018). https://doi.org/10.1016/j.cmpb.2018.08.008

13. Kim, J., Shin, H.S., Shin, K., et al.: Robust algorithm for arrhythmia classification in ECG using extreme learning machine. Biomed. Eng. Online **8**(1), 31 (2009). https://doi.org/10.1186/1475-925x-8-31

14. Chen, G., Hong, Z., Guo, Y., Pang, C.: A cascaded classifier for multi-lead ECG based on feature fusion. Comput. Methods Programs Biomed. **178**, 135–143 (2019). https://doi.org/10.1016/j.cmpb.2019.06.021

15. Sumathi, S., Beaulah, H.L., Vanithamani, R.: A wavelet transform based feature extraction and classification of cardiac disorder. J. Med. Syst. **38**(9), 98 (2014). https://doi.org/10.1007/s10916-014-0098-x

16. Afkhami, R.G., Azarnia, G., Tinati, M.A.: Cardiac arrhythmia classification using statistical and mixture modeling features of ECG signals. Pattern Recogn. Lett. **70**, 45–51 (2016). https://doi.org/10.1016/j.patrec.2015.11.018

17. Yu, S.N., Chou, K.T.: Selection of significant independent components for ECG beat classification. Expert Syst. Appl. **36**(2), 2088–2096 (2019). https://doi.org/10.1016/j.eswa.2007.12.016

18. Yildirim, O., Baloglu, U.B., Tan, R.-S., Ciaccio, E.J., Acharya, U.R.: A new approach for arrhythmia classification using deep coded features and LSTM networks. Comput. Methods Programs Biomed. **176**, 121–133 (2019). https://doi.org/10.1016/j.cmpb.2019.05.004

19. Yang, W., Si, Y., Wang, D., Guo, B.: Automatic recognition of arrhythmia based on principal component analysis network and linear support vector machine. Comput. Biol. Med. **101**, 22–32 (2018). https://doi.org/10.1016/j.compbiomed.2018.08.003

20. Wang, R., Fan, J., Li, Y.: Deep multi-scale fusion neural network for multi-class arrhythmia detection. IEEE J. Biomed. Health Inform. **24**(9), 2461–2472 (2020). https://doi.org/10.1109/JBHI.2020.2981526

21. Irfan, S., Anjum, N., Althobaiti, T., Alotaibi, A.A., Siddiqui, A.B., Ramzan, N.: Heartbeat classification and arrhythmia detection using a multi-model deep-learning technique. Sensors **22**(15), 5606 (2022). https://doi.org/10.3390/s22155606

22. Luo, X., Yang, L., Cai, H., Tang, R., Chen, Y., Li, W.: Multi-classification of arrhythmias using a HCRNet on imbalanced ECG datasets. Comput. Methods Programs Biomed. **208**, 106258 (2021). https://doi.org/10.1016/j.cmpb.2021.106258

23. Ullah, W., Siddique, I., Zulqarnain, R.M., Alam, M.M., Ahmad, I., Raza, U.A.: Classification of arrhythmia in heartbeat detection using deep learning. Comput. Intell. Neurosci. **2021**, 1–13 (2021). https://doi.org/10.1155/2021/2195922

24. Iftene, A., Burlacu, A., Gifu, D.: Atrial fibrillation detection based on deep learning models. Procedia Comput. Sci. **207**, 3752–3760 (2022). https://doi.org/10.1016/j.procs.2022.09.436

25. Anwar, S.M., Gul, M., Majid, M., Alnowami, M.: Arrhythmia classification of ECG signals using hybrid features. Comput. Math. Methods Med. **2018**, 1–8 (2018). https://doi.org/10. 1155/2018/1380348

26. Ma, S., Cui, J., Xiao, W., Liu, L.: Deep learning-based data augmentation and model fusion for automatic arrhythmia identification and classification algorithms. Comput. Intell. Neurosci. **2022**, 1–17 (2022). https://doi.org/10.1155/2022/1577778

27. Wang, J.: Automated detection of premature ventricular contraction based on the improved gated recurrent unit network. Comput. Methods Programs Biomed. **208**, 106284 (2021). https://doi.org/10.1016/j.cmpb.2021.106284

28. Trardi, Y., Ananou, B., Haddi, Z., Ouladsine, M.: Multi-dynamics analysis of QRS complex for atrial fibrillation diagnosis. In: 5th International Conference on Control, Decision and Information Technologies (CoDIT) (2018). https://doi.org/10.1109/codit.2018.8394935

29. Trardi, Y., Ananou, B., Haddi, Z., Ouladsine, M.: A novel method to identify relevant features for automatic detection of atrial fibrillation. In: 26th Mediterranean Conference on Control and Automation (MED) (2018). https://doi.org/10.1109/med.2018.8442479

30. Trardi, Y., Ananou, B., Ouladsine, M.: An advanced arrhythmia recognition methodology based on R-waves time-series derivatives and benchmarking machine-learning algorithms. In: European Control Conference (ECC) (2020). https://doi.org/10.23919/ecc51009.2020. 9143678

31. Trardi, Y., Ananou, B., Ouladsine, M.: Computationally efficient algorithm for atrial fibrillation detection using linear and geometric features of RR time-series derivatives. In: International Conference on Control, Automation and Diagnosis (ICCAD) (2022). https://doi.org/ 10.1109/iccad55197.2022.9853910

32. He, K., Zhang, X., Ren, S., Sun, J.: Deep residual learning for image recognition. IEEE Conf. Comput. Vision Pattern Recogn. (CVPR) (2016). https://doi.org/10.1109/CVPR.2016.90

33. Goldberger, A.L., Amaral, L.A.N., Glass, L., Hausdorff, J.M., Ivanov, P.C., Mark, R.G., Mietus, J.E., Moody, G.B., Peng, C.-K., Stanley, H.E.: PhysioBank, PhysioToolkit, and PhysioNet: components of a new research resource for complex physiologic signals. Circulation **101**(23) (2000). https://doi.org/10.1161/01.cir.101.23.e215

34. Moody, G.B., Mark, R.G.: The impact of the MIT-BIH arrhythmia database. IEEE Eng. Med. Biol. Mag. **20**(3), 45–50 (2001). https://doi.org/10.1109/51.932724

35. Butterworth, S.: On the theory of filter amplifiers. Exper. Wirel. Wirel. Eng. **7**, 536–541 (1930)

36. Oppenheim, A.V., Willsky, A.S., Young, I.T.: Signals and Systems (1983)

37. Chawla, N.V., Bowyer, K.W., Hall, L.O., Kegelmeyer, W.P.: SMOTE: synthetic minority over-sampling technique. J. Artif. Intell. Res. **16**, 321–357 (2002). https://doi.org/10.1613/jai r.953

38. Sakoe, H., Chiba, S.: Dynamic programming algorithm optimization for spoken word recognition. IEEE Trans. Acoust. Speech Signal Process. **26**(1), 43–49 (1978). https://doi.org/10. 1109/tassp.1978.1163055

39. Hochreiter, S., Schmidhuber, J.: Long short-term memory. Neural Comput. **9**(8), 1735–1780 (1997). https://doi.org/10.1162/neco.1997.9.8.1735

40. Yang, J., Li, C., Dai, X., Yuan, L., Gao, J.: Focal modulation networks. Adv. Neural Inf. Process. Syst. **35** (NeurIPS 2022) (2022). https://proceedings.neurips.cc/paper_files/paper/ 2022/hash/1b08f585b0171b74d1401a5195e986f1-Abstract-Conference.html

41. Vaswani, A., Shazeer, N., Parmar, N., Uszkoreit, J., Jones, L., Gomez, A.N., Kaiser, L., Polosukhin, I.: Attention Is All You Need. Advances in Neural Information Processing Systems 30 (NeurIPS 2017) (2017). https://proceedings.neurips.cc/paper_files/paper/2017/hash/ 3f5ee243547dee91fbd053c1c4a845aa-Abstract.html

42. Qin, Q., Li, J., Zhang, L., Yue, Y., Liu, C.: Combining low-dimensional wavelet features and support vector machine for arrhythmia beat classification. Sci. Rep. **7**(1), 6067 (2017). https:// doi.org/10.1038/s41598-017-06596-z

43. Yıldırım, Ö., Pławiak, P., Tan, R.-S., Acharya, U.R.: Arrhythmia detection using deep convolutional neural network with long duration ECG signals. Comput. Biol. Med. **102**, 411–420 (2018). https://doi.org/10.1016/j.compbiomed.2018.09.009

44. Oh, S.L., Ng, E.Y.K., Tan, R.S., Acharya, U.R.: Automated beat-wise arrhythmia diagnosis using modified U-net on extended electrocardiographic recordings with heterogeneous arrhythmia types. Comput. Biol. Med. **105**, 92–101 (2019). https://doi.org/10.1016/j.compbiomed.2018.12.012

45. Kim, J.-H., Seo, S.-Y., Song, C.-G., Kim, K.-S.: Assessment of electrocardiogram rhythms by GoogLeNet deep neural network architecture. J. Healthc. Eng. **2019**, 1 (2019). https://doi.org/10.1155/2019/2826901

46. Zubair, M., Yoon, C.: Cost-sensitive learning for anomaly detection in imbalanced ECG data using convolutional neural networks. Sensors **22**(11), 4075 (2022). https://doi.org/10.3390/s22114075

Graph-Based Data Representation and Prediction in Medical Domain Tasks Using Graph Neural Networks

Vdovkina Sofiia[1]($\boxtimes$) (iD), Derevitskii Ilya[1] (iD), Abramyan Levon[2] (iD),
and Vatian Aleksandra[1] (iD)

[1] ITMO University, Saint-Petersburg, Russia
{sophia.vdovkina,ilyaderevitskii}@niuitmo.ru
[2] World-Class Research Centre for Personalized Medicine, Almazov National Medical
Research Centre, 197341 St. Petersburg, Russia

Abstract. Medical data often presents as a time series, reflecting the disease's progression. This can be captured through longitudinal health records or hospital treatment notes, encompassing diagnoses, health states, medications, and procedures. Understanding disease evolution is critical for effective treatment. Graph embedding of such data is advantageous, as it inherently captures entity relationships, offering significant utility in medicine. Hence, this study aims to develop a graph representation of Electronic Health Records (EHRs) and combine it with a method for predictive analysis of COVID-19 using network-based embedding. Evaluation of Graph Neural Networks (GNNs) against Recurrent Neural Networks (RNNs) reveals superior performance of GNNs, underscoring their potential in medical data analysis and forecasting.

Keywords: Graph neural networks · data representation · electronic health records

1 Introduction

Recent advancements in deep learning have yet to fully integrate into clinical decision-support systems, but the digitization of health records has spurred machine learning research in medicine, particularly in EHRs. While this surge underscores the growing relevance of the field, the utilization of EHRs spans from database establishment to embedding methodologies [1]. Our focus lies specifically on leveraging graph embeddings within medical records, representing a novel approach in this domain.

Medical data often manifests as time series, portraying disease progression over time within longitudinal health records or clinical notes documenting hospital treatments. EHRs serve as rich repositories for various electronic health tasks, including predictive modeling of clinical risks such as in-hospital mortality and readmission rates, disease correlations exploration, classification, and medical decision support. Predictive modeling of future clinical events is thus a vital objective in medical practice. Predicting sequences of clinical events typically involves latent entity and event embedding

L. Franco et al. (Eds.): ICCS 2024, LNCS 14835, pp. 371–378, 2024.
https://doi.org/10.1007/978-3-031-63772-8_32

coupled with neural network models like RNNs. However, medical records pose unique challenges due to their sparsity, irregularity, and heterogeneity, hampering effective predictive model creation. Graph embedding offers inherent advantages by capturing entity relationships, particularly beneficial in the medical domain. Therefore, our study aims to develop a predictive analysis tool for disease progression using network-based embedding methodologies, specifically exploring methods for transforming longitudinal medical record data into graphs to elucidate disease trajectories, followed by the construction and comprehensive analysis of GNNs.

The remainder of this paper is organized as follows. Section 2 reviews related work in the field. Section 3 provides details on the creation of patient graphs. In Sect. 4, the experimental setups with the proposed embedding approach and GNN are explained, along with the presentation of results. Finally, Sect. 5 concludes this paper, summarizing the findings and suggesting avenues for future research.

2 Related Work

2.1 Transformation Methods

The temporal dynamics inherent in time series data can be effectively captured through a graph model representation, akin to a discrete model of the underlying dynamic system. This model, guided by Takens theorem, facilitates the restoration of the system's state space. In [2], various transformation methods for time series data into complex network representations are categorized into three distinct classes: proximity networks, visibility graphs, and transition networks. In the context of personal medicine, prediction, and classification tasks utilizing EHRs, the conversion of temporal EHR data into graph structures has emerged as a pivotal approach for leveraging inherent graph properties in subsequent analyses. Given the heterogeneous nature of EHRs encompassing medications, diagnoses, clinical notes, and lab results, diverse modeling techniques are imperative to accommodate this diversity. Despite the prevalence of articles mentioning "medical records" and "graphs" a significant portion does not delve into graph theory or utilize graphs to represent individual patient data [2]. However, a subset of studies effectively leverages graph structures to encode temporal relationships within time series data.

2.2 Methods of Health Records Transformation to Graph

Patient medical records can be represented in a knowledge model in two primary ways: as a scope depicting common patient features and their interrelations or as temporal graphs for individual patients. Notably, Khademi M. and Nedialkov N. S. [3] constructed a probabilistic graphical model using clinical data, integrating it with deep belief networks for breast cancer prognosis. Chen et al. [4] proposed a graph-based semi-supervised learning algorithm for risk prediction, utilizing Cause of Death information as labels and temporal relationships between examination items as edges. Liu et al. [5] developed a temporal phenotyping approach based on graph representations of EHR events, extracting significant graph bases for interpretability. Esteban et al. [6] applied latent embedding models to clinical data for event sequence prediction, while Zhang et al. [7, 8] proposed integrative medical temporal graph-based prediction approaches, albeit with varied success rates. Notably, Tong et al. [9] introduced an LSTM-GNN model for patient outcome prediction, effectively combining temporal and graph-based features.

Similarly, graph methods for patient clustering were developed [10], emphasizing the construction of network-like structures based on patient similarities. Additionally, Hanzlicek et al. [11] and Kaur et al. [12] described graph-based models for storing medical records, facilitating efficient data retrieval and decision support functionalities. These studies underscore the diverse applications and potential of graph representations in healthcare, from predictive modeling to decision support systems.

From the review, it can be inferred that graphs are not yet widely adopted as a form of health data representation. However, this area of research shows promise and warrants further exploration. For sequence prediction tasks, methods based on individual patient data graphs are recommended, as they effectively capture temporal relationships between health states. Conversely, for classification tasks, scope representation may be more beneficial, as it encapsulates common features and their interrelations. Nevertheless, the most promising avenue for future research lies in the discovery of methods that represent inter-patient connections. Such approaches hold the potential to enhance prognostic capabilities and inform clinical decision-making by leveraging similarities among medical cases. Thus, further investigation into graph-based representations in healthcare is imperative for advancing predictive analytics and decision support systems in medicine.

3 Creating EHR Representation Through Graphs

EHRs serve as comprehensive repositories of patient health information, capturing vital aspects of their medical journey. Transforming this rich temporal data into a structured graph representation holds immense potential for facilitating various healthcare tasks, ranging from predictive modeling to clinical decision support. In this chapter, we present a novel pipeline for creating graph representations of EHRs. Our approach involves gathering patient state data during hospitalization periods, embedding each state, and subsequently constructing a graph where nodes represent patient states and edges encode temporal and proximity-based relationships.

3.1 Data Description

The study utilized a dataset comprising 6188 medical records encompassing 1992 distinct patients who received treatment for COVID-19 at Almazov National Medical Research Centre in St. Petersburg, Russia, spanning from June 2020 to March 2021. Each treatment case is characterized by a comprehensive set of indicators, as detailed in Table 1.

These medical indicators comprise a spectrum of information, spanning past medical conditions, laboratory analysis results, physical measurements, lifestyle factors, and medication types. They collectively provide a comprehensive overview of patients' health profiles, supporting clinical analysis and decision-making. The dataset comprises 40 integer features, 11 floating-point features, and 3 date features.

Table 1. Medical indicators of treatment cases.

Group	Features
Controlled Medications	Omeprazole, nadroparin calcium, esomeprazole, amlodipine, ambroxol, domperidone, mebrofenin, technetium, mometasone, bisoprolol, dexamethasone, hydrochlorothiazide, hydroxychloroquine, rabeprazole, enoxaparin sodium, perindopril, acetylcysteine, azithromycin, valsartan, methylprednisolone, loratadine, chloroquine, sodium chloride, indapamide, prednisolone, atorvastatin, dextran, lisinopril, losartan
Dynamic Factors	Temperature, lymphocytes count, aspartate aminotransferase, heart rate, respiratory rate, total bilirubin, mean platelet volume, platelet crit, lymphocytes percentage, decreased consciousness, severity grade on CT scan, lactate dehydrogenase, platelet distribution width
Static Factors	Age, sex
Controlled Procedures	Blood transfusion, oxygen therapy, non-invasive ventilation, invasive ventilation
Process Variables	Process stages, current process duration
Dates and length of stay	Admission date, end episode, length of observation
Target Variable	Outcome

3.2 Pipeline Description

The pipeline is initiated by collecting data pertaining to patients' states during hospitalization periods. Each state is characterized by a set of features capturing relevant clinical parameters and temporal information. Preprocessing steps involve cleaning the data, handling missing values, and standardizing features to ensure uniformity across the dataset.

To capture the intricate relationships within patient states, TabNet, a state-of-the-art tabular data embedding technique, is employed [13]. This network leverages sequential attention mechanisms to learn informative embeddings from tabular data, effectively capturing both local and global dependencies. By embedding each patient state using TabNet, the multidimensional characteristics are encoded into a compact representation, facilitating downstream graph construction.

The crux of our pipeline lies in the construction of a graph representation that encapsulates the temporal and proximity-based relationships among patient states. A two-pronged approach to edge creation is adopted: temporal connections and proximity-based connections. Incorporating both features is crucial for capturing the sequential order of events and the contextual relationships between patient states in EHRs, our model gains a deeper understanding of disease dynamics and patient interactions. This enriched context improves the model's ability to generalize to new data and adapt to changes in healthcare practices.

Temporal Connections: Patient states are sequentially connected in time, reflecting the temporal progression of their medical journey. Each state is linked to its subsequent state within the same patient's trajectory, forming a temporal sequence of nodes.

Proximity-based Connections: To capture inter-patient relationships, the Euclidean distance between each patient state is computed to identify the closest states of other patients. Two approaches are explored: (1) connecting each state to the closest three states of any other patient (see Fig. 1), or (2) imposing a constraint wherein a state can only be connected to a state of another patient once, thereby fostering diverse inter-patient connections (see Fig. 2). The choice between these approaches can be determined experimentally or tailored to specific healthcare objectives. For visualization purposes, graphs featuring 10 patients are provided. The complete graph comprises 6182 nodes, each with 24 features, wherein each node is labeled with either 0 or 1 based on the outcome, and 22737 edges representing temporal and interpersonal connections.

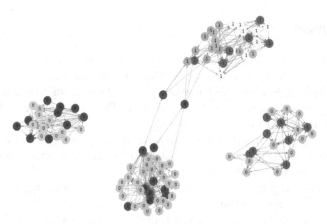

Fig. 1. Graph without constraints on the connections of patients. Colors represent different patients, and node labels are selected by the outcome. (Color figure online)

Constructing a graph without connection constraints reveals discernible clusters within the data. Notably, patients experiencing negative outcomes tend to cluster together, suggesting shared characteristics or medical trajectories. Additionally, the observed clustering patterns hint at the potential grouping of patients with close medical histories, indicating the influence of shared anamnesis factors on cluster formation.

In a graph with connection constraints, distinct groups of patients are still discernible, albeit with more blurred boundaries. However, this method holds promise for classification tasks as it captures more ambiguous connections, allowing for the identification of subtle relationships between patient clusters.

Further experimentation and validation are needed to assess its efficacy and generalizability across healthcare contexts.

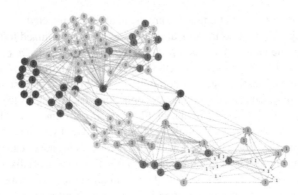

Fig. 2. A graph constructed with diverse inter-patient connections. Colors represent different patients, and node labels are selected by the outcome. (Color figure online)

4 Experiments

The experimental results provide valuable insights into the performance of the suggested representation method used as a source for Graph Convolutional Network (GCN) and Graph Isomorphism Network (GIN) models for mortality prediction tasks based on patient state graphs, performing node classification. This choice is due to their unique ability to effectively capture relational information and structural dependencies inherent in graph data. By aggregating information from neighboring nodes, GCNs can effectively model the complex interactions between patient states, capturing crucial temporal and contextual dependencies. GINs are invariant under permutations of the nodes, so they learn and generalize patterns across different patient trajectories, regardless of their specific ordering or representation.

The proposed GCN architecture consists of three hidden layers of GCNConv, each followed by ReLU activation function. The input to the model is the feature vector representing each patient state, which is passed through the GCN layers along with the edge indices representing the connectivity of the graph. Dropout with a probability of 0.5 is applied after the first GCN layer to prevent overfitting. The output of the final GCN layer is passed through a linear layer with an output dimension of 1, followed by a sigmoid function to produce the final classification probability indicating the likelihood of patient mortality.

The GIN architecture consists of two GINConv layers, each comprising a sequence of linear transformations followed by batch normalization and ReLU activation functions. Similar to GCN, the input to the model is the feature vector representing each patient state, which is passed through the GINConv layers along with the edge indices representing graph connectivity. The aggregated features are then passed through a linear layer with an output dimension of 1, followed by a sigmoid function as well. And for the sake of experiment, along with graph networks RNN model tested with 3 hidden layers was.

The choice of metrics was driven by their relevance to mortality prediction. While accuracy measures overall correctness, precision assesses the proportion of true positive cases among predicted positives, and recall gauges the model's ability to identify all true positive cases. Emphasizing recall ensures the accurate identification of patients at risk of mortality, vital for early intervention.

The observed metrics of the models are presented in Table 2.

Table 2. Observed metrics of the models.

Model	Accuracy	Precision	Recall
GCN without connection constraints	0.8208	0.5341	0.7743
GIN without connection constraints	**0.8722**	0.5771	**0.8329**
GCN with connection constraints	0.8023	0.2875	0.6145
GIN with connection constraints	0.8458	0.3298	0.6875
RNN	0.7832	**0.6680**	0.7352

The first type of graph, without connection constraints, allowed for the formation of clusters, leading to improved performance of both GCN and GIN models. This suggests that the graph structure contributed to more accurate representations and enhanced predictive performance. The absence of constraints likely facilitated the models' ability to capture underlying patterns and relationships within the data. While both models performed well, GIN showed superior ability in capturing true positive cases while minimizing false positives.

In perspective of outcome prediction, these results suggest that incorporating graph-based representations of patient states can enrich predictive modeling.

5   Conclusion

In conclusion, our study introduces a novel approach for Electronic Health Record (EHR) representation using graphs, promising advancements in healthcare analytics and predictive modeling. By employing graph-based representations of patient states, temporal and relational dependencies crucial for accurate predictions were captured. Leveraging advanced machine learning techniques like Graph Convolutional Networks and Graph Isomorphism Networks further enhances predictive capabilities.

Our experiments confirm the effectiveness of our approach. While the incorporation of connection constraints in graph construction didn't improve efficiency, its potential applications warrant further investigation. Analyzing specific cluster characteristics could shed light on their impact on model performance.

In summary, adopting graph-based representations for EHRs revolutionizes healthcare analytics, enabling comprehensive analysis while preserving temporal and relational contexts. Our approach lays the groundwork for future advancements in EHR representation and predictive modeling, driving towards more effective healthcare solutions.

Acknowledgements. This work was supported by Russian Science Foundation, Grant № 23-11-00346.

References

1. Solares, J.R.A., et al.: Deep learning for electronic health records: a comparative review of multiple deep neural architectures. J. Biomed. Inform. **101**, 103337 (2020)
2. Schrodt, J., et al.: Graph-representation of patient data: a systematic literature review. J. Med. Syst. **44**(4), 1–7 (2020)
3. Khademi, M., Nedialkov, N.S.: Probabilistic graphical models and deep belief networks for prognosis of breast cancer. In: 2015 IEEE 14th International Conference on Machine Learning and Applications (ICMLA), pp. 727–732. IEEE (2015)
4. Chen, L., et al.: Mining health examination records—a graph-based approach. IEEE Trans. Knowl. Data Eng. **28**(9), 2423–2437 (2016)
5. Liu, C., et al.: Temporal phenotyping from longitudinal electronic health records: a graph based framework. In: Proceedings of the 21th ACM SIGKDD International Conference on Knowledge Discovery and Data Mining, pp. 705–714 (2015)
6. Esteban, C., et al.: Predicting sequences of clinical events by using a personalized temporal latent embedding model. In: 2015 International Conference on Healthcare Informatics, pp. 130–139. IEEE (2015)
7. Zhang, S., et al.: MTPGraph: a data-driven approach to predict medical risk based on temporal profile graph. In: 2016 IEEE Trustcom/BigDataSE/ISPA, pp. 1174–1181. IEEE (2016)
8. Zhang, J., Gong, J., Barnes, L.: HCNN: heterogeneous convolutional neural networks for comorbid risk prediction with electronic health records. In: 2017 IEEE/ACM International Conference on Connected Health: Applications, Systems and Engineering Technologies (CHASE), pp. 214–221. IEEE (2017)
9. Tong, C., Rocheteau, E., Veličković, P., Lane, N., Liò, P.: Predicting Patient Outcomes with Graph Representation Learning. In: Shaban-Nejad, A., Michalowski, M., Bianco, S. (eds.) AI for Disease Surveillance and Pandemic Intelligence: Intelligent Disease Detection in Action, pp. 281–293. Springer, Cham (2022). https://doi.org/10.1007/978-3-030-93080-6_20
10. Ochoa, J.G.D., Mustafa, F.E.: Graph neural network modelling as a potentially effective method for predicting and analyzing procedures based on patients' diagnoses. Artif. Intell. Med. **131**, 102359 (2022)
11. Hanzlicek, P., et al.: User interface of MUDR electronic health record. Int. J. Med. Inform. **74**(2–4), 221–227 (2005)
12. Kaur, K., Rani, R.: Managing data in healthcare information systems: many models, one solution. Computer **48**(3), 52–59 (2015)
13. Arik, S.Ö., Pfister, T.: TabNet: attentive interpretable tabular learning. In: Proceedings of the AAAI Conference on Artificial Intelligence, vol. 35, no. 8, pp. 6679–6687 (2021)

Global Induction of Oblique Survival Trees

Malgorzata Kretowska$^{(\boxtimes)}$ and Marek Kretowski

Faculty of Computer Science, Bialystok University of Technology, Wiejska 45a,
15-351 Bialystok, Poland
{m.kretowska,m.kretowski}@pb.edu.pl

Abstract. Survival analysis focuses on the prediction of failure time and serves as an important prognostic tool, not solely confined to medicine but also across diverse fields. Machine learning methods, especially decision trees, are increasingly replacing traditional statistical methods which are based on assumptions that are often difficult to meet. The paper presents a new global method for inducing survival trees containing Kaplan–Mayer estimators in leaves. Using a specialized evolutionary algorithm, the method searches for oblique trees in which multivariate tests in internal nodes divide the feature space using hyperplanes. Specific variants of mutation and crossover operators have been developed, making evolution effective and efficient. The fitness function is based on the integrated Brier score and prevents overfitting taking into account the size of the tree. A preliminary experimental verification and comparison with classical univariate trees was carried out on real medical datasets. The evaluation results are promising.

Keywords: survival tree · oblique splits · evolutionary computation

1 Introduction

Is it possible to predict the risk of death after a cancer diagnosis? Can we classify individuals into risk groups for disease relapse? These are some of the questions that survival analysis attempts to answer.

What exactly is survival analysis? It is a set of tools, often statistical, which are able to cope with survival data, in which time of a certain event occurrence is investigated. A characteristic element of this type of data is censoring, which means that for some observations the precise time of the event of interest, called failure, is unknown. Statistical methods often rely on various assumptions that must be met for the results to be accurate [5]. Machine learning methods, on the other hand, are not subject to such limitations. They are constantly developed attempting to successfully address the aforementioned questions.

Tree-based models are among the most commonly used machine learning methods for analyzing censored data. We can differentiate between individual trees and ensembles. In both cases, univariate (with axis-parallel tests in internal nodes) and oblique solutions are available, with the former being predominant.

© The Author(s), under exclusive license to Springer Nature Switzerland AG 2024
L. Franco et al. (Eds.): ICCS 2024, LNCS 14835, pp. 379–386, 2024.
https://doi.org/10.1007/978-3-031-63772-8_33

The construction process for typical survival tree models follows a greedy, top-down approach. This involves two main phases: induction and pruning. During induction, the focus is on recursively minimizing a specified impurity measure [20] or maximizing between-node separation [17]. The pruning step typically involves cost-complexity pruning [3] or its survival extension, split-complexity pruning [17]. Despite pruning efforts, the resulting trees often remain overgrown [13], and adopting more global approaches may be beneficial.

A different approach, called a conditional inference framework, was introduced by Hothorn et al. [8], and next by Kundu and Ghosh [14], where the split importance is assessed during node creation, obviating the need for additional pruning phases. To other univariate survival tree models belong also median regression trees [4] or a non-greedy induction method introduced in [2]. Oblique trees, in which multivariate tests in internal nodes are in a form of a hyperplane, belong to less common solutions. Kretowska [12] proposed dipolar survival tree while oblique random survival forest was introduced by Jaeger et al. [10].

The paper presents a novel method for global inducing survival trees that include Kaplan–Meier estimators in their leaves. Using a specialized evolutionary algorithm, the method searches for a whole oblique tree, simultaneously tree structure and all tests. Specific variants of mutation and crossover operators have been developed to ensure effective and efficient evolution. The fitness function, based on the integrated Brier score, mitigates overfitting by considering the model complexity. A preliminary experimental verification was conducted using five medical datasets. The prediction ability was compared with that of two state-of-the-art survival trees. Additionally, the interpretable model obtained for the follicular cell lymphoma dataset was discussed in detail.

2   Preliminaries

Survival Data. We assume having a learning set, L, which consists of M observations. In survival analysis, the ith observation is described by a set of three values $(\mathbf{x}_i, t_i, \delta_i)$, where $\mathbf{x}_i$ is the N-dimensional feature vector, t_i is the observed time, which for uncensored subjects is equal to its failure time, for censored - it takes values of the follow-up time, δ_i is the failure indicator, which takes one of two values: 0 for censored observations or 1 otherwise.

The distribution of the survival time may be represented by a survival function $S(t) = P(T > t)$, which gives the probability of surviving beyond the time t. Kaplan–Meier (KM) method [11] is one on the most common nonparametric estimators of the survival function. If we assume that the events of interest occur at D distinct times $t_{(1)} < t_{(2)} < \ldots < t_{(D)}$, it is calculated as follows:

$$\hat{S}(t) = \prod_{j|t_{(j)} \leq t} \left(\frac{m_j - d_j}{m_j} \right),$$ (1)

where d_j is the number of events at time $t_{(j)}$ and m_j is the number of patients at risk at $t_{(j)}$ (i.e., who are alive or experience the event of interest at $t_{(j)}$).

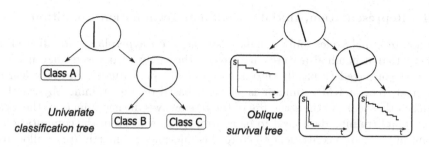

Fig. 1. Univariate decision tree vs. oblique survival tree.

Oblique Survival Trees. A tree structure consists of internal and terminal nodes called leaves. The purpose of trees is to divide a feature space (internal nodes) into homogeneous data regions (leaves) for a given task. In case of survival tree, each terminal node, in our approach, is characterised by the Kaplan–Meier estimator and we aimed at obtaining areas with homogeneous survival experience.

Proposed here, oblique survival trees, divide the feature space by the test of the form of hyperplane $H(\mathbf{w}, \theta)$, where $\mathbf{w} = [w_1, \dots, w_N]^T$, instead of one single variable x_i. Terminal nodes contain the feature vectors corresponding to distinct regions of the feature space (Fig. 1).

Integrated Brier Score. Taking into account censored survival data, the exact failure time is unknown for some of the observations. Therefore, a direct comparison of real and predicted survival times is impossible. One of the most common performance measures is the integrated Brier score [7]. The first step in calculating the IBS is to compute the Brier score as:

$$BS(t) = \frac{1}{M} \sum_{i=1}^{M} (\hat{S}(t|\mathbf{x}_i)^2 I(t_i \leq t \wedge \delta_i = 1)\hat{G}(t_i)^{-1}$$
$$+ (1 - \hat{S}(t|\mathbf{x}_i))^2 I(t_i > t)\hat{G}(t)^{-1}), \qquad (2)$$

here $\hat{S}(t|\mathbf{x}_i)$ is the KM estimator, $\hat{G}(t)$ denotes the KM estimator of the censoring distribution, $I(condition)$ is equal to 1 if the condition is fulfilled and zero otherwise. Integrated Brier score (IBS) is obtained by:

$$IBS = \frac{1}{\max(t_i)} \int_0^{\max(t_i)} BS(t)dt, \qquad (3)$$

3   Evolutionary Induction

The proposed evolutionary algorithm extends the global induction of standard decision trees [13]. In this short paper, we have to concentrate on the elements crucial to survival analysis and omit more general elements.

3.1   Representation, Initialization, and Termination Condition

In non-terminal nodes, only oblique tests based on hyperplanes are allowed. It means that if nominal features are part of the analyzed datasets they need to be first converted (typically into a group of binary features). In every leaf of survival trees, a KM estimator is situated based on the training objects that reached that leaf. As the structure of the tree evolves during induction, the locations of all training data need to be constantly known. From the computational point of view, it is convenient to store this information smartly (once allocated table) and, as the modifications are often local, update only the corresponding areas/subtrees. As a consequence, individuals are not encoded but represented as standard binary trees with additional data structures.

Appropriate population initialization significantly shortens evolution and allows for more efficient use of resources. The initial trees should be diverse and preferably similar in size to the target survival trees, which are usually quite compact. A simple top-down algorithm with a tree height limit is used, which is activated on random small subsets of the training set. Tests in internal nodes are created based on randomly selected dipoles - pairs of observations. Dipoles can be created between uncensored observations and between earlier uncensored observations and later censored observations. When generating tests, longer dipoles are preferred (taking into account the difference between failure times), because their intersection means that these observations will be in two disjoint subtrees (and ultimately leaves). The same mechanism of selecting dipoles and creating tests based on them is also used in the mutation operator.

Evolutionary induction ends when, after a given number of iterations, no individual with a better fitness value is found (default 1000 iterations) or when the limit of the number of iterations is reached (default 5000).

3.2   Genetic Operators

As in a typical evolutionary algorithm [13], two genetic operators are used. Crossover allows the exchange of genetic material between two individuals. In the most typical variant, two nodes (including subtrees) are randomly selected in two trees and the entire subtrees are replaced. It is also possible to: exchange the tests themselves or conduct a crossover with the best individual so far. Since crossing in relation to tree structures can be destructive [6], this operator is applied to trees with a rather low probability (default 0.2). Mutation is the main mechanism for differentiating individuals and is performed on the tree with a high probability (default 0.8). The tree structure can be modified directly, by pruning a randomly chosen subtree to a leaf, or by replacing the leaf with an internal node with a new test. The structure may change indirectly when an existing test is modified (e.g., by randomly changing or resetting one weight in the hyperplane) and the corresponding subtree is changed as a result. The key operation ensuring the efficiency of exploration of the search space is generating a new test. The dipole mechanism described earlier is used here, thanks to which it is possible to direct the search sensibly (by avoiding ineffective tests).

3.3 Fitness Function

The fitness function is the most crucial component of any evolutionary algorithm. In the context of evolutionary machine learning, defining functions directly is not feasible, as the objective of the algorithms is to perform (predict) as effectively as possible not on the training data, but on data that is unavailable during induction. A common approach is to optimize a measure of solution quality on the training data, coupled with an additional factor reflecting model size to prevent overfitting (regularization). In the case of survival trees, we calculate the integrated Brier score (IBS) for the training data and determine the tree size (number of leaves). The fitness function is then defined as follows:

$$Fitness(T) = IBS(T) + \alpha(Size(T) - 1), \tag{4}$$

Adjusting the α value allows for the control of the expected complexity of the resulting tree. In this formulation, all tests (irrespective of the number of features used) hold equal importance. If a preference for simpler tests is desired, the $Size$ term could be made dependent on the number of features used.

4 Preliminary Experimental Validation

The first part of the experiments aims to compare the predictive ability of the evolutionary induced oblique survival tree (EIOST) with two state-of-the-art univariate survival trees: the conditional inference tree (CItree) [8] and the recursive partitioning for survival trees (RPtree) [3]. Both of these solutions are publicly available in R packages `party` and `rpart`, respectively. The parameters of the EIOST (see Sect. 3) were kept constant during all experiments.

The experiments were conducted using five publicly available medical datasets with the percentage of censored cases from 30.4 to 87.3, the number of observations from 418 to 2231, and 4 to 39 attributes. In Table 1, the IBS calculated for RPART, CItree, and EIOST are presented. The EIOSTs were induced with a default value of $\alpha = 0.001$. This indicates that the reported values of IBS may not represent the optimal performance for the specific dataset, leaving potential for further enhancement. For this fixed α value, in three for five datasets the proposed method gives the best IBS values. The number of leaves is similar to other solutions.

The follicular cell lymphoma study (follic) dataset contains information about 541 patients described by four attributes: age, hemoglobin(hgb), clinical stage (stg), and chemotherapy (ch). In Fig. 2, we can observe the impact of the α value on the IBS and the number of leaves obtained for the follic dataset. It is evident that the tree complexity decreases with increasing values of α. We start with approximately 33 leaves for $\alpha = 0.0001$ and gradually decrease to only one leaf at $\alpha = 0.1$. The best IBS value of 19.9 was achieved for $\alpha = 0.005$, with a corresponding number of leaves equals 3. This result clearly outperforms the one reported in Table 1 and is better than the IBS calculated for the univariate trees.

Table 1. Integrated Brier score of three types of survival trees: RPtree, CItree, and EIOST inducted with $\alpha = 0.001$; Size denotes the mean number of leaves and IBS denotes the mean ± standard deviation of IBS calculated over 5 runs of 10-fold cross-validation multiplied by 100.

Dataset	#obs	#at	%cen	RPtree		CItree		EIOST	
				IBS	Size	IBS	Size	IBS	Size
Pbc [5]	418	8	61.5	15.42 ± 0.7	12.5	14.65 ± 0.5	6.9	**14.50 ± 0.9**	8.2
Follic [18]	541	4	49.7	20.88 ± 0.4	3.9	**20.63 ± 0.5**	4	21.43 ± 1.1	9.6
Nwtco [19]	668	5	87.3	10.43 ± 0.2	4.1	10.24 ± 0.2	3.8	**10.10 ± 0.3**	4.1
Mgus2 [15]	1384	6	30.4	14.76 ± 0.3	2	14.14 ± 0.3	12.6	**13.16 ± 0.3**	4.7
Peakv02 [9]	2231	39	67.5	**16.37 ± 0.1**	3.9	16.63 ± 0.1	7.3	16.46 ± 0.6	3.2

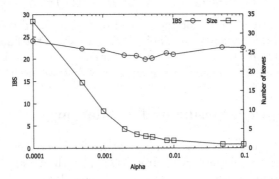

Fig. 2. Accuracy versus interpretability trade-off for follic dataset.

In Fig. 3a, we can see the EIOST induced for the follic dataset with $\alpha = 0.005$. The tree divides the feature space into three distinct regions represented by leaves. L3 represents the region with the worst prognosis, having a median survival time of 6.17 years, while the best prognosis is for L1, where the median survival time cannot be calculated. This is evident in Fig. 3b, where three KM survival functions are depicted. The disparities between estimators are statistically significant (log-rank test, $p < 0.0001$), indicating that the hyperplanes in the internal nodes have split the feature space into areas with varying survival experiences.

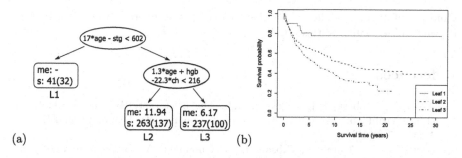

Fig. 3. EIOST inducted for follic dataset (a) with the corresponding KM survival functions (b); *me* denotes the median survival time in the leaf, *s* indicates the number of observations (number of censored cases).

5   Conclusions

In this paper, we propose a novel method for global induction of oblique survival trees tailored for analyzing censored data, which includes observations with unknown exact failure times. An essential aspect of survival analysis tools is the ability to leverage this incomplete information during the induction process. We achieve this through specialized evolutionary algorithm with specific initialization, variants of genetic operators, as well as the fitness function combining the integrated Brier score and tree complexity.

Based on five real datasets, two characteristics of the survival trees, predictive ability and model complexity, were compared with two existing univariate tree models. The preliminary results are encouraging. In three datasets, the predictive ability of EIOST is better than the results obtained for the competitors, while the number of nodes is small facilitating the interpretability of the model. The experiments were conducted with default values of α and the quality of the oblique survival tree may be improved by adjusting it to a given problem.

One of the major tasks of the resulting tree is the ability to distinguish areas in the feature space that would contain patients with varying survival experiences. The example of the tree model induced for the follicular cell lymphoma dataset points out that this objective was achieved. The Kaplan-Meier survival functions calculated for leaves differ significantly.

The proposed solution requires further investigation. A possible path is to replace the integrated Brier score in the fitness function with other measures, such as the likelihood method [16], as the IBS can favor tests with higher specificity [1]. Additionally, we aim to extend the algorithm to accommodate other types of survival data, such as discrete survival data or data with competing risks.

Acknowledgement. This work was supported by Bialystok University of Technology under the grant WZ/WI-IIT/4/2023 founded by Ministry of Science and Higher Education.

Conflict of Interest. The authors have no competing interests to declare that are relevant to the content of this article.

References

1. Assel, M., Sjoberg, D.D., Vickers, A.J.: The brier score does not evaluate the clinical utility of diagnostic tests or prediction models. Diagn. Prognostic Res. 1(1), 1–7 (2017)
2. Bertsimas, D., Dunn, J., Gibson, E., Orfanoudaki, A.: Optimal survival trees. Mach. Learn. 111(8), 2951–3023 (2022)
3. Breiman, L., Friedman, J., Olshen, R., Stone, C.: Classification and Regression Trees. Wadsworth, Belmont, CA (1984)
4. Cho, H.J., Hong, S.M.: Median regression tree for analysis of censored survival data. IEEE Trans. Syst. Man Cybern. Part A Syst. Hum. 38(3) (2008)
5. Fleming, T.R., Harrington, D.P.: Counting Processes and Survival Analysis. Wiley (1991)
6. Freitas, A.A.: Data Mining and Knowledge Discovery with Evolutionary Algorithms. Springer, Heidelberg (2002). https://doi.org/10.1007/978-3-662-04923-5
7. Graf, E., Schmoor, C., Sauerbrei, W., Schumacher, M.: Assessment and comparison of prognostic classification schemes for survival data. Stat. Med. 18, 2529–2545 (1999)
8. Hothorn, T., Hornik, K., Zeileis, A.: Unbiased recursive partitioning: a conditional inference framework. J. Comput. Graph. Stat. 15(3), 651–674 (2006)
9. Hsich, E., Gorodeski, E.Z., Blackstone, E.H., Ishwaran, H., Lauer, M.S.: Identifying important risk factors for survival in patient with systolic heart failure using random survival forests. Circ. Cardiovasc. Qual. Outcomes 4(1), 39–45 (2011)
10. Jaeger, B.C., et al.: Oblique random survival forests. Ann. Appl. Stat. 13(3), 1847–1883 (2019)
11. Kaplan, E.L., Meier, P.: Nonparametric estimation from incomplete observations. J. Am. Stat. Assoc. 53, 457–481 (1958)
12. Kretowska, M.: Piecewise-linear criterion functions in oblique survival trees induction. Arif. Intell. Med. 75, 32–39 (2017)
13. Kretowski, M.: Evolutionary Decision Trees in Large-Scale Data Mining. SBD, vol. 59. Springer, Cham (2019). https://doi.org/10.1007/978-3-030-21851-5
14. Kundu, M.G., Ghosh, S.: Survival trees based on heterogeneity in time-to-event and censoring distributions using parameter instability test. Stat. Anal. Data Mining ASA Data Sci. J. 14(5), 466–483 (2021)
15. Kyle, R.A., et al.: A long-term study of prognosis in monoclonal gammopathy of undetermined significance. N. Engl. J. Med. 346(8), 564–569 (2002)
16. LeBlanc, M., Crowley, J.: Relative risk trees for censored survival data. Biometrics 48, 411–425 (1992)
17. LeBlanc, M., Crowley, J.: Survival trees by goodness of split. J. Am. Stat. Assoc. 88(422), 457–467 (1993)
18. Pintilie, M.: Competing Risks: A Practical Perspective, vol. 58. Wiley (2006)
19. Therneau, T.M.: Survival: Survival Analysis (2016). http://CRAN.R-project.org/package=survival. R package version 2.39
20. Therneau, T.M., Grambsch, P.M., Fleming, T.R.: Martingale-based residuals for survival models. Biometrika 77(1), 147–160 (1990)

Development of a VTE Prediction Model Based on Automatically Selected Features in Glioma Patients

Sergei Leontev[1]([✉])(iD), Maria Simakova[2](iD), Vitaly Lukinov[3](iD),
Konstantin Pishchulov[2](iD), Ilia Derevitskii[1](iD), Levon Abramyan[2]([✉])(iD),
and Alexandra Vatian[1](iD)

[1] ITMO University, Kronverksky Pr. 49, bldg. A, St. Petersburg 197101, Russia
`international@itmo.ru`
[2] Personalized Medicine Centre, Almazov National Medical Research Centre, Saint Petersburg,
St. Petersburg, Russian Federation
`Abramyan_LK@almazovcentre.ru`
[3] Institute of Computational Mathematics and Mathematical Geophysics SB RAS, Novosibirsk,
Russia

Abstract. Venous thromboembolism (VTE) poses a significant risk to patients undergoing cancer treatment, particularly in the context of advanced and metastatic disease. In the realm of neuro-oncology, the incidence of VTE varies depending on tumor location and stage, with certain primary and secondary brain tumors exhibiting a higher propensity for thrombotic events. In this study, we employ advanced machine learning techniques, specifically XGBoost, to develop identifying models for predictors searching associated with VTE risk in patients with gliomas. By comparing the diagnosis testing accuracy of our XGBoost models with traditional logistic regression approaches, we aim to enhance our understanding of VTE prediction in this population. Our findings contribute to the growing body of literature on thrombosis risk assessment in cancer patients and may inform the development of personalized prevention and treatment strategies to mitigate the burden of VTE in individuals with gliomas at the hospital term.

Keywords: VTE · Predictive modeling · Machine learning · Clinical diagnosis

1 Introduction

Venous thromboembolism (VTE) is a condition that encompasses superficial vein thrombosis, deep vein thrombosis (DVT), venous gangrene, and pulmonary embolism (PE). In the field of cardio-oncology, cancer-associated thrombosis is a significant concern and is strongly associated with increased early all-cause mortality during cancer chemotherapy and surgery [1–3].

The construction of VTE prediction models remains an urgent issue to this day [4–8].

It has been observed that certain cancer sites such as the pancreas, kidneys, ovaries, lungs, gastrointestinal tract, and brain tumors have a higher tendency to cause blood clots.

L. Franco et al. (Eds.): ICCS 2024, LNCS 14835, pp. 387–395, 2024.
https://doi.org/10.1007/978-3-031-63772-8_34

These tumors are categorized as either primary or secondary tumors that are linked with metastasis. The two most prevalent primary brain tumors are meningioma and glial tumors, which account for 35.6% and 35.5% of cases, respectively [9].

The purpose of this study was to compare XGBoost and logistic regression methods as tools for creating a risk stratification model for venous thromboembolic events in patients with gliomas.

2   Materials and Methods

A study was conducted at the Almazov National Medical Research Center from January 2021 to May 2023, which enrolled 286 consecutive patients with histologically verified glioma who underwent surgery. The group consisted of 133 (51.2%) men and 132 (49.8%) women, with an average age of 54 [41; 63] years. The diagnosis of pulmonary embolism and deep vein thrombosis was made in accordance with current clinical recommendations [10].

The study determined the frequency of binary variables indicating the occurrence of VTE, clinical manifestations of neoplasms, and concomitant cardiovascular pathology. 95% confidence intervals (95% CI) were estimated using the Wilson formula and compared by the Fisher's exact test.

The models were compared by the areas under the ROC curves (AUC) using the DeLong test.

Prognostic characteristics related to VTE development were also compared. Sensitivity and specificity were assessed using McNemar's test, while positive and negative predictive values (PPV and NPV) were compared using a weighted generalized test (WSG test).

To address the issue of multiple comparisons, p-values were adjusted using the Benjamini-Hochberg method. Statistical hypotheses were tested at a significance level of $p = 0.05$. Differences were considered statistically significant if $p < 0.05$.

3   Model Building

3.1   Metrics

The parameters chosen for maximizing in model construction are specificity and precision (sensitivity). This means that models with higher sum of specificity and precision will be considered of higher quality.

3.2 Feature Selection for the Final Model

XGBoost models were chosen both as the final model and the feature selection model. These models were partitioned into test and training sets based on the rules described in previous chapters.

Subsequently, the remaining parameters underwent sequential inclusion into the model. Those parameters which led to the greatest increase in specificity were chosen. The process involved starting with one parameter and then adding another, continuing until sets of parameters ranging from 5 to 10 pieces were studied. The final model exhibited the most favorable results with 7 parameters.

The selected parameters were subsequently automatically transferred to the main model for its training.

3.3 Model Building

Next, the final model was built, which dynamically receives as input the parameters obtained in previous stage. The model itself is also an XGBoost, but with parameters different from those that were used when selecting parameters.

3.4 Results

The parameter selection stage for the transmitted data resulted in the following 7 parameters: ['D-dimer', 'BMI', 'bed rest (more than 3 days), prolonged lying position', 'PulmonaryDis', 'varicose veins', 'Hypertension', 'Dyslipidemia'].

The model was able to achieve the following indicators according to the main metrics (Table 1).

Table 1. Model result metrics.

Metric	Values [95% CI]
Specificity	93% [87%; 99%]
Precision	77% [58%; 94%]
Accuracy	84% [76%; 90%]
Recall	61% [42%; 77%]
F1-score	68% [51%; 81%]

Diagnostic testing accuracy table is shown in Fig. 1.

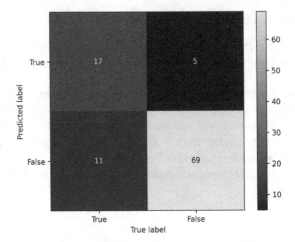

Fig. 1. Diagnostic testing accuracy table.

The area under ROC curve is 0.77 (Fig. 2).

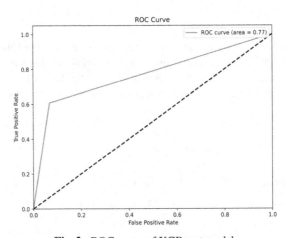

Fig. 2. ROC curve of XGBoost model.

3.5 Comparison with Logistic Regression

Significant predictors of VTE were identified through the construction of single-factor logistic regression models. Independent predictors of VTE development were identified through the construction of a multi-factor logistic regression model. The data from both single-factor and multi-factor regression analyses are presented in Table 2.

The summary characteristic of the model based on ROC analysis data is presented in the Table 3 and on Fig. 3.

Table 2. The data from the multi-factor and single-factor regression analyses.

Covariates	Single-factor models	Multi-factor models
	p	p
Bed rest (more than 3 days), prolonged lying position	< 0.001*	< 0.001*
D-dimer	< 0.001*	0.006*
PLT	0.006*	0.099
Age at the time of inclusion	0.010*	0.067
Radiation therapy	0.044*	0.047*

Table 3. The summary characteristic of the logistic regression.

Parameter	Values [95% CI]
Specificity	78.6% [49.2%; 95.3%]
Sensitivity	93.5% [88.4%; 96.8%]
Positive predictive value	52.4% [29.8%; 74.3%]
Negative predictive value	98% [94.2%; 99.6%]
Positive likelihood ratio	12.1 [6.3; 23.4]
Negative likelihood ratio	0.2 [0.1; 0.6]

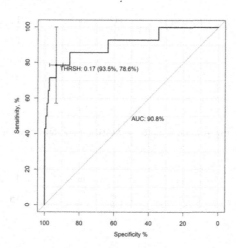

Fig. 3. Roc curve of multifactor logistic regression model.

The next step involved validating the model on a prospective sample of 100 patients with CNS gliomas who underwent treatment at the V. A. Almazov National Medical

Research Center of the Ministry of Health of Russia during the period from 2022 to 2023. The validation data is presented in Table 4.

Table 4. The internal validation data of the model on the prospective sample.

Parameter	Values [95% CI]
Specificity	95% [87%; 99%]
Sensitivity	47% [23%; 72%]
Positive predictive value	73% [39%; 94%]
Negative predictive value	85% [77%; 94%]
Positive likelihood ratio	10.2 [3.03; 34.35]
Negative likelihood ratio	0.56 [0.35; 0.87]

4   Model Comparison

A comparison of the models in the prospective sample is presented in Tables 5, 6, 7, 8 and Fig. 4. The difference in the total number of patients is due to incomplete data.

Table 5. Diagnostic testing accuracy table of XGBoost model for all data

	Outcome+	Outcome−	Total
Test+	11	5	16
Test−	7	69	76
Total	18	74	92

Table 6. Diagnostic testing accuracy table of XGBoost model for adjusted with multi-factor logistic regression data

	Outcome+	Outcome−	Total
Test+	10	5	15
Test−	7	60	67
Total	17	65	82

No statistically significant difference in the characteristics of the models was found.

Table 7. Diagnostic testing accuracy table of multifactor logistic regression model

	Outcome+	Outcome−	Total
Test+	8	3	11
Test−	9	65	71
Total	17	65	82

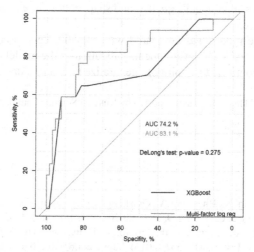

Fig. 4. Comparing ROC curves of models on prospective data.

Table 8. Comparing diagnostic models accuracy

Name	1. XGBoost model on primary prospective data value [95% CI]	2. XGBoost model on adjusted data value [95% CI]	3. Multi-factor logistic regression on adjusted data value [95% CI]	Compare 1–3 2–3 P (the same)
Apparent prevalence	0.17[0.10; 0.27]	0.18[0.11; 0.28]	0.13[0.07; 0.23]	0.423
True prevalence	0.20[0.12; 0.29]	0.21[0.13; 0.31]	0.21[0.13; 0.31]	–
Sensitivity	0.61[0.36; 0.83]	0.59[0.33; 0.82]	0.47[0.23; 0.72]	0.414
Specificity	0.93[0.85; 0.98]	0.92[0.83; 0.97]	0.95[0.87; 0.99]	0.480
Positive predictive value	0.69[0.41; 0.89]	0.67[0.38; 0.88]	0.73[0.39; 0.94]	0.711
Negative predictive value	0.91[0.82; 0.96]	–	0.87[0.77; 0.94]	0.483

5   Conclusions

In this study, a XGBoost model was constructed to predict the development of venous thromboembolism (VTE) in glioma patients. This model was based on the analysis of automatically selected parameters. An XGBoost algorithm was employed to build the model, optimized based on the obtained parameters.

The XGBoost model demonstrated good performance according to key metrics including specificity, precision, recall, and F1-score. Error analysis and ROC curve were utilized to assess the model's quality, and shap values were generated to illustrate parameter importance.

Additionally, the model was compared with logistic regression model, revealing significant predictors of VTE development. Validation of the model on an independent sample of patients confirmed its ability to generalize to external data.

Acknowledgements. This work was supported by Russian Science Foundation, Grant № 23-11-00346.

References

1. Nicholson, M., Chan, N., Bhagirath, V., Ginsberg, J.: Prevention of venous thromboembolism in 2020 and beyond. J. Clin. Med. **9**, 1–27 (2020). https://doi.org/10.3390/jcm9082467

2. Kearon, C., et al.: Antithrombotic therapy for VTE disease: CHEST guideline and expert panel report. Chest **149**(2), 315–352 (2016). https://doi.org/10.1016/j.chest.2015.11.026

3. Connors, J.M., Levy, J.H.: COVID-19 and its implications for thrombosis and anticoagulation. Blood **135**, 2033–2040 (2020). https://doi.org/10.1182/BLOOD.2020006000

4. Xu, Q., Lei, H., Li, X., Li, F., Shi, H., Wang, G., Sun, A., Wang, Y., Peng, B.: Machine learning predicts cancer-associated venous thromboembolism using clinically available variables in gastric cancer patients. Heliyon **9**(1) (2023). https://doi.org/10.1016/j.heliyon.2022.e12681

5. He, L., Luo, L., Hou, X., Liao, D., Liu, R., Ouyang, C., Wang, G.: Predicting venous thromboembolism in hospitalized trauma patients: a combination of the Caprini score and data-driven machine learning model. BMC Emergency Med. **21**(1) (2021). https://doi.org/10.1186/s12873-021-00447-x

6. Lin, C.C., et al.: Derivation and validation of a clinical prediction model for risks of venous thromboembolism in diabetic and general populations. Medicine **100**(39), E27367 (2021). https://doi.org/10.1097/MD.0000000000027367

7. Gerotziafas, G.T., Papageorgiou, L., Salta, S., Nikolopoulou, K., Elalamy, I.: Updated clinical models for VTE prediction in hospitalized medical patients. Thromb. Res. **164**, S62–S69 (2018). https://doi.org/10.1016/j.thromres.2018.02.004

8. Beal, E.W., Tumin, D., Chakedis, J., Porter, E., Moris, D., Zhang, X. feng, Abdel-Misih, S., Dillhoff, M., Manilchuk, A., Cloyd, J., et al.: Identification of patients at high risk for post-discharge venous thromboembolism after hepato-pancreato-biliary surgery: which patients benefit from extended thromboprophylaxis? HPB **20**(7), 621–630 (2018). https://doi.org/10.1016/j.hpb.2018.01.004

9. Lee, E.J., Chang, C.H., Wang, L.C., Hung, Y.C., Chen, H.H.: Two primary brain tumors, meningioma and glioblastoma multiforme, in opposite hemispheres of the same patient. J. Clin. Neurosci. **9**(5), 589–591 (2002). https://doi.org/10.1054/jocn.2002.1086. PMID: 12383424

10. Farge, D., Frere, C., Connors, J.M., Khorana, A.A., Kakkar, A., Ay, C., Muñoz, A., et al.: 2022 international clinical practice guidelines for the treatment and prophylaxis of venous thromboembolism in patients with cancer, including patients with COVID-19. Lancet Oncol. **23**, e334–e347 (2022). https://doi.org/10.1016/S1470-2045(22)00160-7

Author Index

Printed in the United States
by Baker & Taylor Publisher Services

Printed in the United States
by Baker & Taylor Publisher Services